VIDEO WORKBOOK WITH THE MATH COACH

BEGINNING ALGEBRA: EARLY GRAPHING

FOURTH EDITION

John Tobey
North Shore Community College
Danvers, Massachusetts

Jeffrey Slater
North Shore Community College
Danvers, Massachusetts

Jamie Blair
Orange Coast College
Costa Mesa, California

Jennifer Crawford
Normandale Community College
Bloomington, Minnesota

P

Pearson

Pearson

ISBN-13: 978-0-13-459201-5
ISBN-10: 0-13-459201-8

Contents

Contents

Guided Learning Video Worksheets

Preface

This *Video Workbook with the Math Coach* provides a convenient, ready-to-use format with ample work space to help your students to stay organized and take notes. The workbook provides an effective way to check to see if students understand the skills and concepts from each section of their textbook. The *Video Workbook with the Math Coach* is made up of two sections:

1. Worksheets with the Math Coach and
2. Guided Learning Video Worksheets

The worksheets with the Math Coach are the Tobey series' popular extra-practice student supplement that correlates with the Math Coach feature in the textbook itself. The worksheets with the Math Coach include

- Key vocabulary terms and vocabulary practice problems
- Step-by-step worked examples
- Extra Student Practice problems (modeled after each guided example)
- Concept Check questions from each section of the book
- Additional exercises with space for students to show their work
- Math Coach problems for students to work along with the Math Coach videos

The second section makes up the latest addition to the Tobey series: the Guided Learning Video Worksheets. This video note-taking guide encourages students to follow along with the objective-level Guided Learning Videos in their MyMathLab course, prompting them to fill in the blanks as they go. Once completed, instructors have a way to verify that students have watched the video and understood the key concepts, and students have a study tool for future reference. The Guided Learning Video Worksheets include

- Understanding the Big Picture vocabulary terms and concept explanations
- Worked-out examples that follow along step-by-step with the Guided Learning Video
- Paused Student Practice with ample space for student work
- Two Active Video Lessons that prompt students to complete examples by themselves:
  - Active Video Lesson 1 supplies a worked-out solution
  - Active Video Lesson 2 supplies a static answer

The two sections of the *Video Workbook with the Math Coach* are flexible and can be used independently of each other or integrated with the Guided Learning Video Worksheets being completed at point of use. Both the Math Coach exercises and Guided Learning Videos are assignable in MyMathLab and can be integrated into your homework plan.

Name: _____   Date: _____

Instructor: _____   Section: _____

Chapter 0 Prealgebra Review
0.1 Simplifying Fractions

Vocabulary
whole numbers • fractions • numerator • denominator • numerals
simplest form • reduced form • simplifying • reducing • natural numbers
counting numbers • factor • prime numbers • lowest terms • multiplicative identity
proper fraction • improper fraction • mixed number

1. The whole numbers 1, 2, 3, 4, 5, 6, 7, 8, and 9 are called _____ or counting numbers.

2. In the fraction $\dfrac{1}{25}$, 1 is the numerator and 25 is the _____.

3. When we obtain an equivalent fraction in simplest form (or reduced form), the new simplified fraction is said to be in _____.

4. The natural number factors of _____ are 1 and themselves.

Example	**Student Practice**
1. Simplify each fraction.	**2.** Simplify each fraction.
(a) $\dfrac{14}{21}$	(a) $\dfrac{33}{39}$
First factor 14 and 21, $\dfrac{14}{21} = \dfrac{7 \times 2}{7 \times 3}$.	
Now divide the numerator and denominator by 7.	
$\dfrac{14}{21} = \dfrac{\cancel{7} \times 2}{\cancel{7} \times 3} = \dfrac{2}{3}$	
(b) $\dfrac{20}{70}$	(b) $\dfrac{126}{210}$
Factor 20 and 70, $\dfrac{20}{70} = \dfrac{2 \times 2 \times 5}{7 \times 2 \times 5}$.	
Now divide the numerator and denominator by both 2 and 5.	
$\dfrac{20}{70} = \dfrac{2 \times \cancel{2} \times \cancel{5}}{7 \times \cancel{2} \times \cancel{5}} = \dfrac{2}{7}$	

Example	Student Practice
3. Simplify the fraction $\dfrac{7}{21}$. First factor 7 and 21, $\dfrac{7}{21} = \dfrac{7\times1}{7\times3}$. Now divide the numerator and denominator by 7. $\dfrac{7}{21} = \dfrac{\cancel{7}\times1}{\cancel{7}\times3} = \dfrac{1}{3}$ Notice that all the prime numbers in the numerator divided out. When this happens, we must remember that 1 is left in the numerator.	**4.** Simplify the fraction $\dfrac{13}{169}$.
5. Simplify the fraction $\dfrac{70}{10}$. $\dfrac{70}{10} = \dfrac{7\times\cancel{5}\times\cancel{2}}{\cancel{5}\times\cancel{2}\times1} = 7$ Notice that all the prime numbers in the denominator divided out. When this happens, we do not need to leave 1 in the denominator because the answer is a whole number.	**6.** Simplify the fraction $\dfrac{423}{47}$.
7. Cindy got 48 out of 56 questions correct on a test. Write this as a fraction in simplest form. Express as a fraction the number of correct responses out of the total number of questions on the test. 48 out of 56 $\rightarrow \dfrac{48}{56}$ Express this fraction in simplest form. Factor 48 and 56, then divide out common factors. $\dfrac{48}{56} = \dfrac{6\times\cancel{8}}{7\times\cancel{8}} = \dfrac{6}{7}$ Cindy answers the questions correctly $\dfrac{6}{7}$ of the time.	**8.** Last June, Jacob went to the gym 18 out of the 30 days. Write this as a fraction in simplest form.

2

Example	Student Practice
9. Change $\dfrac{7}{4}$ to a mixed number or to a whole number.	**10.** Change $\dfrac{63}{7}$ to a mixed number or to a whole number.

Divide the denominator into the numerator.

$$\dfrac{7}{4} = 7 \div 4$$

$$\begin{array}{r} 1 \\ 4\overline{)7} \\ \underline{4} \\ 3 \end{array} \quad \text{Remainder}$$

The quotient, 1, is the whole-number part of the mixed number. The remainder from the division, 3, is the numerator of the fraction. The denominator of the fraction remains unchanged.

Thus, $\dfrac{7}{4} = 1\dfrac{3}{4}$.

11. Change $3\dfrac{1}{7}$ to an improper fraction.	**12.** Change $6\dfrac{5}{12}$ to an improper fraction

To change a mixed number to an improper fraction, multiply the whole number by the denominator. Add this to the numerator. The result is the new numerator. The denominator does not change.

$$3\dfrac{1}{7} = \dfrac{(3 \times 7)+1}{7} = \dfrac{21+1}{7} = \dfrac{22}{7}$$

Thus, $3\dfrac{1}{7} = \dfrac{22}{7}$.

Example	Student Practice
13. Find the missing number.	**14.** Find the missing number.

13. (a) $\dfrac{3}{5} = \dfrac{?}{25}$

A fraction can be changed to an equivalent fraction with a different denominator by multiplying both numerator and denominator by the same number. Observe that we need to multiply the denominator by 5 to obtain 25. So we multiply the numerator 3 by 5 also.

$$\frac{3 \times 5}{5 \times 5} = \frac{15}{25}$$

The desired number is 15.

(b) $\dfrac{2}{9} = \dfrac{?}{36}$

Observe that $9 \times 4 = 36$. We need to multiply the numerator by 4 to get the new numerator.

$$\frac{2 \times 4}{9 \times 4} = \frac{8}{36}$$

The desired number is 8.

14. (a) $\dfrac{1}{9} = \dfrac{?}{99}$

(b) $\dfrac{4}{7} = \dfrac{?}{35}$

Extra Practice

1. Simplify the fraction $\dfrac{60}{15}$.

2. Change $\dfrac{556}{10}$ to a mixed number.

3. Change $1\dfrac{12}{17}$ to an improper fraction.

4. Find the missing numerator, $\dfrac{8}{15} = \dfrac{?}{120}$.

Concept Check

Explain in your own words how to change a mixed number to an improper fraction.

Chapter 0 Prealgebra Review
0.2 Adding and Subtracting Fractions

Vocabulary
fraction • numerator • denominator • common denominator
least common denominator • prime numbers • mixed numbers • perimeter

1. The _____ of two or more fractions is the smallest whole number that is exactly divisible by each denominator of the fractions.

2. Before you can add or subtract fractions, they must have the same _____.

3. The distance around a polygon is called the _____.

4. When adding or subtracting _____ first change them to improper fractions.

Example	Student Practice
1. Add the fractions. Simplify your answer whenever possible.	**2.** Add the fractions. Simplify your answer whenever possible.

Example

1. Add the fractions. Simplify your answer whenever possible.

(a) $\dfrac{5}{7}+\dfrac{1}{7}$

Add the numerators and keep the denominator the same.

$$\dfrac{5}{7}+\dfrac{1}{7}=\dfrac{5+1}{7}=\dfrac{6}{7}$$

This fraction cannot be simplified further.

(b) $\dfrac{1}{8}+\dfrac{3}{8}+\dfrac{2}{8}$

Add the numerators and keep the denominator the same.

$$\dfrac{1}{8}+\dfrac{3}{8}+\dfrac{2}{8}=\dfrac{1+3+2}{8}=\dfrac{6}{8}$$

Now simplify.

$$\dfrac{6}{8}=\dfrac{3}{4}$$

Student Practice

2. Add the fractions. Simplify your answer whenever possible.

(a) $\dfrac{1}{4}+\dfrac{3}{4}$

(b) $\dfrac{7}{10}+\dfrac{5}{10}+\dfrac{3}{10}$

Example	Student Practice
3. Subtract the fractions $\dfrac{9}{11} - \dfrac{2}{11}$. Simplify your answer if possible. Subtract the numerators and keep the denominator the same. $$\dfrac{9}{11} - \dfrac{2}{11} = \dfrac{9-2}{11} = \dfrac{7}{11}$$ This fraction cannot be simplified further.	**4.** Subtract the fractions $\dfrac{7}{8} - \dfrac{5}{8}$. Simplify your answer if possible.
5. Find the LCD of $\dfrac{5}{6}$ and $\dfrac{1}{15}$ using the prime factor method. Write each denominator as the product of prime factors. The LCD is a product containing each different prime factor. $6 = 2 \cdot 3$ $15 = \downarrow\ 3 \cdot 5$ $\quad\ \ \downarrow\ \downarrow\ \downarrow$ $LCD = 2 \cdot 3 \cdot 5$ The different factors are 2, 3, and 5, and each factor appears at most once in any one denominator. Multiply the factors to find the LCD. $LCD = 2 \cdot 3 \cdot 5 = 30$	**6.** Find the LCD of $\dfrac{1}{10}$ and $\dfrac{6}{35}$ using the prime factor method.
7. Find the LCD of $\dfrac{5}{12}$, $\dfrac{1}{15}$, and $\dfrac{7}{30}$. Write each denominator as the product of prime factors. The LCD is a product containing each different factor, with the factor 2 appearing twice since it occurs twice in the factorization of 12. $12 = 2 \cdot 2 \cdot 3$ $15 = \downarrow\quad\ 3 \cdot 5$ $\quad\ \ \downarrow$ $30 = \downarrow\ 2 \cdot 3 \cdot 5$ $\quad\ \ \downarrow\ \downarrow\ \downarrow\ \downarrow$ $LCD = 2 \cdot 2 \cdot 3 \cdot 5 = 60$	**8.** Find the LCD of $\dfrac{11}{18}$, $\dfrac{2}{21}$, and $\dfrac{13}{42}$.

6

Example	Student Practice
9. Combine. $\dfrac{1}{5} + \dfrac{1}{6} - \dfrac{3}{10}$	**10.** Combine. $\dfrac{1}{4} + \dfrac{13}{18} - \dfrac{5}{54}$

First find the LCD.

$$5 = 5$$
$$6 = \quad 2 \cdot 3$$
$$10 = 5 \cdot 2 \quad \downarrow$$
$$\downarrow \quad \downarrow \quad \downarrow$$
$$\text{LCD} = 5 \cdot 2 \cdot 3 = 30$$

Now we change $\dfrac{1}{5}$, $\dfrac{1}{6}$, and $\dfrac{3}{10}$ to equivalent fractions that have the LCD for a denominator.

$$\dfrac{1}{5} = \dfrac{?}{30} \qquad \dfrac{1 \times 6}{5 \times 6} = \dfrac{6}{30}$$
$$\dfrac{1}{6} = \dfrac{?}{30} \qquad \dfrac{1 \times 5}{6 \times 5} = \dfrac{5}{30}$$
$$\dfrac{3}{10} = \dfrac{?}{30} \qquad \dfrac{3 \times 3}{10 \times 3} = \dfrac{9}{30}$$

Combine the three fractions.

$$\dfrac{1}{5} + \dfrac{1}{6} - \dfrac{3}{10} = \dfrac{6}{30} + \dfrac{5}{30} - \dfrac{9}{30}$$
$$= \dfrac{6 + 5 - 9}{30}$$
$$= \dfrac{2}{30}$$

Now simplify.

$$\dfrac{2}{30} = \dfrac{1}{15}$$

Thus, $\dfrac{1}{5} + \dfrac{1}{6} - \dfrac{3}{10} = \dfrac{1}{15}$.

Example	Student Practice
11. Manuel is enclosing a triangular shaped exercise yard for his new dog. He wants to determine how many feet of fencing he will need. The sides of the yard measure $20\frac{3}{4}$ feet, $15\frac{1}{2}$ feet, and $18\frac{1}{8}$ feet. What is the perimeter of (the total distance around) the triangle?	**12.** Find the perimeter of the parallelogram shaped park below.

Begin by drawing a picture. Find the perimeter by adding up the lengths of the three sides of the triangle. To add mixed numbers, change them to improper fractions then add.

$$20\frac{3}{4}+15\frac{1}{2}+18\frac{1}{8}=\frac{83}{4}+\frac{31}{2}+\frac{145}{8}$$

$$=\frac{166}{8}+\frac{124}{8}+\frac{145}{8}$$

$$=\frac{435}{8}=54\frac{3}{8}$$

He will need $54\frac{3}{8}$ feet of fencing.

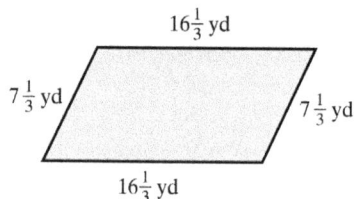

Extra Practice

1. Find the LCD of $\frac{6}{25}$ and $\frac{99}{100}$. Do not combine the fractions; only find the LCD.

2. Combine $\frac{31}{45}-\frac{2}{15}$. Be sure to simplify your answer if possible.

3. Combine $\frac{7}{9}+2\frac{5}{12}$. Be sure to simplify your answer if possible.

4. Combine $6\frac{3}{11}-2\frac{17}{22}$. Be sure to simplify your answer if possible.

Concept Check

Explain how you would find the LCD of the fractions $\frac{4}{21}$ and $\frac{5}{18}$.

Chapter 0 Prealgebra Review
0.3 Multiplying and Dividing Fractions

Vocabulary

fractions • numerators • denominators • invert and multiply method
common factors • prime numbers • mixed numbers • complex fraction

1. A(n) _____ has one fraction in the numerator and one fraction in the denominator.

2. To multiply two fractions, multiply the _____ and multiply the denominators.

3. Use the _____ to divide two fractions.

4. When multiplying or dividing _____, first change them to improper fractions.

Example	**Student Practice**
1. Multiply.	2. Multiply.
(a) $\dfrac{3}{5} \times \dfrac{5}{7}$	(a) $\dfrac{2}{7} \times \dfrac{7}{17}$
Multiply the numerators and multiply the denominators. $$\dfrac{3}{5} \times \dfrac{5}{7} = \dfrac{3 \cdot 5}{5 \cdot 7}$$ Divide the numerator and denominator by 5 and simplify. $$\dfrac{3}{5} \times \dfrac{5}{7} = \dfrac{3 \cdot 5}{5 \cdot 7} = \dfrac{3 \cdot \overset{1}{\cancel{5}}}{\cancel{5} \cdot 7} = \dfrac{3}{7}$$	
(b) $\dfrac{15}{8} \times \dfrac{10}{27}$	(b) $\dfrac{55}{12} \times \dfrac{4}{45}$
If we factor each number, we can see the common factors. Remove common factors and multiply. $$\dfrac{15}{8} \times \dfrac{10}{27} = \dfrac{\overset{1}{\cancel{3}} \cdot 5}{2 \cdot 2 \cdot \underset{1}{\cancel{2}}} \times \dfrac{5 \cdot \overset{1}{\cancel{2}}}{\underset{1}{\cancel{3}} \cdot 3 \cdot 3} = \dfrac{25}{36}$$	

Example	Student Practice

3. Multiply $7 \times \dfrac{3}{5}$.

Write the whole number as a fraction whose denominator is 1.

$$7 \times \dfrac{3}{5} = \dfrac{7}{1} \times \dfrac{3}{5}$$

Follow the multiplication rule for fractions. Multiply numerators and multiply denominators.

$$7 \times \dfrac{3}{5} = \dfrac{7}{1} \times \dfrac{3}{5} = \dfrac{21}{5} \text{ or } 4\dfrac{1}{5}$$

4. Multiply $25 \times \dfrac{4}{5}$.

---

5. How do we find the area of a rectangular field $3\dfrac{1}{3}$ miles long and $2\dfrac{1}{2}$ miles wide?

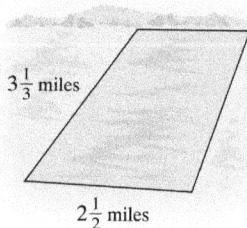

$3\dfrac{1}{3}$ miles

$2\dfrac{1}{2}$ miles

To find the area, we multiply length times width, $3\dfrac{1}{3} \times 2\dfrac{1}{2}$.

To multiply two mixed numbers, change them to improper fractions then follow the multiplication rule for fractions.

$$3\dfrac{1}{3} \times 2\dfrac{1}{2} = \dfrac{10}{3} \times \dfrac{5}{2} = \dfrac{\cancel{2} \cdot 5}{3} \times \dfrac{5}{\cancel{2}} = \dfrac{25}{3}$$

Write the answer as a mixed number.

The area is $8\dfrac{1}{3}$ square miles.

6. Find the area of a rectangular pool that is $7\dfrac{1}{12}$ meters long and $6\dfrac{3}{10}$ meters wide.

Example	Student Practice
7. Divide $\dfrac{2}{5} \div \dfrac{3}{10}$. To divide two fractions, invert the second fraction (the divisor) and multiply. $$\dfrac{2}{5} \div \dfrac{3}{10} = \dfrac{2}{5} \times \dfrac{10}{3}$$ Remove common factors and simplify. $$\dfrac{2}{5} \div \dfrac{3}{10} = \dfrac{2}{5} \times \dfrac{10}{3}$$ $$= \dfrac{2}{\cancel{5}} \times \dfrac{\cancel{5} \cdot 2}{3} = \dfrac{4}{3} \text{ or } 1\dfrac{1}{3}$$	**8.** Divide $\dfrac{1}{9} \div \dfrac{5}{6}$.
9. Divide $\dfrac{1}{3} \div 2$. Write the whole number as a fraction. Then invert the second fraction (the divisor) and multiply. $$\dfrac{1}{3} \div 2 = \dfrac{1}{3} \div \dfrac{2}{1} = \dfrac{1}{3} \times \dfrac{1}{2} = \dfrac{1}{6}$$	**10.** Divide $7 \div \dfrac{1}{5}$.
11. Divide $\dfrac{\dfrac{3}{7}}{\dfrac{3}{5}}$. This division is in the form of a complex fraction. Write this problem as a standard division first, then complete the problem using the rule for division. $$\dfrac{\dfrac{3}{7}}{\dfrac{3}{5}} = \dfrac{3}{7} \div \dfrac{3}{5} = \dfrac{\cancel{3}}{7} \times \dfrac{5}{\cancel{3}} = \dfrac{5}{7}$$	**12.** Divide $\dfrac{\dfrac{7}{20}}{\dfrac{14}{15}}$.

Example	Student Practice
13. Divide $2\frac{1}{3} \div 3\frac{2}{3}$. To divide mixed numbers, change the mixed numbers to improper fractions and then use the rule for dividing fractions. $2\frac{1}{3} \div 3\frac{2}{3} = \frac{7}{3} \div \frac{11}{3} = \frac{7}{\cancel{3}} \times \frac{\cancel{3}}{11} = \frac{7}{11}$	**14.** Divide $\dfrac{10}{2\frac{4}{5}}$.
15. A chemist has 96 fluid ounces of a solution. She pours the solution into test tubes. Each test tube holds $\frac{3}{4}$ fluid ounces. How many test tubes can she fill? Divide the total number of ounces by the number of ounces in each test tube. $96 \div \frac{3}{4} = \frac{96}{1} \div \frac{3}{4}$ $= \frac{96}{1} \times \frac{4}{3} = \frac{\cancel{3} \cdot 32}{1} \times \frac{4}{\cancel{3}} = 128$ She will be able to fill 128 test tubes.	**16.** A graphic designer created 5 images for the evening news program in $6\frac{1}{4}$ hours. How many images did the designer create per hour?

Extra Practice

1. Multiply $\frac{3}{7} \times \frac{5}{12} \times \frac{8}{35}$. Simplify your answer if possible.

2. Multiply $7\frac{1}{2} \times 2\frac{2}{5}$. Simplify your answer if possible.

3. Divide $\frac{2}{3} \div \frac{4}{9}$. Simplify your answer if possible.

4. Mario cuts $59\frac{1}{2}$ inches of string into pieces that are exactly $4\frac{1}{4}$ inches long. How many pieces of string does Mario cut?

Concept Check

Explain the steps you would take to perform the calculation $3\frac{1}{4} \div 2\frac{1}{2}$.

Chapter 0 Prealgebra Review
0.4 Using Decimals

Vocabulary

decimal • decimal points • decimal places • terminal zeros • divisor
dividend • quotient • caret • fraction

1. A fraction whose denominator is 10, 100, and so on can be expressed as a _____.

2. Divide the dividend by the _____.

3. The number of digits to the right of the decimal point is also the number of _____.

4. To add or subtract decimals, write in column form and line up the _____.

Example	Student Practice
1. Write each of the following decimals as a fraction or mixed number. State the number of decimal places. Write out in words the way the number would be spoken. **(a)** 0.6 In fraction form, $0.6 = \dfrac{6}{10}$. 0.6 has 1 decimal place because it has 1 digit to the right of the decimal point. Use a place value chart if necessary to determine the word name. Written out, 0.6 is six-tenths.	**2.** Write each of the following decimals as a fraction or mixed number. State the number of decimal places. Write out in words the way the number would be spoken. **(a)** 0.00009
(b) 1.38 In fraction form, $1.38 = 1\dfrac{38}{100}$. 1.38 has 2 decimal places because it has 2 digits to the right of the decimal point. Use a place value chart if necessary to determine the word name. Written out, 1.38 is one and thirty-eight hundredths.	**(b)** 4.025

Example	Student Practice
3. Write $\dfrac{2}{11}$ as a decimal.	**4.** Write $\dfrac{13}{15}$ as a decimal.

$$\begin{array}{r} 0.1818 \\ 11\overline{)2.0000} \\ \underline{1\,1} \\ 90 \\ \underline{88} \\ 20 \\ \underline{11} \\ 90 \\ \underline{88} \\ 2 \end{array}$$

Thus, $\dfrac{2}{11} = 0.1818\ldots = 0.\overline{18}$.

Example	Student Practice
5. Write 0.138 as a fraction and simplify if possible.	**6.** Write 0.72 as a fraction and simplify if possible.

To convert from a decimal to a fraction, write the decimal as a fraction with a denominator of 10, 100, 1000, and so on, and simplify the result.

Write 0.138 as a fraction with numerator 138 and denominator 1000. Then simplify.

$$0.138 = \dfrac{138}{1000} = \dfrac{69}{500}$$

Example	Student Practice
7. Subtract $127.32 - 38.48$.	**8.** Add $21.16 + 2.04 + 13.91$.

To add or subtract decimals, write in column form and line up the decimal points. Then add or subtract the digits.

$$\begin{array}{r} 127.32 \\ -\ \ 38.48 \\ \hline 88.84 \end{array}$$

Example	Student Practice
9. Multiply 2.56×0.003. To multiply decimals, first multiply as with whole numbers. To determine the position of the decimal point, count the total number of decimal places in the two numbers being multiplied. This will determine the number of decimal places that should appear in the answer. $\quad\quad 2.56$ (two decimal places) $\underline{\times\quad 0.003}$ (three decimal places) $\quad 0.00768$ (five decimal places) Thus, $2.56 \times 0.003 = 0.00768$.	**10.** Multiply 0.85×0.0005.
11. Divide $16.2 \div 0.027$. There are three decimal places in the divisor, so we move the decimal point three places to the right in the divisor and dividend and mark the new position by a caret. Note that we must add two zeros to 16.2 in order to do this. $\quad\quad\quad 600.$ $0.027_\wedge)\overline{16.200_\wedge}$ Now perform the division as with whole numbers. The decimal point in the answer is directly above the caret. $\quad\quad\quad\ 600.$ $0.027_\wedge)\overline{16.200_\wedge}$ $\quad\quad\ \underline{16\ 2}$ $\quad\quad\quad\ 000$ Thus, $16.2 \div 0.027 = 600$.	**12.** Divide $25.2 \div 0.0003$.

Example	Student Practice
13. Multiply 0.0026×1000. When multiplying by 10, 100, 1000, and so on, for every zero in the multiplier, move the decimal point one place to the right. Since 1000 has 3 zeros, move the decimal point 3 places to the right. $0.0026 \times 1000 = 2.6$	**14.** Multiply 0.93×100.
15. Divide $0.0038 \div 100$. When dividing by 10, 100, 1000, and so on, for every zero in the divisor, move the decimal point one place to the left. Since 100 has 2 zeros, move the decimal point 2 places to the left. $0.0038 \div 100 = 0.000038$	**16.** Divide $7138.3 \div 10$.

Extra Practice

1. Write 32.082 as a fraction in simplified from. Write the value in words.

2. Subtract $14.03 - 7.8932$.

3. Divide $1.1592 \div 0.06$.

4. Multiply 16.785×100 by moving the decimal point.

Concept Check

Explain how you would place the decimal points when performing the calculation $0.252 \div 0.0035$.

Chapter 0 Prealgebra Review
0.5 Percents, Rounding, and Estimating

Vocabulary
decimal • decimal point • percent • % symbol • estimation • nonzero digit

1. A _____ is a fraction that has a denominator of 100.

2. To estimate by rounding, first round each number so that there is one _____ .

3. _____ is the process of finding an approximate answer.

4. To change a percent to a decimal, move the _____ two places to the left and remove the % symbol.

Example	**Student Practice**
1. Change to a percent. **(a)** 0.0364 We move the decimal point two places to the right and add the % symbol. 0.0364 = 3.64% **(b)** 0.4 0.4 = 40%	**2.** Change to a percent. **(a)** 0.00019 **(b)** 0.6
3. Change to a percent. **(a)** 2.938 We move the decimal point two places to the right and add the % symbol. 2.938 = 293.8% **(b)** 4.5 4.5 = 450%	**4.** Change to a percent. **(a)** 5.95 **(b)** 7.9

Example	Student Practice
5. Change to a decimal. (a) 4% First we move the decimal point two places to the left. Then we remove the % symbol. $$4\% = 4 \underset{\uparrow}{\,.\,} \% = 0.04$$ The unwritten decimal point is understood to be here. (b) 254.8% $254.8\% = 2.548$	**6.** Change to a decimal. (a) 0.8% (b) 101.2%
7. Find 182% of 12. To find the percent of a number, change the percent to a decimal and multiply the number by the decimal. 182% of $12 = 1.82 \times 12 = 21.84$	**8.** Find 5% of 639.
9. A store is having a sale of 35% off the retail price of all sofas. Melissa wants to buy a particular sofa that normally sells for $595. (a) How much will Melissa save if she buys the sofa on sale? Find 35% of $595. $0.35 \times 595 = 208.25$ Thus, Melissa will save $208.25. (b) What will the purchase price be if Melissa buys the sofa on sale? The purchase price is the difference between the original price and the amount saved. $595.00 - 208.25 = 386.75$ If Melissa buys the sofa on sale, she will pay $386.75.	**10.** A store is having a sale of 65% off the retail price of all boots. Marcus wants to buy a particular pair of boots that normally sells for $123. (a) How much will Marcus save if he buys the boots on sale? (b) What will the purchase price be if Marcus buys the boots on sale?

Example	Student Practice
11. What percent of 24 is 15?	**12.** What percent of 72 is 63?

11. (continued)

First, write a fraction with the two numbers. The number after the word of is always the denominator, and the other number is the numerator.

$$\frac{15}{24}$$

Simplify the fraction.

$$\frac{15}{24} = \frac{5}{8}$$

Change the simplified fraction to a decimal.

$$\frac{5}{8} = 0.625$$

Express the decimal as a percent.
$$0.625 = 62.5\%$$

Thus, 62.5% of 24 is 15.

---

13. Marcia made 29 shots on goal during the last high school field hockey game. She actually scored a goal 8 times. What percent of her total shots were goals? Round your answer to the nearest whole percent.

We want to know what percent of 29 is 8. This relationship expressed as a fraction is $\frac{8}{29}$.

This fraction cannot be reduced. The next step is to write the fraction as a decimal.

$$\frac{8}{29} = 0.2758...$$

Change the decimal to a percent and round to the nearest whole number.
$$0.2758... = 27.58...\% \approx 28\%$$

Thus, Marcia scored a goal about 28% of the time she made a shot on goal.

14. In a shipment of 600 widgets, 38 widgets are warped. What percent of the total widgets are warped? Round your answer to the nearest whole percent.

19

Example	Student Practice
15. The four walls of a college classroom are $22\frac{1}{4}$ feet long and $8\frac{3}{4}$ feet high. A painter needs to know the area of these four walls in square feet. Since paint is sold in gallons, an estimate will do. Use estimation by rounding to approximate the area of the four walls.	**16.** A landscaper is to seed a lawn that is $47\frac{1}{2}$ meters long and $18\frac{1}{4}$ meters wide. The landscaper needs to know the area of the lawn in square meters to determine how much grass seed to buy. Use estimation by rounding to approximate the area of the lawn.

To estimate by rounding, round each number so that there is one nonzero digit. Round $22\frac{1}{4}$ feet to 20 feet.

Round $8\frac{3}{4}$ feet to 9 feet.

Multiply 20×9 to obtain an estimate of the area of one wall. Multiply $20 \times 9 \times 4$ to obtain an estimate of the area of all four walls.

$20 \times 9 \times 4 = 720$

Our estimate for the painter is 720 square feet of wall space.

Extra Practice

1. Change 0.00576 to a percent.

2. Change 100% to a decimal.

3. 65 is what percent of 25?

4. Follow the principles of estimation to find an approximate value of $\dfrac{804}{39,500}$.

Round each number so that there is one nonzero digit. Do not find the exact value.

Concept Check

Explain how you would change 0.0078 to a percent.

Name: _____   Date: _____

Instructor: _____   Section: _____

Chapter 0 Prealgebra Review
0.6 Using the Mathematics Blueprint for Problem Solving

Vocabulary
mathematics blueprint　•　area　•　percent　•　check　•　estimation

1. When solving real-life problems, use a(n) _____.

2. You can _____ your answer to real-life problems using estimation.

Example	Student Practice
1. Nancy and John want to install wall-to-wall carpeting in their living room. The floor of the rectangular living room is $11\frac{2}{3}$ feet wide and $19\frac{1}{2}$ feet long. How much will it cost if the carpet is $18.00 per square yard?	**2.** Kris wants to install wall-to-wall carpet in a bedroom. The floor of this rectangular room is $12\frac{1}{2}$ feet wide and $17\frac{1}{3}$ feet long. How much will it cost if the carpet is $27.00 per square yard?

Read the problem carefully and create a Mathematics Blueprint. Draw a picture if it will help.

Find the area of the floor.
$$11\frac{2}{3}\times19\frac{1}{2}=\frac{35}{3}\times\frac{39}{2}=\frac{455}{2}=227\frac{1}{2}$$

A minimum of $227\frac{1}{2}$ square feet of carpet is needed. Dividing this value by 9 gives the area in square yards,

$227\frac{1}{2}\div9=25\frac{5}{18}$. Multiply $25\frac{5}{18}$ square yards by $18.00 per square yard to find the cost.

$$25\frac{5}{18}\times18=\frac{455}{18}\times\frac{18}{1}=455$$

The carpet will cost a minimum of $455.00 for this room. The check is left to the student.

Example	Student Practice

Example

3. The following chart shows the 2012 sales of Micropower Computer Software for each of the four regions of the United States. Use the chart to answer the following questions (round all answers to the nearest whole percent).

Region of the U.S.	Number of Sales Personnel	Dollar Volume of Sales
Northeast	12	1,560,000
Southeast	18	4,300,000
Northwest	10	3,660,000
Southwest	15	3,720,000
Total	55	13,240,000

(a) What percent of the sales personnel are assigned to the Northeast?

Read the problem carefully and create a Mathematics Blueprint. You must calculate, "12 is what percent of 55?" Divide 12 by 55. Change the decimal to a percent and round.

$$\frac{12}{55} = 0.21818... = 21.818...\% \approx 22\%$$

Thus, about 22% of sales personnel are assigned to the Northeast.

(b) What percent of the volume of sales is attributed to the Northeast?

You must calculate, "1,560,000 is what percent of 13,240,000?"

$$\frac{1,560,000}{13,240,000} = \frac{156}{1324} \approx 0.1178 \approx 12\%$$

Thus, about 12% of the volume of sales is attributed to the Northeast.

Student Practice

4. Use the chart in example **3** to answer the following.

(a) What percent of sales personnel are assigned to the Northwest?

(b) What percent of the volume of sales is attributed to the Northwest?

(c) Of the Northeast and Northwest, which region has sales personnel that appear to be more effective in terms of the volume of sales?

Extra Practice

1. In a recent mayoral election in the town of Oceanside, Ms. Dempsey received 44% of the vote, Mr. Chan received 36%, and Ms. Tobin received 18%. A total of 6150 people voted. How many more people voted for Ms. Dempsey than Mr. Chan?

2. Ally would like to install linoleum on her kitchen floor. The floor of the rectangular kitchen is $16\frac{1}{2}$ feet wide and $22\frac{1}{2}$ feet long. How much will it cost to replace the floor if the linoleum costs $17.00 per square yard?

3. Nadine ran 2.2 kilometers on Monday. On each of the next four days, she increased her distance by 40% compared with the previous day. How many kilometers further did Nadine run on Thursday than on Wednesday? Round to the nearest tenth of a kilometer.

4. Karen earns $5200 per month, of which 35% is taken out in taxes and other withholding. How much does Karen take home every month after taxes and withholding?

Concept Check

Hank knows that 1 kilometer ≈ 0.62 mile. Explain how Hank could find out how many miles he traveled on a 12-kilometer trip to Mexico.

MATH COACH

Mastering the skills you need to do well on the test.

Watch the **MATH COACH** videos in MyMathLab° or on YouTube™ while you work the problems below. These helpful hints will help you avoid making common errors on test problems.

Subtracting Mixed Numbers—Problem 7

Subtract $3\frac{2}{3} - 2\frac{5}{6}$.

> **Helpful Hint:** First change the mixed numbers to improper fractions. Next find the LCD of the two denominators. Then change the fractions to an equivalent form with the LCD as the common denominator before subtracting.

Did you change $3\frac{2}{3}$ to $\frac{11}{3}$ and $2\frac{5}{6}$ to $\frac{17}{6}$?

Yes _____ No _____

If you answered No, stop and change the two mixed numbers to improper fractions.

Did you find the LCD to be 6? Yes _____ No _____

If you answered No, consider how to find the LCD of the two denominators. Once the fractions are written as equivalent fractions with 6 as the denominator, the two like fractions can be subtracted.

If you answered Problem 7 incorrectly, go back and rework the problem using these suggestions.

Dividing Mixed Numbers—Problem 10   Divide $5\frac{3}{8} \div 2\frac{3}{4}$.

> **Helpful Hint:** Be sure to change the mixed numbers to improper fractions before dividing.

Did you change $5\frac{3}{8}$ to $\frac{43}{8}$ and $2\frac{3}{4}$ to $\frac{11}{4}$ before doing any other steps? Yes _____ No _____

If you answered No, stop and change the two mixed numbers to improper fractions.

Next did you change the division to multiplication to obtain $\frac{43}{8} \times \frac{4}{11}$? Yes _____ No _____

If you answered No, stop and make this change.

Did you simplify the product? Yes _____ No _____

If you answered No, try dividing a 4 from the second numerator and first denominator before multiplying. The product will be an improper fraction that can be converted to a mixed number.

Now go back and rework the problem using these suggestions.

Dividing Decimals—Problem 17   Divide $12.88 \div 0.056$.

> **Helpful Hint:** Be careful as you move the decimal point in the divisor to the right. Make sure that the resulting divisor is an integer. Then move the decimal point in the dividend the same number of places to the right. Add zeros if necessary.

Did you move the decimal point in the divisor three places to the right to get 56?

Yes _____   No _____

If you answered No, stop and perform this step first.

Did you move the decimal point in the dividend three places to the right and add one zero to get 12880?

Yes _____   No _____

If you answered No, perform this step now. Be careful of calculation errors as you perform the division.

If you answered problem 17 incorrectly, go back and rework the problem using these suggestions.

Find the Missing Percent—Problem 22   39 is what percent of 650?

> **Helpful Hint:** Write a fraction with the two numbers. The number after the word "of" is always the denominator, and the other number is the numerator.

Did you write the fraction $\dfrac{39}{650}$?

Yes _____   No _____

If you answered No, stop and perform this step.

Did you simplify the fraction to $\dfrac{3}{50}$ before changing the fraction to a decimal?

Yes _____   No _____

If you answered No, consider that simplifying the fraction first makes the division step a little easier. Be sure to place the decimal point correctly in your quotient.

Did you change the quotient from a decimal to a percent?

Yes _____   No _____

If you answered No, stop and perform this final step.

Now go back and rework the problem using these suggestions.

Chapter 1 Real Numbers and Variables
1.1 Adding Real Numbers

Vocabulary
whole numbers • integers • rational numbers • irrational numbers • real numbers
number line • positive numbers • negative numbers • opposite numbers • absolute value
additive inverses

1. The _____ of a number is the distance between that number and zero on a number line.

2. _____, also called additive inverses, have the same magnitude but different signs and can be represented on a number line.

3. _____ are all the rational numbers and all the irrational numbers.

4. _____ are to the left of 0 on a number line.

Example	**Student Practice**
1. Classify as an integer, a rational number, an irrational number, and/or a real number.	**2.** Classify as an integer, a rational number, an irrational number, and/or a real number.
(a) 5	**(a)** $\sqrt{10}$
5 is an integer, a rational number, and a real number.	
(b) $-\dfrac{1}{3}$	**(b)** $\dfrac{1}{9}$
$-\dfrac{1}{3}$ is a rational number and a real number.	
(c) $\sqrt{2}$ is an irrational number and a real number.	**(c)** -4.25

Example	Student Practice
3. Use a real number to represent each situation.	**4.** Use a real number to represent each situation.
(a) A temperature of 128.6° F below zero is recorded at Vostok, Antarctica.	**(a)** A stock loss of 6.23 points
"below" is a keyword indicating that the number is negative.	
128.6° F below zero is -128.6.	**(b)** A temperature of 109.4° F in a desert
(b) The Himalayan peak K2 rises 29,064 feet above sea level.	
"above" is a keyword indicating that the number is positive.	**(c)** A population loss of 345
29,064 feet above sea level is $+29,064$.	
(c) The Dow gains 10.24 points.	**(d)** A 15 yard gain in a football game
A gain of 10.24 points is $+10.24$.	
(d) An oil drilling platform extends 328 feet below sea level.	
328 feet below sea level is -328.	
5. Find the additive inverse (that is, the opposite).	**6.** Find the additive inverse (that is, the opposite).
(a) -7	**(a)** $-\dfrac{7}{8}$
The opposite of -7 is $+7$.	
(b) $\dfrac{1}{4}$	**(b)** 30 feet below sea level
The opposite of $\dfrac{1}{4}$ is $-\dfrac{1}{4}$.	

Example	Student Practice
7. Find the absolute value.	**8.** Find the absolute value.
(a) $\left|-4.62\right|$	**(a)** $\left|5.34\right|$
$\left|-4.62\right| = 4.62$	
(b) $\left|\dfrac{3}{7}\right|$	**(b)** $\left|-\sqrt{5}\right|$
$\left|\dfrac{3}{7}\right| = \dfrac{3}{7}$	
(c) $\left|0\right|$	**(c)** $\left|\dfrac{0}{6}\right|$
$\left|0\right| = 0$	
9. Add. $14+16$	**10.** Add. $-9+(-5)$
To add two numbers with the same sign, add the absolute values. Then use the common sign, $+$, in the answer.	
$14+16 = 30$ $14+16 = +30$	
11. Add. $8+(-7)$	**12.** Add. $3+(-9)$
To add two numbers with different signs, find the difference between the two absolute values. The answer will have the sign of the number with the larger absolute value.	
$8-7 = 1$ $+8+(-7) = +1 \text{ or } 1$	

Example	Student Practice
13. Add. $-1.8 + 1.4 + (-2.6)$	**14.** Add. $5.6 + (-2.34) + (-3.16)$
We take the difference of 1.8 and 1.4 and use the sign of the number with the larger absolute value. Then add the result to -2.6. $-0.4 + (-2.6) = -3.0$	
15. Add. $-8 + 3 + (-5) + (-2) + 6 + 5$	**16.** Add. $-7 + 2 + 5 + (-6) + 1 + (-3)$
Add the three negative numbers and the three positive numbers separately, then add the two results. $$\begin{array}{cc} -8 & +3 \\ -5 & +6 \\ \underline{-2} & \underline{+5} \\ -15 & +14 \end{array}$$ Add the two results, $-15 + 14 = -1$.	

Extra Practice

1. Identify the following as a whole number, rational number, irrational number, integer, or real number. Remember, a number can be identified as more than one type.

2. Find the absolute value of $-\dfrac{7}{8}$.

3. Use a real number to represent the situation.

You owe your best friend $25.

4. Add. $57 + (-32) + 90 + (-100)$

Concept Check

Explain why when you add two negative numbers, you always obtain a negative number, but when you add one negative number and one positive number, you may obtain zero, a positive number, or a negative number.

Chapter 1 Real Numbers and Variables
1.2 Subtracting Real Numbers

Vocabulary
additive inverse property   •   subtract

1. To _____ real numbers, add the opposite of the second number to the first.

2. The _____ says that when you add two real numbers that are opposites of each other, you will obtain zero.

Example	**Student Practice**
1. Subtract. $6-(-2)$ Change the subtraction to addition and write the opposite of the second number. $6-(-2)=6+(+2)$ Then add the two real numbers with the same sign. $6+(+2)=8$	**2.** Subtract. $4-(-8)$
3. Subtract. $-8-(-6)$ Change the subtraction to addition and write the opposite of the second number. $-8-(-6)=-8+(+6)$ Then add the two real numbers with different signs. $-8+(+6)=-2$	**4.** Subtract. $-15-(-9)$

Example	Student Practice
5. Subtract.	**6.** Subtract.

5. Subtract.

(a) $\dfrac{3}{7} - \dfrac{6}{7}$

Note that the problem has two fractions with the same denominator.

$$\dfrac{3}{7} - \dfrac{6}{7} = \dfrac{3}{7} + \left(-\dfrac{6}{7}\right)$$

$$= -\dfrac{3}{7}$$

(b) $-\dfrac{7}{18} - \left(-\dfrac{1}{9}\right)$

Change subtracting to adding the opposite and change $\dfrac{1}{9}$ to $\dfrac{2}{18}$ since the LCD $= 18$.

$$-\dfrac{7}{18} - \left(-\dfrac{1}{9}\right) = -\dfrac{7}{18} + \dfrac{1}{9}$$

$$= -\dfrac{7}{18} + \dfrac{2}{18}$$

$$= -\dfrac{5}{18}$$

6. Subtract.

(a) $\dfrac{2}{9} - \dfrac{7}{9}$

(b) $-\dfrac{2}{5} - \left(-\dfrac{1}{4}\right)$

---

7. Subtract. $-5.2 - (-5.2)$

Change the subtraction problem to one of adding the opposite of the second number. Then, add two numbers with different signs.

$$-5.2 - (-5.2) = -5.2 + 5.2$$

$$= 0$$

8. Subtract. $-4.2 - (-4.2)$

Example	Student Practice
9. Subtract.	**10.** Subtract.
(a) $-8-2$	**(a)** $-12-8$
To subtract, we add the opposite of the second number to the first.	
$-8-2=-8+(-2)=-10$	
(b) $23-28$	**(b)** $15-20$
In a similar fashion, we have the following.	
$23-28=23+(-28)=-5$	
(c) $5-(-3)$	**(c)** $14-(-11)$
$5-(-3)=5+3=8$	
(d) $\dfrac{1}{4}-8$	**(d)** $\dfrac{4}{5}-6$
Write -8 as $-\dfrac{32}{4}$ since the LCD$=4$.	
$\dfrac{1}{4}-8=\dfrac{1}{4}+(-8)$	
$=\dfrac{1}{4}+\left(-\dfrac{32}{4}\right)$	
$=-\dfrac{31}{4}$ or $-7\dfrac{3}{4}$	

Example	Student Practice
11. A satellite is recording radioactive emissions from nuclear waste buried 3 miles below sea level. The satellite orbits Earth at 98 miles above sea level. How far is the satellite from the nuclear waste? We want to find the difference between +98 miles and −3 miles. This means we must subtract −3 from 98. $98 - (-3) = 93 + 3 = 101$ The satellite is 101 miles from the nuclear waste.	**12.** A helicopter flies over a deep sea diver. The helicopter is 500 feet above sea level. The diver is 129 feet below sea level. How far is the helicopter from the diver?

Extra Practice

1. Subtract by adding the opposite.

$(-80.09) - (-76.8)$

2. Subtract by adding the opposite.

$-\dfrac{11}{15} - \left(-\dfrac{1}{2}\right)$

3. Change each subtraction operation to "adding the opposite." Then combine the numbers.

$-30 + 12 - (-15) - 8$

4. In the morning, Maxine began hiking uphill from a valley that is 22 feet below sea level. She ate her lunch on top of a hill whose altitude is 179 feet above sea level. Write an expression to represent the difference in altitude. How many feet did Maxine climb during her hike?

Concept Check

Explain the different results that are possible when you start with a negative number and then subtract a negative number.

Name: _____   Date: _____

Instructor: _____   Section: _____

Chapter 1 Real Numbers and Variables
1.3 Multiplying and Dividing Real Numbers

Vocabulary

multiplication • division • positive • negative • undefined

1. Division by zero is _____.

2. When you multiply or divide two numbers with different signs, you obtain a _____ number.

3. _____ can be indicated by the symbol ÷ or by the fraction bar —.

4. _____ is commutative and associative.

Example	**Student Practice**
1. Multiply.	**2.** Multiply.
(a) $\left(-\dfrac{5}{7}\right)\left(-\dfrac{2}{9}\right)$	**(a)** $(5)(4)$
When multiplying two numbers with the same sign, the result is a positive number.	
$\left(-\dfrac{5}{7}\right)\left(-\dfrac{2}{9}\right)=\dfrac{10}{63}$	
(b) $-4(8)$	**(b)** $\left(-\dfrac{5}{12}\right)(5)$
When multiplying two numbers with different signs, the result is a negative number.	
$-4(8)=-32$	

Example	Student Practice
3. Multiply.	**4.** Multiply.

3. Multiply.

(a) $\left(-\dfrac{1}{2}\right)(-1)(-4)$

Begin by multiplying the first two numbers. The signs are the same, so the answer is positive. Then multiply the remaining two numbers. The signs are different, so the answer is negative.

$$\left(-\dfrac{1}{2}\right)(-1)(-4) = +\dfrac{1}{2}(-4) = -2$$

(b) $-2(-2)(-2)(-2)$

$$-2(-2)(-2)(-2) = +4(-2)(-2)$$
$$= -8(-2)$$
$$= +16 \text{ or } 16$$

4. Multiply.

(a) $-6(-5.4)$

(b) $-3(-2)(-5)(-1)(-2)$

---

5. Divide.

(a) $12 \div 4$

When dividing two numbers with the same sign, the result is a positive number.

$$12 \div 4 = 3$$

(b) $\dfrac{-36}{18}$

When dividing two numbers with different signs, the result is a negative number.

$$-\dfrac{36}{18} = -2$$

6. Divide.

(a) $(-32) \div (-4)$

(b) $\dfrac{45}{-9}$

Example	Student Practice
7. Divide.	**8.** Divide.

7. Divide.

(a) $-36 \div 0.12$

When dividing two numbers with different signs, the result is a negative number. We then divide the absolute values.

$$0.12_\wedge \overline{)36.00_\wedge} \quad \begin{array}{r} 3\ 00. \\ \end{array}$$
$$\underline{36}$$
$$00$$

Thus $-36 \div 0.12 = -300$.

(b) $-2.4 \div (-0.6)$

$$0.6_\wedge \overline{)2.4_\wedge} \quad \begin{array}{r} 4. \\ \end{array}$$
$$\underline{2\ 4}$$

Thus $-2.4 \div (-0.6) = 4$.

8. Divide.

(a) $-121 \div (-0.11)$

(b) $-0.54 \div 0.9$

---

9. Divide. $-\dfrac{12}{5} \div \dfrac{2}{3}$

We invert the second fraction and multiply by the first fraction. The answer is negative since the two numbers divided have different signs.

$$-\frac{12}{5} \div \frac{2}{3} = \left(-\frac{12}{5}\right)\left(\frac{3}{2}\right)$$

$$= \left(-\frac{\overset{6}{\cancel{12}}}{5}\right)\left(\frac{3}{\underset{1}{\cancel{2}}}\right)$$

$$= -\frac{18}{5} \text{ or } -3\frac{3}{5}$$

10. Divide. $-\dfrac{2}{9} \div \left(-\dfrac{8}{15}\right)$

Example	Student Practice
11. Divide. $\dfrac{-\dfrac{2}{3}}{-\dfrac{7}{13}}$	**12.** Divide. $\dfrac{\dfrac{32}{4}}{-\dfrac{4}{5}}$

$$\dfrac{-\dfrac{2}{3}}{-\dfrac{7}{13}} = -\dfrac{2}{3} \div \left(-\dfrac{7}{13}\right)$$

$$= -\dfrac{2}{3}\left(-\dfrac{13}{7}\right)$$

$$= \dfrac{26}{21} \text{ or } 1\dfrac{5}{21}$$

Extra Practice

1. Multiply. Be sure to write your answer in the simplest form.

$$(15.8)(-29.3)$$

2. Multiply. You may want to determine the sign of the product before you multiply.

$$(2.5)(-0.4)(-3.2)(5)$$

3. Divide.

$$87.5 \div (-0.5)$$

4. Divide.

$$-\dfrac{11}{16} \div \dfrac{33}{40}$$

Concept Check

Explain how you can determine the sign of the answer if you multiply several negative numbers.

Chapter 1 Real Numbers and Variables
1.4 Exponents

Vocabulary
base • exponent • variable • squared • cubed • to the (exponent)-th power

1. The _____ tells you how many times the base is used as a factor.

2. If the base has an exponent of 3, we say the base is _____.

3. The _____ tells you what number is being multiplied in exponent form.

4. If we do not know the value of a number, we use a letter, called a(n) _____, to represent the unknown number.

Example	**Student Practice**
1. Write in exponent form.	**2.** Write in exponent form.
(a) $9(9)(9)$	**(a)** $-3(-3)(-3)(-3)$
The base is 9 and it is used as a factor three times. So, the exponent is 3.	
$9(9)(9) = 9^3$	
(b) $-7(-7)(-7)(-7)(-7)$	**(b)** $-6(-6)(-6)(-6)(-6)(-6)(-6)(-6)$
The base is -7 and it is used as a factor five times. The answer must contain parentheses.	
$-7(-7)(-7)(-7)(-7) = (-7)^5$	**(c)** $(n)(n)(n)(n)(n)(n)$
(c) $(y)(y)(y)$	
$(y)(y)(y) = y^3$	

Example	Student Practice
3. Evaluate.	**4.** Evaluate.
(a) 2^5	**(a)** 4^3
$2^5 = (2)(2)(2)(2)(2) = 32$	
(b) $2^3 + 4^4$	
First evaluate each power. Then add.	**(b)** $1^7 + 6^2$
$2^3 + 4^4 = 8 + 256$ $ = 264$	
5. Evaluate.	**6.** Evaluate.
(a) $(-2)^3$	**(a)** $(-4)^3$
The answer is negative since the base is negative and the exponent is odd, $(-2)^3 = -8$.	
(b) $(-4)^6$	**(b)** $(-4)^4$
The answer is positive since the exponent 6 is even, $(-4)^6 = +4096$.	
(c) -3^6	**(c)** -4^4
The negative sign is not contained within parentheses. Thus, find 3^6 and take the opposite of that value, $-3^6 = -729$.	
(d) $-\left(5^4\right)$	**(d)** $-\left(4^4\right)$
The negative sign is outside the parentheses, $-\left(5^4\right) = -625$.	

Example	Student Practice
7. Evaluate.	**8.** Evaluate.

7. Evaluate.

(a) $\left(\dfrac{1}{2}\right)^4$

$$\left(\dfrac{1}{2}\right)^4 = \left(\dfrac{1}{2}\right)\left(\dfrac{1}{2}\right)\left(\dfrac{1}{2}\right)\left(\dfrac{1}{2}\right)$$
$$= \dfrac{1}{16}$$

(b) $\left(0.2\right)^4$

$$\left(0.2\right)^4 = (0.2)(0.2)(0.2)(0.2)$$
$$= 0.0016$$

(c) $\left(\dfrac{2}{5}\right)^3$

$$\left(\dfrac{2}{5}\right)^3 = \left(\dfrac{2}{5}\right)\left(\dfrac{2}{5}\right)\left(\dfrac{2}{5}\right)$$
$$= \dfrac{8}{125}$$

(d) $\left(3\right)^3 \left(2\right)^5$

First we evaluate each power, then we multiply.

$$\left(3\right)^3 \left(2\right)^5 = (27)(32)$$
$$= 864$$

(e) $2^3 - 3^4$

$$2^3 - 3^4 = 8 - 81$$
$$= -73$$

8. Evaluate.

(a) $\left(\dfrac{1}{5}\right)^3$

(b) $\left(0.3\right)^3$

(c) $\left(\dfrac{4}{7}\right)^3$

(d) $\left(2\right)^4 \left(5\right)^2$

(e) $10^3 - 2^{10}$

Extra Practice

1. Write the product in exponent form. Do not evaluate.

$$(-ab)(-ab)$$

2. Evaluate.

$$\left(-\frac{1}{2}\right)^3$$

3. Evaluate.

$$-6^2 - (-2)^2$$

4. Evaluate.

$$7^2 - (-2)^3$$

Concept Check

Explain the difference between $(-2)^6$ and -2^6. How do you decide if the answers are positive or negative?

Chapter 1 Real Numbers and Variables
1.5 The Order of Operations

Vocabulary
order of operations   •   parentheses   •   multiply and divide   •   add and subtract

1. When simplifying an expression, the last priority is to _____ numbers from left to right.

2. When simplifying an expression, the first priority is to do all operations inside _____ .

3. The list of operations to do first is called the _____ .

4. When simplifying an expression, the third priority is to _____ numbers from left to right.

Example	**Student Practice**
1. Evaluate. $8 \div 2 \cdot 3 + 4^2$ Evaluate $4^2 = 16$ because the highest priority in this problem is raising to a power. $8 \div 2 \cdot 3 + 4^2 = 8 \div 2 \cdot 3 + 16$ Next, multiply and divide from left to right. $8 \div 2 \cdot 3 + 16 = 4 \cdot 3 + 16$ $\qquad\qquad\qquad = 12 + 16$ Finally, add. $12 + 16 = 28$	**2.** Evaluate. $36 \div 3 \cdot 2 + 6^2$

Example	Student Practice
3. Evaluate. $(-3)^3 - 2^4$	**4.** Evaluate. $(-5)^2 - 1^{10}$

3. Evaluate. $(-3)^3 - 2^4$

The highest priority is to raise the expressions to the appropriate powers.

In $(-3)^3$ we are cubing the number -3 to obtain -27.

$$(-3)^3 - 2^4 = -27 - 2^4$$

Be careful; -2^4 is not $(-2)^4$. We raise 2 to the fourth power. Then, add and subtract from left to right.

$$(-3)^3 - 2^4 = -27 - 16$$
$$= -43$$

4. Evaluate. $(-5)^2 - 1^{10}$

---

5. Evaluate. $2 \cdot (2-3)^3 + 6 \div 3 + (8-5)^2$

Combine the numbers inside the parentheses.

$$2 \cdot (2-3)^3 + 6 \div 3 + (8-5)^2$$
$$= 2 \cdot (-1)^3 + 6 \div 3 + 3^2$$

We need parentheses for -1 because of the negative sign, but they are not needed for 3. Next, raise to a power.

$$2 \cdot (-1)^3 + 6 \div 3 + 3^2 = 2 \cdot (-1) + 6 \div 3 + 9$$

Next, multiply and divide from left to right, $2 \cdot (-1) + 6 \div 3 + 9 = -2 + 2 + 9$.

Finally, add and subtract from left to right, $-2 + 2 + 9 = 9$.

6. Evaluate. $3 \cdot (3-5)^2 + 15 \div 5 + (7-5)^3$

Example	Student Practice
7. Evaluate. $\left(-\dfrac{1}{5}\right)\left(\dfrac{1}{2}\right)-\left(\dfrac{3}{2}\right)^{2}$	**8.** Evaluate. $\left(\dfrac{3}{4}\right)^{2}+\dfrac{14}{15}\left(-\dfrac{5}{7}\right)$

The highest priority is to raise $\dfrac{3}{2}$ to the second power, $\left(\dfrac{3}{2}\right)^{2}=\left(\dfrac{3}{2}\right)\left(\dfrac{3}{2}\right)=\dfrac{9}{4}.$

$$\left(-\dfrac{1}{5}\right)\left(\dfrac{1}{2}\right)-\left(\dfrac{3}{2}\right)^{2}=\left(-\dfrac{1}{5}\right)\left(\dfrac{1}{2}\right)-\dfrac{9}{4}$$

Next we multiply.

$$\left(-\dfrac{1}{5}\right)\left(\dfrac{1}{2}\right)-\dfrac{9}{4}=-\dfrac{1}{10}-\dfrac{9}{4}$$

We need to write each fraction as an equivalent fraction with the LCD of 20.

$$\left(-\dfrac{1}{5}\right)\left(\dfrac{1}{2}\right)-\dfrac{9}{4}=-\dfrac{1\cdot 2}{10\cdot 2}-\dfrac{9\cdot 5}{4\cdot 5}$$
$$=-\dfrac{2}{20}-\dfrac{45}{20}$$

Finally, subtract.

$$-\dfrac{2}{20}-\dfrac{45}{20}=-\dfrac{47}{20} \text{ or } -2\dfrac{7}{20}$$

Extra Practice

1. Evaluate. Simplify all fractions.

$$1 + 2 - 3 \times 4 \div 6$$

2. Evaluate. Simplify all fractions.

$$50 \div 5 \times 2 + (11 - 8)^2$$

3. Evaluate. Simplify all fractions.

$$\left(\frac{1}{2}\right)^2 \div \left(\frac{1}{2}\right)^3$$

4. Evaluate. Simplify all fractions.

$$\left(\frac{2}{3}\right)^2 (-18) + \frac{3}{7} \div \frac{5}{21}$$

Concept Check

Explain in what order you would perform the calculations to evaluate the expression $4 - (-3 + 4)^3 + 12 \div (-3)$.

Chapter 1 Real Numbers and Variables
1.6 Using the Distributive Property to Simplify Algebraic Expressions

Vocabulary
algebraic expression   •   term   •   distributive property   •   factors

1. The _____ states that for all real numbers a, b, and c, $a(b+c) = ab + ac$.

2. A(n) _____ is a quantity that contains numbers and variables.

3. Two or more algebraic expressions that are multiplied are called _____.

4. A(n) _____ is a number, variable, or a product of numbers and variables.

Example	Student Practice
1. Multiply.	**2.** Multiply.
(a) $5(a+b)$	**(a)** $6(x+3y)$
Multiply the factor $(a+b)$ by the factor 5 using the distributive property.	
$5(a+b) = 5a + 5b$	
(b) $-3(3x+2y)$	**(b)** $-4(2m+5n)$
$-3(3x+2y) = -3(3x) + (-3)(2y)$ $= -9x - 6y$	
3. Multiply. $-(a-2b)$	**4.** Multiply. $-(m-6n)$
$-(a-2b) = (-1)(a-2b)$ $= (-1)(a) + (-1)(-2b)$ $= -a + 2b$	

Example	Student Practice
5. Multiply.	**6.** Multiply.

5. Multiply.

(a) $\frac{2}{3}\left(x^2 - 6x + 8\right)$

Apply the distributive property.

$\frac{2}{3}\left(x^2 - 6x + 8\right)$

$= \left(\frac{2}{3}\right)\left(1x^2\right) + \left(\frac{2}{3}\right)\left(-6x\right) + \left(\frac{2}{3}\right)\left(8\right)$

$= \frac{2}{3}x^2 + \left(-4x\right) + \frac{16}{3}$

$= \frac{2}{3}x^2 - 4x + \frac{16}{3}$

(b) $1.4\left(a^2 + 2.5a + 1.8\right)$

$1.4\left(a^2 + 2.5a + 1.8\right)$

$= 1.4\left(1a^2\right) + \left(1.4\right)\left(2.5a\right) + \left(1.4\right)\left(1.8\right)$

$= 1.4a^2 + 3.5a + 2.52$

6. Multiply.

(a) $\frac{3}{4}\left(m^2 - 8m + 5\right)$

(b) $1.3\left(x^2 + 1.3x + 0.7\right)$

---

7. Multiply. $-2x\left(3x + y - 4\right)$

$-2x\left(3x + y - 4\right)$

$= -2\left(x\right)\left(3\right)\left(x\right) - 2\left(x\right)\left(y\right) - 2\left(x\right)\left(-4\right)$

$= -2\left(3\right)\left(x\right)\left(x\right) - 2\left(xy\right) - 2\left(-4\right)\left(x\right)$

$= -6x^2 - 2xy + 8x$

8. Multiply. $-5x\left(x + 3y - 6\right)$

---

9. Multiply. $\left(2x^2 - x\right)\left(-3\right)$

$\left(2x^2 - x\right)\left(-3\right) = 2x^2\left(-3\right) + \left(-x\right)\left(-3\right)$

$= -6x^2 + 3x$

10. Multiply. $\left(4y - 3y^2\right)\left(-2\right)$

Example	Student Practice
11. A farmer has a rectangular field that is 300 feet wide. One portion of the field is $2x$ feet long. The other portion of the field is $3y$ long. Use the distributive property to find an expression for the area of this field.	**12.** A convenience store is 500 feet wide. One portion of the store is $7x$ feet long. The other portion of the store is $8y$ feet long. Use the distributive property to find an expression for the area of this store.

First we draw a picture of a field that is 300 feet wide and $2x+3y$ feet long.

To find the area of the field, we multiply the width times the length.

$$300(2x+3y)$$

Then, apply the distributive property.

$$300(2x+3y)=300(2x)+300(3y)$$

Finally, simplify.

$$300(2x)+300(3y)=600x+900y$$

Thus the area of the field in square feet is $600x+900y$.

Extra Practice

1. Multiply. Use the distributive property.

$$-2(6x+3)$$

2. Multiply. Use the distributive property.

$$\frac{1}{2}\left(\frac{1}{3}x+\frac{2}{5}y\right)$$

3. Multiply. Use the distributive property.

$$(6x-4y+2z)(-3)$$

4. Mr. Jorgensen's backyard is a rectangle that is 30 feet wide. He recently built a fence that extends across his backyard. The distance from his house to the fence is 15 feet. The distance from the fence to the back edge of his property is $7n$. Write an expression that represents the area of Mr. Jorgensen's property in square feet, then simplify this expression.

Concept Check

Multiply. Use the distributive property and explain how you would multiply to obtain the answer for $\left(-\frac{3}{7}\right)(21x^2-14x+3)$.

Chapter 1 Real Numbers and Variables
1.7 Combining Like Terms

Vocabulary
combine like terms   •   like terms   •   combining   •   term

1. A _____ is a number, a variable, or a product of numbers and variables separated by plus or minus signs in an algebraic expression.

2. We can add or subtract quantities that are like quantities, which is called _____ like quantities.

3. To _____ is to add or subtract like terms in an algebraic expression.

4. _____ are terms that have identical variables and exponents.

Example	Student Practice
1. List the like terms of each expression.	**2.** List the like terms of each expression.
(a) $5x - 2y + 6x$	**(a)** $4x - 6y + 2x + 7$
$5x$ and $6x$ are like terms.	
(b) $2x^2 - 3x - 5x^2 - 8x$	**(b)** $4x^2 + 5x - 6x^2 + 9x$
$2x^2$ and $-5x^2$ are like terms. $-3x$ and $-8x$ are like terms. Note that x^2 and x are not like terms.	
3. Combine like terms.	**4.** Combine like terms.
(a) $-4x^2 + 8x^2$	**(a)** $10z^5 - 3z^5$
Each term contains the factor x^2.	
$-4x^2 + 8x^2 = (-4 + 8)x^2 = 4x^2$	
	(b) $5z - 7z + 4z$
(b) $5x + 3x + 2x$	
$5x + 3x + 2x = (5 + 3 + 2)x = 10x$	

Example	Student Practice
5. Simplify. $5a^2 - 2a^2 + 6a^2$ $5a^2 - 2a^2 + 6a^2 = (5 - 2 + 6)a^2$ $\qquad\qquad\qquad = 9a^2$	**6.** Simplify. $x^3 - 7x^3 + 9x^3$
7. Simplify. **(a)** $5.6a + 2b + 7.3a - 6b$ We combine the a terms and the b terms separately. $5.6a + 2b + 7.3a - 6b = 12.9a - 4b$ **(b)** $3x^2y - 2xy^2 + 6x^2y$ Note that x^2y and xy^2 are not like terms because of different powers. $3x^2y - 2xy^2 + 6x^2y = 9x^2y - 2xy^2$ **(c)** $2a^2b + 3ab^2 - 6a^2b^2 - 8ab$ These terms cannot be combined; there are no like terms in this expression. $2a^2b + 3ab^2 - 6a^2b^2 - 8ab$	**8.** Simplify. **(a)** $4.1m + 6n + 3.5m + 1.2n$ **(b)** $6a^2b - 4ab^3 + 2a^2b$ **(c)** $6xy + 2x^2y - 9xy^2 + 5x^2y^2$
9. Simplify. $3a - 2b + 5a^2 + 6a - 8b - 12a^2$ There are three pairs of like terms. You can rearrange the terms so that like terms are together. $\underbrace{3a + 6a}_{a\ \text{terms}} - \underbrace{2b - 8b}_{b\ \text{terms}} + \underbrace{5a^2 - 12a^2}_{a^2\ \text{terms}}$ $= 9a - 10b - 7a^2$	**10.** Simplify. $9x - 9c + 2t + 4c - 4t + 5x$

Example	Student Practice
11. Simplify. $\dfrac{3}{4}x^2 - 5y - \dfrac{1}{8}x^2 + \dfrac{1}{3}y$	**12.** Simplify. $\dfrac{2}{5}x - \dfrac{6}{7}y + 3x + \dfrac{1}{2}y$

We need the least common denominator for the x^2 terms, which is 8.

$$\dfrac{3}{4}x^2 - \dfrac{1}{8}x^2 = \dfrac{3 \cdot 2}{4 \cdot 2}x^2 - \dfrac{1}{8}x^2$$

$$= \dfrac{6}{8}x^2 - \dfrac{1}{8}x^2$$

$$= \dfrac{5}{8}x^2$$

The least common denominator for the y terms is 3.

$$-\dfrac{5}{1}y + \dfrac{1}{3}y = \dfrac{-5 \cdot 3}{1 \cdot 3}y + \dfrac{1}{3}y$$

$$= \dfrac{-15}{3}y + \dfrac{1}{3}y$$

$$= -\dfrac{14}{3}y$$

Thus, our solution is $\dfrac{5}{8}x^2 - \dfrac{14}{3}y$.

13. Simplify. $6(2x + 3xy) - 8x(3 - 4y)$	**14.** Simplify. $5(2c + 7cd) - 3c(9 - 4d)$

First remove the parentheses using the distributive property. Then combine like terms.

$$6(2x + 3xy) - 8x(3 - 4y)$$
$$= 12x + 18xy - 24x + 32xy$$
$$= -12x + 50xy$$

Extra Practice

1. Combine like terms.

$$9a + 10 - 2a^2 - 11a + 2 + 6a^2$$

2. Combine like terms.

$$\frac{3}{7}x^2 - \frac{1}{2}y + \frac{3}{14}x^2 - \frac{1}{8}y$$

3. Simplify. Use the distributive property to remove parentheses; then combine like terms.

$$-2(7a - 3b) + 4(-2a + 4b)$$

4. Simplify. Use the distributive property to remove parentheses; then combine like terms.

$$2a(b - 4c) - 3c(-7a + b - 10d)$$

Concept Check

Explain how you would remove parentheses and then combine like terms to obtain the answer for $1.2(3.5x - 2.2y) - 4.5(2.0x + 1.5y)$.

Chapter 1 Real Numbers and Variables
1.8 Using Substitution to Evaluate Algebraic Expressions and Formulas

Vocabulary
evaluate • substituted • perimeter • area • right angles • altitude • rectangle
parallelogram • square • trapezoid • triangle • circle • circumference

1. A(n) _____ is a rectangle with all four sides equal.

2. _____ is the distance around a circle.

3. _____ is a measure of the amount of surface in a region.

4. You will use the order of operations to _____ variable expressions.

Example	**Student Practice**
1. Evaluate $\frac{2}{3}x - 5$ for $x = -6$. Substitute -6 for x. $\frac{2}{3}x - 5 = \frac{2}{3}(-6) - 5$ $\quad\quad = -4 - 5$ $\quad\quad = -9$	**2.** Evaluate $\frac{3}{4}x - 7$ for $x = -24$.
3. Evaluate for $x = -3$. **(a)** $2x^2$ Here the value x is squared. $2x^2 = 2(-3)^2$ $\quad\quad = 2(9)$ $\quad\quad = 18$ **(b)** $(2x)^2$ Here the value $(2x)$ is squared. $(2x)^2 = [2(-3)]^2$ $\quad\quad = (-6)^2$ $\quad\quad = 36$	**4.** Evaluate for $x = -5$. **(a)** $3y^2$ **(b)** $(3y)^2$

Example	Student Practice
5. Evaluate $x^2 + 3x$ for $x = -4$. Replace each x by -4. $$x^2 + 3x = (-4)^2 + 3(-4)$$ $$= 16 + 3(-4)$$ $$= 16 - 12$$ $$= 4$$	**6.** Evaluate $5x^2 + 4x$ for $x = -3$.
7. Evaluate $x^3 + 2xy - 3x + 1$ for $x = 2$ and $y = -\dfrac{1}{4}$. Replace x with 2 and y with $-\dfrac{1}{4}$. $x^3 + 2xy - 3x + 1$ $$= (2)^3 + 2(2)\left(-\dfrac{1}{4}\right) - 3(2) + 1$$ $$= 8 + (-1) - 6 + 1$$ $$= 8 + (-1) + (-6) + 1$$ $$= 9 + (-7)$$ $$= 2$$	**8.** Evaluate $7x^2 - 10xy - 50y^2$ for $x = -4$ and $y = \dfrac{1}{5}$.
9. Find the area of a triangle with a base of 16 centimeters (cm) and a height of 12 centimeters (cm). Substitute $a = 12$ cm and $b = 16$ cm in $A = \dfrac{1}{2}ab$. $$A = \dfrac{1}{2}(12 \text{ cm})(16 \text{ cm})$$ $$= (6)(16)(\text{cm})^2$$ $$= 96 \text{ square centimeters}$$ The area of the triangle is 96 square centimeters or 96 cm^2.	**10.** Find the area of a triangle with a height of 9 yards and a base of 3 yards.

Example	Student Practice
11. Find the area of a circle if the radius is 2 inches. Write the formula and substitute the given values for the letters. $$A = \pi r^2 \approx (3.14)(2 \text{ inches})^2$$ Raise to a power. Then multiply. $$(3.14)(2 \text{ inches})^2 = (3.14)(4)(\text{in.})^2$$ $$= 12.56 \text{ in.}^2$$ Thus the area is 12.56 square inches or 12.56 in.^{2}	**12.** Find the area of a circle if the radius is 4 yards.
13. What is the Celsius temperature when the Fahrenheit temperature is $F = -22°$? Use the formula. Substitute -22 for F in the formula. $$C = \frac{5}{9}(F - 32)$$ $$= \frac{5}{9}\left[(-22) - 32\right]$$ Combine the numbers inside the brackets. Then simplify and multiply. $$\frac{5}{9}\left[(-22) - 32\right] = \frac{5}{9}(-54)$$ $$= (5)(-6)$$ $$= -30$$ The temperature is $-30°$ Celsius or $-30° \text{C}$.	**14.** What is the Celsius temperature when the Fahrenheit temperature is $F = -40°$? Use the formula $C = \frac{5}{9}(F - 32)$.

Example	Student Practice
15. You are driving on a highway in Mexico. It has a posted maximum speed of 100 kilometers per hour. You are driving at 61 miles per hour. Are you exceeding the speed limit?	**16.** You are driving on a Canadian highway. The posted maximum speed limit is 110 kilometers per hour. You are driving at 70 miles per hour. Are you exceeding the speed limit?

Use the formula. Replace r by 61. Then multiply the numbers.

$$k \approx 1.61r$$
$$= (1.61)(61)$$
$$= 98.21$$

You are driving approximately 98 kilometers per hour. You are not exceeding the speed limit.

Extra Practice

1. Evaluate $\frac{2}{7}y - 7$ for $y = 14$.

2. Evaluate $2x^2 + 3x - 8$ for $x = 3$.

3. Evaluate $\frac{2a^2 - 3b}{b^2}$ for $a = -1$ and $b = 2$.

4. A circular patch of grass has a radius of 6 meters. What is the area of the patch of grass, rounded to the nearest square meter?

Concept Check
Explain how you would find the area of a circle if you know its diameter is 12 meters.

Chapter 1 Real Numbers and Variables
1.9 Grouping Symbols

Vocabulary
grouping symbols   •   fraction bars   •   distributive property   •   like terms

1. _____ are also considered grouping symbols.

2. Use the _____ and the rules for real numbers to remove grouping symbols

3. Many expressions in algebra use _____ such as parentheses, brackets, and braces.

Example	Student Practice
1. Simplify. $3\left[6-2(x+y)\right]$	**2.** Simplify. $4\left[6x-5(x+3)\right]$
We want to remove the innermost parentheses first. Therefore, we first use the distributive property to simplify.	
$3\left[6-2(x+y)\right]=3\left[6-2x-2y\right]$	
Use the distributive property again.	
$3\left[6-2x-2y\right]=18-6x-6y$	
3. Simplify. $-2\left[3a-(b+2c)+(d-3e)\right]$	**4.** Simplify. $-4\left[5x-(2y-7)+(3x-9)\right]$
Remove both inner sets of parentheses.	
$-2\left[3a-(b+2c)+(d-3e)\right]$ $=-2\left[3a-b-2c+d-3e\right]$	
Now remove the brackets by multiplying each term by -2.	
$-2\left[3a-b-2c+d-3e\right]$ $=-6a+2b+4c-2d+6e$	

Example	Student Practice
5. Simplify. $$2\big[3x-(y+w)\big]-3\big[2x+2(3y-2w)\big]$$ $$=2\big[3x-y-w\big]-3\big[2x+6y-4w\big]$$ $$=6x-2y-2w-6x-18y+12w$$ $$=-20y+10w \text{ or } 10w-20y$$	**6.** Simplify. $$5\big[x-3(4-2x)\big]+(-3)\big[7x-6(x+1)\big]$$

7. Simplify. $-3\big\{7x-2\big[x-(2x-1)\big]\big\}$

Remember to remove the innermost grouping symbols first.

$$-3\big\{7x-2\big[x-(2x-1)\big]\big\}$$
$$=-3\big\{7x-2\big[x-2x+1\big]\big\}$$
$$=-3\big\{7x-2\big[-x+1\big]\big\}$$
$$=-3\big\{7x+2x-2\big\}$$
$$=-3\big\{9x-2\big\}$$
$$=-27x+6$$

8. Simplify. $-4\big\{6x-\big[3x-(2-x)\big]\big\}$

Extra Practice

1. Simplify. Remove grouping symbols and combine like terms. $-2a-3(a+3b)$

2. Simplify. Remove grouping symbols and combine like terms.
$$-2\big[3(x-y)-2(x+y)\big]$$

3. Simplify. Remove grouping symbols and combine like terms.
$$2b-\big\{3a+4\big[2a-(-2b+3a)\big]\big\}$$

4. Simplify. Remove grouping symbols and combine like terms.
$$-2\big\{5x^2+2\big[3x-(5-x)\big]\big\}$$

Concept Check

Explain how you would simplify the following expression to combine like terms whenever possible. $3\big\{2-3\big[4x-2(x+3)+5x\big]\big\}$

MATH COACH

Mastering the skills you need to do well on the test.

Watch the **MATH COACH** videos in MyMathLab°or on You Tube
while you work the problems below. These helpful hints will
help you avoid making common errors on test problems.

Evaluating Exponential Expressions When the Base Is

a Fraction—Problem 9   Simplify $\left(\dfrac{2}{3}\right)^4$.

> **Helpful Hint:** Always write out the repeated multiplication.
> - If a number is raised to the fourth power, then we write out the multiplication with that number appearing as a factor a total of four times.
> - When the base is a fraction, this means that we must multiply the numerators and then multiply the denominators.

Did you rewrite the problem as $\left(\dfrac{2}{3}\right)\left(\dfrac{2}{3}\right)\left(\dfrac{2}{3}\right)\left(\dfrac{2}{3}\right)$?

Yes _____ No _____

If you answered No to this question, please complete this step now.

Did you multiply $2\times2\times2\times2$ in the numerator and $3\times3\times3\times3$ in the denominator? Yes _____ No _____

If you answered No, make this correction and complete the calculations.

If you answered Problem 9 incorrectly, go back and rework the problem using these suggestions.

Using the Order of Operations with Numerical Expressions—Problem 11

Simplify $3(4-6)^3 +12\div(-4)+2$.

> **Helpful Hint:** First, do all operations inside parentheses. Second, raise numbers to a power. Then multiply and divide numbers from left to right. As the last step, add and subtract numbers from left to right.

Did you first combine $4-6$ to obtain -2?
Yes _____ No _____

If you answered No, perform the operation inside the parentheses first.

Next, did you calculate $(-2)^3 = -8$ before multiplying by 3?
Yes _____ No _____

If you answered No, go back and evaluate the exponential expression.

Did you multiply and divide before adding?
Yes _____ No _____

If you answered No, remember that multiplication and division must be performed before addition and subtraction once the operations inside the parentheses are done and exponents are evaluated.

Evaluate Algebraic Expressions for a Specified Value—Problem 19

Evaluate $3x^2 - 7x - 11$ for $x = -3$.

Helpful Hint: When you replace a variable by a particular value, place parentheses around that value. Then use the order of operations to evaluate the expression.

Did you rewrite the expression as $3(-3)^2 - 7(-3) - 11$, using parentheses to complete the substitution?

Yes _____ No _____

If you answered No, please go back and perform that substitution step using parentheses around the specified value.

Did you next raise -3 to the second power to obtain $3(9) - 7(-3) - 11$?

Yes _____ No _____

If you answered No, review the order of operations and complete this step.

Remember that multiplication must be performed before addition and subtraction.

If you answered Problem 19 incorrectly, go back and rework this problem using these suggestions.

Simplifying Algebraic Expressions with Many Grouping Symbols—Problem 26

Simplify $-3\{a + b[3a - b(1 - a)]\}$.

Helpful Hint: Work from the inside out. Remove the innermost symbols, ( ), first. Then remove the next level of innermost symbols, [ ]. Finally remove the outermost symbols, { }. Be careful to avoid sign errors.

Did you first obtain the expression $-3\{a + b[3a - b + ab]\}$ when you removed the innermost parentheses?

Yes _____ No _____

If you answered No, go back to the original problem and use the distributive property to remove the innermost grouping symbol, the parentheses.

Did you next obtain the expression $-3\{a + 3ab - b^2 + ab^2\}$ when you removed the brackets?

Yes _____ No _____

If you answered No, use the distributive property to distribute b on the outside of the bracket, [ ], to each term inside the bracket.

Finally, use the distributive property in the last step to distribute the -3 across all of the terms inside the outermost brackets, { }. Be careful with the $+/-$ signs.

Now go back and rework the problem using these suggestions.

Chapter 2 Equations, Inequalities, and Applications
2.1 The Addition Principle of Equality

Vocabulary

Equation • solution • equivalent equations • solving the equation • identity
satisfies • checking • addition principle • opposite in sign • additive inverse

1. A number that is opposite in sign to another number is called its _____.

2. A(n) _____ is a statement in which the equals sign $(=)$ is used to indicate that
    two expressions are equal.

3. If the same value appears on both sides of the equals sign, the equation is called a(n)
    _____.

4. If the same number is added to both sides of an equation, the _____ states that
    the results on both sides are equal.

Example	Student Practice
1. Solve for x. $x+16=20$	**2.** Solve for x. $x+8=42$

Example

1. Solve for x. $x+16=20$

Use the addition principle to add -16 to both sides and simplify.

$$x+16=20$$
$$x+16+(-16)=20+(-16)$$
$$x+0=4$$
$$x=4$$

Substitute the value found for x into the original equation to verify the solution.

$$x+16=20$$
$$4+16\overset{?}{=}20$$
$$20=20$$

The solution checks.

Example	Student Practice
3. Solve for x. $1.5 + 0.2 = 0.3 + x + 0.2$	**4.** Solve for x. $2.7 - 0.2 = 0.4 + x - 0.6$

3. Solve for x. $1.5 + 0.2 = 0.3 + x + 0.2$

Simplify by adding.

$$1.5 + 0.2 = 0.3 + x + 0.2$$
$$1.7 = x + 0.5$$

Add -0.5 to both sides.

$$1.7 + (-0.5) = x + 0.5 + (-0.5)$$

Simplify.

$$1.2 = x$$

The check is left to the student.

---

5. Is 10 the solution to the equation $-15 + 2 = x - 3$? If not, find the solution.

Substitute 10 for x in the equation.

$$-15 + 2 = x - 3$$
$$-15 + 2 \overset{?}{=} 10 - 3$$
$$-13 \neq 7$$

The values are not equal. Thus, 10 is not the solution. Solve the original equation to find the solution. Start by simplifying.

$$-15 + 2 = x - 3$$
$$-13 = x - 3$$

Add 3 to both sides.

$$-13 = x - 3$$
$$-13 + 3 = x - 3 + 3$$
$$-10 = x$$

The check is left to the student.

6. Is 26 the solution to the equation $x - 12 = 28 - 4$? If not, find the solution.

Example	Student Practice
7. Find the value of x that satisfies the equation. $$\frac{1}{5} + x = -\frac{1}{10} + \frac{1}{2}$$	**8.** Find the value of x that satisfies the equation. $$\frac{1}{3} + x = -\frac{1}{21} + \frac{1}{7}$$

To be combined, the fractions must have common denominators. Rewrite each fraction as an equivalent fraction with a denominator of 10 and simplify.

$$\frac{1}{5} \cdot \frac{2}{2} + x = -\frac{1}{10} + \frac{1}{2} \cdot \frac{5}{5}$$

$$\frac{2}{10} + x = -\frac{1}{10} + \frac{5}{10}$$

$$\frac{2}{10} + x = \frac{4}{10}$$

Add $-\dfrac{2}{10}$ to each side.

$$\frac{2}{10} + \left(-\frac{2}{10}\right) + x = \frac{4}{10} + \left(-\frac{2}{10}\right)$$

$$x = \frac{2}{10} = \frac{1}{5}$$

Check.

$$\frac{1}{5} + x = -\frac{1}{10} + \frac{1}{2}$$

$$\frac{1}{5} + \left(\frac{1}{5}\right) \overset{?}{=} -\frac{1}{10} + \frac{1}{2}$$

$$\frac{2}{5} \overset{?}{=} -\frac{1}{10} + \frac{5}{10}$$

$$\frac{2}{5} \overset{?}{=} \frac{4}{10}$$

$$\frac{2}{5} = \frac{2}{5}$$

The solution checks.

Extra Practice

1. Solve for x. Check your answers.

$$14 = 8 + x$$

2. Solve for x. Check your answers.

$$1.7 + 4.2 + x = 19.23 - 9.8$$

3. Is 6 the solution to the equation $x + 7 = 14 - 19 + 6$? If it is not, find the solution.

4. Is 28 the solution to the equation $-17 + x - 4 = 12 - 8 + 3$? If it is not, find the solution.

Concept Check

Explain how you would check to verify whether $x = 3.8$ is the solution to $-1.3 + 1.6 + 3x = -6.7 + 4x + 3.2$.

Chapter 2 Equations, Inequalities, and Applications
2.2 The Multiplication Principle of Equality

Vocabulary
terminating decimal • multiplicative inverse • multiplication principle • division principle

1. If both sides of an equation are multiplied by the same nonzero number, the
   _____ states that the results on both sides are equal.

2. The _____ states that dividing both sides of an equation by the same nonzero
   number results in an equivalent equation.

3. For any nonzero number a, the _____ is $\dfrac{1}{a}$.

4. A _____ is a decimal with a definite number of digits.

Example	Student Practice
1. Solve for x. $\dfrac{1}{3}x = -15$	**2.** Solve for x. $-\dfrac{1}{7}x = -3$

We know that $(3)\left(\dfrac{1}{3}\right) = 1$. Multiply each side of the equation by 3 to isolate x.

$$3\left(\dfrac{1}{3}x\right) = 3(-15)$$

$$\left(\dfrac{3}{1}\right)\left(\dfrac{1}{3}\right)x = -45$$

$$x = -45$$

Check.

$$\dfrac{1}{3}(-45) \overset{?}{=} -15$$

$$-15 = -15$$

The solution checks.

Example	Student Practice
3. Solve for x. $5x = 125$ Divide both sides by 5. $$\frac{5x}{5} = \frac{125}{5}$$ $$x = 25$$ Check. $$5x = 125$$ $$5(25) \overset{?}{=} 125$$ $$125 = 125$$ The solution checks.	**4.** Solve for x. $6x = 42$
5. Solve for x. $9x = 60$ Divide both sides by 9 and simplify. $$9x = 60$$ $$\frac{9x}{9} = \frac{60}{9}$$ $$x = \frac{20}{3}$$ The check is left to the student.	**6.** Solve for x. $16x = 30$
7. Solve for x. $-3x = 48$ Divide both sides by -3. $$\frac{-3x}{-3} = \frac{48}{-3}$$ $$x = -16$$ The check is left to the student.	**8.** Solve for x. $-9x = 63$

Example	Student Practice
9. Solve for x. $-x = -24$	**10.** Solve for x. $-x = 15$

9. Solve for x. $-x = -24$

Rewrite the equation. Note that $-1x$ is the same as $-x$.

$-1x = -24$

Divide both sides by the coefficient -1.

$$\frac{-1x}{-1} = \frac{-24}{-1}$$
$$x = 24$$

The check is left to the student.

---

11. Solve for x. $-78 = 5x - 8x$

12. Solve for x. $24 = 4x - 8x$

11. Solve for x. $-78 = 5x - 8x$

Combine the like terms on the right side.

$-78 = 5x - 8x$
$-78 = -3x$

Divide both sides by the coefficient -3.

$$\frac{-78}{-3} = \frac{-3x}{-3}$$
$$26 = x$$

The check is left to the student.

---

13. Solve for x. $31.2 = 6.0x - 0.8x$

14. Solve for x. $15.4 = 4.8x - 2.6x$

13. Solve for x. $31.2 = 6.0x - 0.8x$

Combine the like terms on the right side.

$31.2 = 6.0x - 0.8x$
$31.2 = 5.2x$

Divide both sides by 5.2.
$$\frac{31.2}{5.2} = \frac{5.2x}{5.2}$$
$$6 = x$$

Extra Practice

1. Solve for x. Check your answers.

   $-13 = -x$

2. Solve for x. Check your answers.

   $-5 = \dfrac{x}{5}$

3. Is 7 the solution to the equation $\dfrac{3}{7}x = 3$? If it is not, find the correct solution.

4. Is 0.12 the solution to the equation $3x = -0.36$? If it is not, find the correct solution.

Concept Check

Explain how you would check to verify whether $x = 36\dfrac{2}{3}$ is the solution to $-22 = -\dfrac{3}{5}x$.

Name: _____   Date: _____

Instructor: _____   Section: _____

Chapter 2 Equations, Inequalities, and Applications
2.3 Using the Addition and Multiplication Principles Together

Vocabulary

distributive property   •   like terms   •   multiplication principle   •   addition principle

1. To solve an equation of the form $ax + b = c$, we must use both the addition principle and the _____.

2. If the variable appears on both sides of the equation, apply the _____ to the variable term to collect the variable terms on one side of the equation.

3. If an equation contains parentheses, first apply the _____ to remove the parentheses.

4. Where it is possible, first collect _____ on one or both sides of the equation.

Example	**Student Practice**
1. Solve for x. $5x + 3 = 18$	**2.** Solve for x. $6x - 8 = 4$

1. Solve for x. $5x + 3 = 18$

Use the addition principle to add -3 to both sides of the equation and simplify.

$$5x + 3 = 18$$
$$5x + 3 + (-3) = 18 + (-3)$$
$$5x = 15$$

Use the division principle to divide both sides by 5 and simplify.

$$5x = 15$$
$$\frac{5x}{5} = \frac{15}{5}$$
$$x = 3$$

Check. $5(3) + 3 \overset{?}{=} 18$

$$15 + 3 \overset{?}{=} 18$$
$$18 = 18$$

Example	Student Practice
3. Solve for x and check your solution. $9x + 3 = 7x - 2$	**4.** Solve for x and check your solution. $4x + 6 = x + 4$

Add $-7x$ to both sides of the equation to get all variable terms on one side and combine like terms.

$$9x + 3 = 7x - 2$$
$$9x + (-7x) + 3 = 7x + (-7x) - 2$$
$$2x + 3 = -2$$

Add -3 to both sides and simplify.

$$2x + 3 = -2$$
$$2x + 3 + (-3) = -2 + (-3)$$
$$2x = -5$$

Divide both sides by 2 and simplify.

$$2x = -5$$
$$\frac{2x}{2} = -\frac{5}{2}$$
$$x = -\frac{5}{2}$$

Check.

$$9x + 3 = 7x - 2$$
$$9\left(-\frac{5}{2}\right) + 3 \overset{?}{=} 7\left(-\frac{5}{2}\right) - 2$$
$$-\frac{45}{2} + 3 \overset{?}{=} -\frac{35}{2} - 2$$
$$-\frac{45}{2} + \frac{6}{2} \overset{?}{=} -\frac{35}{2} - \frac{4}{2}$$
$$-\frac{39}{2} = -\frac{39}{2}$$

The solution checks.

Example	Student Practice
5. Solve for x and check your solution. $5x + 26 - 6 = 9x + 12x$ Combine like terms. Then add $(-5x)$ to both sides. $$5x + 26 - 6 = 9x + 12x$$ $$5x + 20 = 21x$$ $$5x + (-5x) + 20 = 21x + (-5x)$$ $$20 = 16x$$ Divide both sides by 16. $$\frac{20}{16} = \frac{16x}{16}$$ $$\frac{5}{4} = x$$ The check is left for the student.	**6.** Solve for x and check your solution. $3x + 22 - 5 = 8x - x$
7. Solve for x and check your solution. $4(x+1) - 3(x-3) = 25$ Multiply by 4 and -3 to remove the parentheses. Then combine like terms. Be careful of the signs. $$4(x+1) - 3(x-3) = 25$$ $$4x + 4 - 3x + 9 = 25$$ $$x + 13 = 25$$ Subtract 13 from both sides to isolate the variable. $$x + 13 = 25$$ $$x + 13 - 13 = 25 - 13$$ $$x = 12$$ The check is left to the student.	**8.** Solve for x and check your solution. $-6(x-3) = 4(2x+5)$

Example	Student Practice
9. Solve for x and check your solution. $0.3(1.2x-3.6)=4.2x-16.44$	**10.** Solve for z and check your solution. $0.6(z-2)=0.4(z+7)$

Remove the parentheses.

$$0.3(1.2x-3.6)=4.2x-16.44$$
$$0.36x-1.08=4.2x-16.44$$

Solve.

$$0.36x-1.08=4.2x-16.44$$
$$0.36x+(-0.36x)-1.08=4.2x+(-0.36x)-16.44$$
$$-1.08+16.44=3.84x-16.44+16.44$$
$$15.36=3.84x$$
$$\frac{15.36}{3.84}=\frac{3.84x}{3.84}$$
$$4=x$$

The check is left to the student.

Extra Practice

1. Solve for x and check your solution.
$3x+15=30$

2. Solve for x and check your solution.
$2x-3=x+9$

3. Solve for x and check your solution.
$18-3=16x-4x+12$

4. Solve for z and check your solution.
$3(2z-3)-3=3z-2(2z-1)$

Concept Check

Explain how you would solve the equation $3(x-2)+2=2(x-4)$.

Chapter 2 Equations, Inequalities, and Applications
2.4 Solving Equations with Fractions

Vocabulary

LCD • multiplication principle • infinite number of solutions • no solution • decimal

1. A(n) _____ is a fraction written a special way.

2. If there is no value of x for which an equation is true, the equation has _____.

3. To eliminate the fractions in an equation, multiply each term by the _____.

Example	Student Practice
1. Solve for x. $\dfrac{1}{4}x - \dfrac{2}{3} = \dfrac{5}{12}x$	**2.** Solve for x. $\dfrac{1}{18}x + \dfrac{2}{3} = \dfrac{5}{6}x$

To eliminate the fractions, multiply both sides by the LCD, 12, and apply the distributive property to simplify.

$$\frac{1}{4}x - \frac{2}{3} = \frac{5}{12}x$$

$$12\left(\frac{1}{4}x - \frac{2}{3}\right) = 12\left(\frac{5}{12}x\right)$$

$$\left(\frac{12}{1}\right)\left(\frac{1}{4}\right)x - \left(\frac{12}{1}\right)\left(\frac{2}{3}\right) = \left(\frac{12}{1}\right)\left(\frac{5}{12}\right)x$$

$$3x - 8 = 5x$$

Add $-3x$ to both sides, then divide by 2.

$$3x - 8 = 5x$$

$$3x + (-3x) - 8 = 5x + (-3x)$$

$$-8 = 2x$$

$$-\frac{8}{2} = \frac{2x}{2}$$

$$-4 = x$$

The check is left to the student.

Example	Student Practice
3. Solve for x and check your solution. $$\frac{x}{3}+3=\frac{x}{5}-\frac{1}{3}$$	**4.** Solve for x and check your solution. $$\frac{x}{2}+2=\frac{x}{3}-\frac{1}{2}$$

Multiply each term by the LCD, 15.

$$15\left(\frac{x}{3}\right)+15(3)=15\left(\frac{x}{5}\right)-15\left(\frac{1}{3}\right)$$
$$5x+45=3x-5$$

Subtract $3x$ and 45 from both sides.

$$5x+45=3x-5$$
$$5x-3x+45-45=3x-3x-5-45$$
$$2x=-50$$

Divide by 2.

$$\frac{2x}{2}=\frac{-50}{2}$$
$$x=-25$$

Check.

$$\frac{(-25)}{3}+3\overset{?}{=}\frac{(-25)}{5}-\frac{1}{3}$$
$$-\frac{25}{3}+\frac{9}{3}\overset{?}{=}-\frac{5}{1}-\frac{1}{3}$$
$$-\frac{16}{3}\overset{?}{=}-\frac{15}{3}-\frac{1}{3}$$
$$-\frac{16}{3}=-\frac{16}{3}$$

The solution checks.

Example	Student Practice
5. Solve for x and check your solution. $\dfrac{1}{3}(x-2)=\dfrac{1}{5}(x+4)+2$ Remove the parentheses. $\dfrac{x}{3}-\dfrac{2}{3}=\dfrac{x}{5}+\dfrac{4}{5}+2$ Multiply all terms by the LCD, 15. Then, solve the resulting equation. $15\left(\dfrac{x}{3}\right)-15\left(\dfrac{2}{3}\right)=15\left(\dfrac{x}{5}\right)+15\left(\dfrac{4}{5}\right)+15(2)$ $5x-10=3x+12+30$ $5x-10=3x+42$ $2x=52$ $x=26$ The check is left to the student.	**6.** Solve for x and check your solution. $\dfrac{1}{4}(x-4)=\dfrac{1}{6}(x+2)+8$
7. Solve for x. $0.2(1-8x)+1.1=-5(0.4x-0.3)$ Remove parentheses. Then, multiply each term by 10 to get integer coefficients and solve. $0.2-1.6x+1.1=-2.0x+1.5$ $2-16x+11=-20x+15$ $4x+13=15$ $4x=2$ $x=\dfrac{1}{2}$ or 0.5 The decimal form of the solution should only be given if it is a terminating decimal. The check is left to the student.	**8.** Solve for x. $0.3(1-6x)+1.3=3(-0.8x+0.6)$

Extra Practice

1. Solve for p. Check your solution.

$$\frac{1}{3}p = p - 1$$

2. Solve for x. Check your solution.

$$\frac{x+2}{3} = \frac{x}{18} + \frac{1}{9}$$

3. Solve for x. Check your solution.

$$\frac{3}{4}(2x+6) - 3 = 3(x+3)$$

4. Solve for x. Check your solution.

$$0.4(x-6) = 0.7(2x+3) + 0.5$$

Concept Check

Explain how you would solve $\dfrac{x+5}{6} = \dfrac{x}{2} + \dfrac{3}{4}$.

Chapter 2 Equations, Inequalities, and Applications
2.5 Translating English Phrases into Algebraic Expressions

Vocabulary
more than • sum of • increased by • added to • greater than • plus • minus
decreased by • less than • subtracted from • smaller than • fewer than • of
diminished by • difference between • double • twice • product of • times
divided by • quotient of

1. "The _____ a number and three" translates to $x+3$.

2. "The _____ a number and four" translates to $x-4$.

3. "The _____ a two and a number" translates to $2x$.

4. "The _____ a number and five" translates to $\dfrac{x}{5}$.

Example	Student Practice
1. Write each English phrase as an algebraic expression.	**2.** Write each English phrase as an algebraic expression.
(a) A quantity is increased by five. $x+5$	**(a)** Five less than a quantity
(b) Double the value. $2x$	**(b)** Triple the discount.
(c) One-third of the weight $\dfrac{x}{3}$ or $\dfrac{1}{3}x$	**(c)** One-fifth of the height
(d) Seven less than a number $x-7$ Note that the variable or expression that follows the words "less than" always comes first.	**(d)** Ten more than a number

Example	Student Practice
3. Write each English phrase as an algebraic expression.	**4.** Write each English phrase as an algebraic expression.

3. Write each English phrase as an algebraic expression.

(a) Seven more than double a number

$2x+7$

(b) The value of the number is increased by seven and then doubled.

Note that the word "then" tells us to add x and 7 before doubling.

$2(x+7)$

Note that this is not the same as $2x+7$.

(c) One-half of the sum of a number and 3

$\dfrac{1}{2}(x+3)$

4. Write each English phrase as an algebraic expression.

(a) Eight more than triple a number

(b) The value of the number is increased by eight and then tripled.

(c) One-fourth of the sum of a number and 6

---

5. Use a variable and an algebraic expression to describe two quantities in the English sentence "Mike's salary is $2000 more than Fred's salary."

The two quantities that are being compared are Mike's and Fred's salaries. Since Mike's salary is being compared to Fred's salary, we let the variable represent Fred's salary. The choice of the letter f helps us to remember that the variable represents Fred's salary.

Let $f =$ Fred's salary.

Then, $f + \$2000 =$ Mike's salary, since Mike's salary is $2000 more than Fred's.

6. Use a variable and an algebraic expression to describe two quantities in the English sentence "Tom's car has 12,500 more miles on it than Chuck's truck."

Example	Student Practice
7. The length of a rectangle is 3 meters shorter than twice the width. Use a variable and an algebraic expression to describe the length and the width. Draw a picture of the rectangle and label the length and width.	**8.** The length of a rectangle is 7 inches shorter than triple the width. Use a variable and an algebraic expression to describe the length and the width. Draw a picture of the rectangle and label the length and width.

The length of the rectangle is being compared to the width. Use the letter w for width. Let $w =$ the width. Express the length in terms of the width. The length is 3 meters shorter than twice the width. Then $2w - 3 =$ the length. A picture of the rectangle is shown.

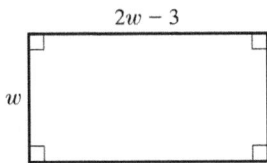

$2w - 3$

w

9. The first angle of a triangle is triple the second angle. The third angle of the triangle is 12° more than the second angle. Describe each angle algebraically. Draw a diagram of the triangle and label its parts.

10. The first angle of a triangle is five times the second angle. The third angle of the triangle is 40° more than the second angle. Describe each angle algebraically. Draw a diagram of the triangle and label its parts.

Since the first and third angles are described in terms of the second angle, we let the variable represent the number of degrees in the second angle.

Let $s =$ the number of degrees in the second angle. Then $3s =$ the number of degrees in the first angle and $s + 12 =$ the number of degrees in the third angle.

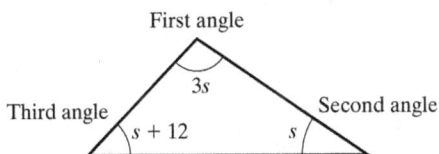

First angle

$3s$

Third angle

Second angle

$s + 12$ s

Extra Practice

1. Write an algebraic expression for the quantity. Let x represent the unknown value.

   a value increased by 12

2. Write an algebraic expression for the quantity. Let x represent the unknown value.

   one-fourth of a number, decreased by one

3. Write an algebraic expression for the quantities being compared.

   The value of Alicia's car is $2300 more than the value of Allison's car.

4. Write an algebraic expression for the quantities being compared.

   Patricia owns 17 more comic books than Scott, Walter owns three times as many as Scott, and Adrienne owns five fewer than four times as many as Scott.

Concept Check

Explain how you would decide whether to use $\frac{1}{3}(x+7)$ or $\frac{1}{3}x+7$ as an algebraic expression for the phrase "one-third of the sum of a number and seven."

Chapter 2 Equations, Inequalities, and Applications
2.6 Using Equations to Solve Word Problems

Vocabulary

understand the problem • write an equation • solve and state the answer • check

1. When solving word problems, the last step is to _____, which involves checking the solution in the original equation and determining if the answer is reasonable.

2. When solving word problems, the third step is to _____, which involves solving the equation to determine the answer to the problem.

3. When solving word problems, the second step is to _____, which involves looking for key words to help you to translate the words into algebraic symbols and expressions.

Example	**Student Practice**
1. Two-thirds of a number is eighty-four. What is the number?	2. Four-fifths of a number is sixty-eight. What is the number?
Understand the problem. Draw a sketch. Let x = the unknown number.	
Write an equation. The word "of" translates to multiplication and the word "is" translates to equals.	
$$\frac{2}{3}x = 84$$	
Solve and state the answer.	
$$\frac{2}{3}x = 84$$ $$2x = 252$$ $$x = 126$$	
The check is left to the student.	

Example	Student Practice
3. Five more than six times a quantity is three hundred five. Find the number.	**4.** Nineteen more than four times a quantity is two hundred seventy-nine. Find the number.

3. Five more than six times a quantity is three hundred five. Find the number.

Understand the problem. Read the problem carefully. You may not need to draw a sketch. Let $x =$ the unknown quantity.

Write an equation. "More than" translates to addition, "times" translates to multiplication, and "is" translates to equals.

$$5 + 6x = 305$$

Solve and state the answer.

$$6x + 5 = 305$$
$$6x + 5 - 5 = 305 - 5$$
$$6x = 300$$
$$\frac{6x}{6} = \frac{300}{6}$$
$$x = 50$$

The quantity, or number, is 50.

Check. Is five more than six times 50 three hundred five?

$$6(50) + 5 \overset{?}{=} 305$$
$$300 + 5 \overset{?}{=} 305$$
$$305 = 305$$

The answer checks.

Example	Student Practice

5. The smallest angle of an isosceles triangle measures $24°$. The other two angles are larger. What are the measurements of the other two angles?

Let $x =$ the measure in degrees of each of the larger angles. Draw a sketch.

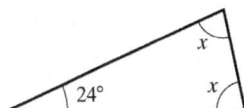

Write an equation showing the sum of all three angles is $180°$, solve for x.

$$24° + x + x = 180°$$
$$x = 78°$$

6. The largest angle of an isosceles triangle measures $116°$. The other two angles are smaller. What are the measurements of the other two angles?

---

7. Two people travel in separate cars. They each travel a distance of 330 miles on an interstate highway. To maximize fuel economy, Fred travels at exactly 50 mph. Sam travels at exactly 55 mph. How much time did the trip take each person?

Read the problem carefully and create a Mathematics Blueprint. Write an equation using the formula

$$\text{distance} = (\text{rate})(\text{time}) \text{ or } d = rt.$$

Substitute the known values into the formula and solve for t.

$$d = rt \qquad\qquad d = rt$$
$$330 = 50t_f \text{ and } 330 = 55t_s$$
$$6.6 = t_f \qquad\qquad 6.6 = 6t_s$$

It took Fred 6.6 hours. It took Sam 6 hours. The check is left to the student.

8. Two people travel in separate cars. They each travel a distance of 780 miles on an interstate highway. To maximize fuel economy, John travels at exactly 65 mph. Yuri travels at exactly 60 mph. How much time did the trip take each person?

Extra Practice

1. What number minus 312 gives 234? Check your solution.

2. A number is tripled and then increased by 27. The result is 72. What is the original number? Check your solution.

3. The local health food store has six times as many energy drinks as energy bars. There are 108 energy drinks in stock. How many energy bars are in stock? Check to see if your answer is reasonable.

4. Amy's cell phone company charges $10.50 per month for 200 minutes of use and $0.15 for each additional minute. Last month Amy's cell phone bill was $55.50. How many additional minutes was she charged for? Check to see if your answer is reasonable.

Concept Check

Explain how you would set up an equation to solve the following problem.
Phil purchased two shirts for $23 each and then purchased several pairs of socks. The socks were priced at $0.75 per pair. How many pairs of socks did he purchase if the total cost was $60.25?

Chapter 2 Equations, Inequalities, and Applications
2.7 Solving Word Problems: The Value of Money and Percents

Vocabulary
percent   •   simple interest   •   compound interest   •   interest

1. _____ is a charge for borrowing money or an income from investing money.

2. _____ is computed by multiplying the amount of money borrowed or invested times the rate of interest times the period of time over which it is borrowed or invested.

3. Applied situations often require finding a _____ of an unknown number.

Example	Student Practice
1. A business executive rented a car. The Supreme Car Rental Agency charged $39 per day and $0.28 per mile. The executive rented the car for two days and the total rental cost was computed to be $176. How many miles did the executive drive the rented car?	2. A business woman rented a room at a motel. The motel charged $52 per day and $2.50 per hour of internet use. The business woman stayed in the room for four days and the total charge was computed to be $248. How many hours of internet use did the business woman accrue?

Understand the problem. It is known that it costs $176 to rent the car for two days. It is necessary to find the number of miles the car was driven. Let $m =$ the number of miles driven in the rented car. Write an equation. Use the relationship for calculating the total cost.

per-day cost + mileage cost = total cost

$$(39)(2) + (0.28)m = 176$$

Solve and state the answer.
$$78 + 0.28m = 176$$
$$0.28m = 98$$
$$m = 350$$

The executive drove 350 miles. The check is left to the student.

Example	Student Practice
3. A sofa was marked with the following sign: "The price of this sofa has been reduced by 23%. You save $138 if you buy now." What was the original price of the sofa? Understand the problem. Let $s =$ the original price of the sofa. Then $0.23s =$ the amount of the price reduction, which is $138. Write an equation and solve. $$0.23s = 138$$ $$\frac{0.23s}{0.23} = \frac{138}{0.23}$$ $$s = 600$$ The original price of the sofa was $600. The check is left to the student.	**4.** A refrigerator was marked with the following sign: "The price of this refrigerator has been reduced by 33%. You save $264 if you buy now." What was the original price of the refrigerator?
5. A woman invested an amount of money in two accounts for one year. She invested some at 8% simple interest and the rest at 6% simple interest. Her total amount invested was $1250. At the end of the year she had earned $86 in interest. How much money had she invested in each account? The simple interest formula is $I = prt$. Let $x =$ amount invested at 8%. Then $\$1250 - x =$ amount invested at 6%. Write an equation and solve for x. $$0.08x + 0.06(1250 - x) = 86$$ $$x = 550$$ The amount invested at 8% is $550. The amount invested at 6% is $700.	**6.** A man invested an amount of money in two accounts for one year. He invested some at 6% simple interest and the rest at 5% simple interest. His total amount invested was $3500. At the end of the year he had earned $195 in interest. How much money had he invested in each account?

Example	Student Practice
7. When Bob got out of math class, he had to make a long-distance call. He had exactly enough dimes and quarters to make a phone call that would cost $2.55. He had one fewer quarter than he had dimes. How many coins of each type did he have?	**8.** When Rebecca got out of chemistry class, she went to the vending machines for a snack. She had exactly enough nickels and dimes to get a combination of items costing $2.75. She had four fewer dimes than she had nickels. How many coins of each type did she have?

Let $d =$ the number of dimes. Then $d-1=$ the number of quarters. The total value of the coins was $2.55. Each dime is worth $0.10 and each quarter is worth $0.25. So, d dimes are worth $0.10d$ dollars and $(d-1)$ quarters are worth $0.25(d-1)$ dollars. Now write an equation for the total value, and solve.

$$0.10d + 0.25(d-1) = 2.55$$
$$0.10d + 0.25d - 0.25 = 2.55$$
$$0.35d - 0.25 = 2.55$$
$$0.35d = 2.80$$
$$d = 8$$

Find the number of quarters Bob has, $d-1=8-1=7$. Thus, Bob has eight dimes and seven quarters.

Check the answer. Bob has $8-7=1$ less quarter than he has dimes. Check that eight dimes and seven quarters are worth $2.55.

$$8(\$0.10) + 7(\$0.25) \overset{?}{=} \$2.55$$
$$\$0.80 + \$1.75 \overset{?}{=} \$2.55$$
$$\$2.55 = \$2.55$$

The solution checks.

Extra Practice

1. Angelina received a pay raise this year. The raise was 5% of last year's salary. This year, Angelina earned $17,640. What was her salary before the raise?

2. Find the simple interest on $6,500 borrowed at 15% for one year.

3. Randall is due to receive a 7% raise, which in dollars will be $3,780 per year. What is his current salary?

4. Mr. Finch keeps money in his pillowcase. Right now, he has equal numbers of five, ten, and twenty-dollar bills, with no other denominations. He has exactly $1505. How many bills does he have all together?

Concept Check

Explain how you would set up an equation to solve the following problem.
Robert has $2.55 in change consisting of nickels, dimes, and quarters. He has twice as many dimes as quarters. He has one more nickel than he has quarters. How many of each coin does he have?

Chapter 2 Equations, Inequalities and Applications
2.8 Solving Inequalities in One Variable

Vocabulary
inequalities • is less than • is greater than • solution • solution set • graph

1. The set of all numbers that make the inequality true is called a(n) _____.

2. Comparisons of values, such as one value being greater than or less than another value, are called _____.

3. A(n) _____ is a visual representation of the solution set.

4. One number _____ another if it is to the right of the other on the number line.

Example	**Student Practice**
1. In each statement, replace the question mark with the symbol < or >.	**2.** In each statement, replace the question mark with the symbol < or >.
(a) 3 ? −1	**(a)** −4 ? −10
$3 > -1$ because 3 is to the right of −1 on the number line.	
(b) −2 ? 1	**(b)** 2 ? −2
$-2 < 1$ because −2 is to the left of 1.	
(c) −3 ? −4	**(c)** −1 ? 4
$-3 > -4$ because −3 is to the right of −4.	**(d)** 0 ? −7
(d) 0 ? 3	
$0 < 3$ because 0 is to the left of 3.	**(e)** 5 ? 8
(e) −3 ? 0	
$-3 < 0$ because −3 is left of 0.	

Example	Student Practice

3. State each mathematical relationship in words and then graph it.

   (a) $x < -2$

   We state "x is less than -2."

   (b) $-3 < x$

   We can state that "-3 is less than x" or, equivalently, that "x is greater than -3." Be sure you see that $-3 < x$ is equivalent to $x > -3$. Although both statements are correct, we usually write the variable first in a simple inequality containing a variable and a numerical value.

4. State each mathematical relationship in words and then graph it.

   (a) $x < 3$

   (b) $\dfrac{5}{2} \geq x$

---

5. Translate each English sentence into an algebraic statement.

   (a) The police on the scene said that the car was traveling more than 80 miles per hour. (Use the variable s for speed.)

   Since the speed must be greater than 80, we have $s > 80$.

   (b) The owner of the trucking company said that the payload of a truck must never exceed 4500 pounds. (Use the variable p for payload.)

   If the payload of the truck can never exceed 4500 pounds, then the payload must always be less than or equal to 4500 pounds. Thus we write $p \leq 4500$.

6. Translate each English sentence into an algebraic statement.

   (a) A student's budget is very tight, so when shopping for a car, her car payment must be less than 125 dollars per month.

   (b) The maximum number of people allowed on the boat at one time is 35.

Example	Student Practice
7. Solve and graph. $3x + 7 \geq 13$	**8.** Solve and graph. $6x - 4 > 5$

7. Solve and graph. $3x + 7 \geq 13$

Subtract 7 from both sides.

$$3x + 7 \geq 13$$
$$3x + 7 - 7 \geq 13 - 7$$
$$3x \geq 6$$

Divide both sides by 3.

$$3x \geq 6$$
$$\frac{3x}{3} \geq \frac{6}{3}$$
$$x \geq 2$$

8. Solve and graph. $6x - 4 > 5$

9. Solve and graph. $5 - 3x > 7$

Subtract 5 from both sides and simplify.

$$5 - 5 - 3x > 7 - 5$$
$$-3x > 2$$

Divide by -3. When dividing by a negative number, the inequality is reversed.

$$-3x > 2$$
$$\frac{-3x}{-3} < \frac{2}{-3}$$
$$x < -\frac{2}{3}$$

Note the direction of the inequality. The graph is as follows.

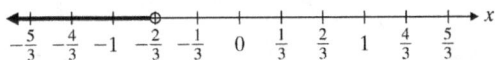

10. Solve and graph. $8 - 3x < 9$

Example	Student Practice
11. Solve and graph. $-\dfrac{13x}{2} \le \dfrac{x}{2} - \dfrac{15}{8}$	**12.** Solve and graph. $-\dfrac{7x}{3} < \dfrac{x}{3} + \dfrac{23}{27}$

Multiply by the LCD, 8, and simplify.

$$8\left(-\frac{13x}{2}\right) \le 8\left(\frac{x}{2}\right) - 8\left(\frac{15}{8}\right)$$

$$-52x \le 4x - 15$$

Subtract both $4x$ from both sides and combine like terms. Then, solve the inequality. Be sure to reverse the direction of the inequality symbol.

$$-52x - 4x \le 4x - 4x - 15$$

$$-56x \le -15$$

$$x \ge \frac{15}{56}$$

The graph is as follows.

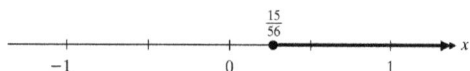

Extra Practice

1. In the statement, replace the question mark with the symbol $<$ or $>$.

$\dfrac{1}{3}$? 5

2. Graph. $x < -4$

3. Translate the English sentence into an algebraic statement.

1,000,000 is greater than or equal to the total weight w.

4. Solve and graph. $3x + 5 < 14$

Concept Check

Explain the difference between $12 < x$ and $x > 12$. Would the graphs of these inequalities be the same or different?

MATH COACH

Mastering the skills you need to do well on the test.

Watch the **MATH COACH** videos in MyMathLab® or on You Tube™ while you work the problems below. These helpful hints will help you avoid making common errors on test problems.

Solving Equations with Both Parentheses and Decimals—

Problem 6   Solve for the variable. $0.8x + 0.18 - 0.4x = 0.3(x + 0.2)$

> **Helpful Hint:** After removing parentheses, it might be most helpful for you to multiply both sides of the equation by 100 in order to obtain a simpler, equivalent equation without decimals. Check to make sure that you did not make any errors in calculations before solving the equation.

Did you remove the parentheses to get the equation $0.8x + 0.18 - 0.4x = 0.3x + 0.06$?
Yes _____   No _____

If you answered No, go back and use the distributive property to remove the parentheses. Be careful to place the decimal point in the correct location when multiplying 0.3 and 0.2 together.

Did you multiply each term of the equation by 100 to move the decimal point two places to the right to get the equivalent equation $80x + 18 - 40x = 30x + 6$?
Yes _____   No _____

If you answered No, stop and carefully complete this step before solving the equation. Remember that you may need to

add a 0 to the end of a term in order to move the decimal point two places to the right.

If you answered Problem 6 incorrectly, go back and rework the problem using these suggestions.

Solving Equations with More Than One Set of Parentheses—Problem 12

Solve for the variable. $20 - (2x + 6) = 5(2 - x) + 2x$

> **Helpful Hint:** Slowly complete the necessary steps to remove each set of parentheses before doing any other steps. Be careful to avoid sign errors.

Did you obtain the equation $20 - 2x - 6 = 10 - 5x + 2x$ after removing each set of parentheses?
Yes _____   No _____

If you answered No, go back and carefully use the distributive property to remove each set of parentheses. Locate any mistakes you have made and make a note of the type of error discovered.

Did you combine like terms to get the equation $14 - 2x = 10 - 3x$?
Yes _____   No _____

If you answered No, stop and perform that step correctly.

Now go back and rework the problem using these suggestions.

Solving Equations with Both Fractions and Parentheses—Problem 17

Solve for x. $\dfrac{2}{3}(x+8)+\dfrac{3}{5}=\dfrac{1}{5}(11-6x)$

> **Helpful Hint:** Remove the parentheses first. This is the most likely place to make a mistake. Next, carefully show every step of your work as you multiply each fraction by the LCD. Be sure to check your work.

Did you remove each set of parentheses to obtain the equation $\dfrac{2}{3}x+\dfrac{16}{3}+\dfrac{3}{5}=\dfrac{11}{5}-\dfrac{6}{5}x$?

Yes _____   No _____

If you answered No, stop and carefully redo your steps of multiplication, showing every part of your work.

Did you identify the LCD as 15 and then multiply each term by 15 to get $10x+80+9=33-18x$?

Yes _____   No _____

If you answered No, stop and write out your steps slowly.

If you answered Problem 17 incorrectly, go back and rework the problem using these suggestions.

Solving and Graphing Inequalities on a Number Line—Problem 19

Solve and graph the inequality. $2-7(x+1)-5(x+2)<0$

> **Helpful Hint:** Be sure to remove parentheses and combine any like terms on each side of the inequality before solving for the variable. Always verify the following:
> 1) Did you multiply or divide by a negative number? If so, did you reverse the inequality symbol?
> 2) In the graph, is your choice of an open circle or closed circle correct?

Did you remove parentheses to get $2-7x-7-5x-10<0$? Did you combine like terms to obtain the inequality $-15-12x<0$? Next, did you add 15 to both sides of the inequality? Yes _____   No _____

If you answered No to any of these questions, stop now and perform those steps.

Did you remember to reverse the inequality symbol in the last step? Yes _____   No _____

If you answered No, please review the rules for when to reverse the inequality symbol and then go back and perform this step.

Did you use an open circle in your number line graph? Is your arrow pointing to the right? Yes _____   No _____

If you answered No to either question, please review the rules for how to graph an inequality involving the < or > inequality symbols.

Now go back and rework the problem using these suggestions.

Chapter 3 Graphing and Functions
3.1 The Rectangular Coordinate System

Vocabulary

graphs • rectangular coordinate system • origin • x-axis • y-axis
ordered pair • coordinates • x-coordinate • y-coordinate • solution

1. The numbers in an ordered pair are often referred to as the _____ of the point.

2. We can illustrate algebraic relationships with drawings called _____.

3. The vertical number line above the origin is often called the _____.

4. The first number in an ordered pair is called the _____ and it represents the distance from the origin measured along the horizontal axis.

Example	Student Practice
1. Plot the point $(5,2)$ on a rectangular coordinate system. Label this point as A.	**2.** Plot the point $(2,5)$ on the preceding rectangular coordinate system. Label this point as B.

Since the x-coordinate is 5, we first count 5 units to the right on the x-axis. Then, because the y-coordinate is 2, we count 2 units up from the point where we stopped on the x-axis. This locates the point corresponding to $(5,2)$. We mark this point with a dot and label it A.

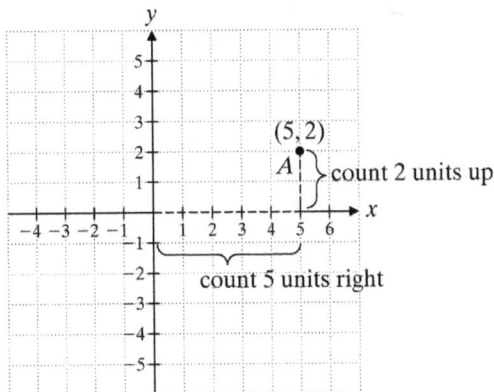

Example	Student Practice
3. Use the rectangular coordinate system to plot each point. Label the points F and G respectively.	**4.** Use the rectangular coordinate system below to plot each point. Label the points I, J, and K, respectively.

3. Use the rectangular coordinate system to plot each point. Label the points F and G respectively.

(a) $(-5, -3)$

Notice that the x-coordinate, -5, is negative. On the coordinate grid, negative x-values appear to the left of the origin. Thus we will begin by counting 5 squares to the left, starting at the origin. Since the y-coordinate, -3, is negative, we will count 3 units down from the point where we stopped on the x-axis.

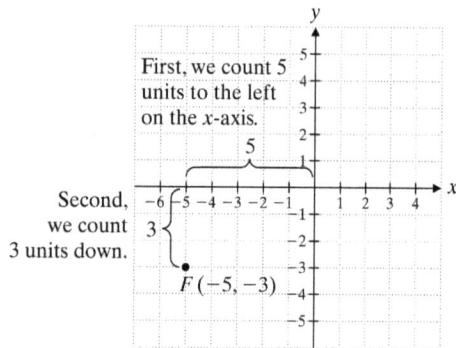

(b) $(2, -6)$

The x-coordinate is positive. Begin by counting 2 squares to the right of the origin. Then count down because the y-coordinate is negative.

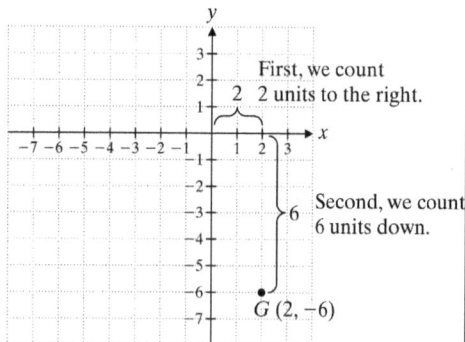

4. Use the rectangular coordinate system below to plot each point. Label the points I, J, and K, respectively.

(a) $(-1, -4)$

(b) $(-4, 3)$

(c) $(3, -5)$

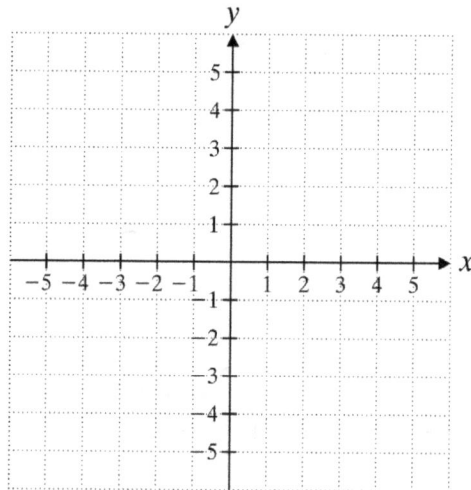

Example	Student Practice
5. What ordered pairs of numbers represent point A and point B on the graph?	**6.** What ordered pair of numbers represents point C on the graph in Example **5**?

5. To find point A, move along the x-axis until you get as close as possible to A, ending at 5. Thus obtaining 5 as the first number of the ordered pair. Then count 4 units upward on a line parallel to the y-axis to reach A. So you obtain 4 as the second number of the ordered pair. Thus point A is $(5,4)$. Use the same approach to find point B: $(-5,-3)$.

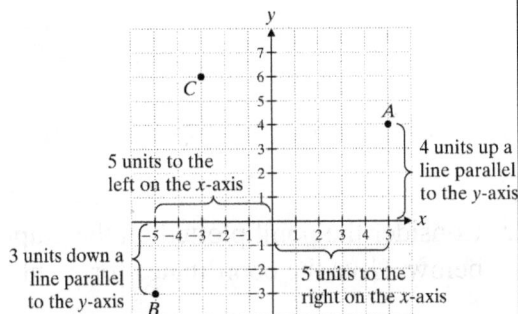

---

7. Is the ordered pair $(-1,4)$ a solution to the equation $3x+2y=5$?

We replace the values for x and y to see if we obtain a true statement.
Replace x with -1 and y with 4.

$$3x+2y=5$$

$$3(-1)+2(4)\overset{?}{=}5$$

$$-3+8\overset{?}{=}5$$

$$5=5$$

The ordered pair $(-1,4)$ is a solution to $3x+2y=5$ because when we replace x with -1 and y with 4, we obtain a true statement.

8. Is the ordered pair $(3,-3)$ a solution to the equation $3x+2y=5$?

Example	Student Practice
9. Find the missing coordinate to complete the ordered-pair solution $(0,?)$ to the equation $2x+3y=15$.	**10.** Find the missing coordinate to complete the ordered-pair solution $(?,3)$ to the equation $6x+5y=3$.

For the ordered pair $(0,?)$, we know that $x=0$. Replace x with 0 in the equation and solve for y.

$$2x+3y=15$$
$$2(0)+3y=15$$
$$0+3y=15$$
$$y=5$$

Thus we have the ordered pair $(0,5)$.

Extra Practice

1. Plot the point $D:(-1,-3)$.

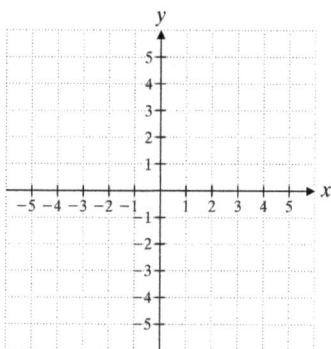

2. Consider the point plotted on the graph below. Give the coordinates for point A.

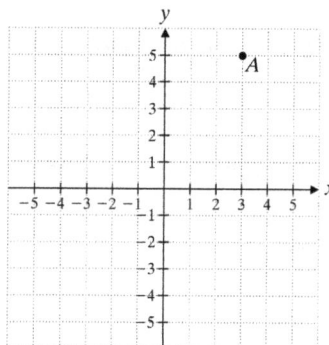

3. Find the missing coordinate to complete the ordered-pair solution to $y=-2x+3$.

(a) $(-1,?)$

(b) $(3,?)$

4. Find the missing coordinate to complete the ordered-pair solution to $4x-2y=12$.

(a) $(-1,?)$

(b) $(4,?)$

Concept Check

Explain how you would find the missing coordinate to complete the ordered-pair solution to the equation $2.5x+3y=12$ if the ordered pair was of the form $(?,-6)$.

Chapter 3 Graphing and Functions
3.2 Graphing Linear Equations

Vocabulary
linear equation • *x*-intercept • *y*-intercept • horizontal • vertical

1. The _____ of a line is the point where the line crosses the *y*-axis.

2. The graph of any _____ in two variables is a straight line.

3. The _____ of a line is the point where the line crosses the *x*-axis.

Example	**Student Practice**
1. Find three ordered pairs that satisfy $y = -2x + 4$. Then graph the resulting straight line.	**2.** Find three ordered pairs that satisfy $y = 2x - 3$. Then graph the resulting straight line. Use the given coordinate system.

Let $x = 0$, $x = 1$, and $x = 3$. For each *x*-value, find the corresponding *y*-value in the equation. Place each *y*-value in the table next to its *x*-value.

x	y
0	4
1	2
3	−2

If we plot these ordered pairs and connect the three points, we get a straight line that is the graph of the equation.

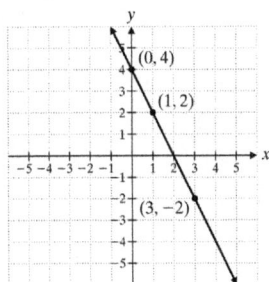

Example	Student Practice

3. Graph $5x - 4y + 2 = 2$.

First, we simplify the equation by subtracting 2 from each side.

$$5x - 4y + 2 = 2$$
$$5x - 4y + 2 - 2 = 2 - 2$$
$$5x - 4y = 0$$

Since we are free to choose any value of x, $x = 0$ is a natural choice. Calculate the value of y when $x = 0$.

$$5(0) - 4y = 0$$
$$-4y = 0$$
$$y = 0$$

A convenient choice for a replacement of x is a number that is divisible by 4. Let $x = 4$ and $x = -4$. Follow the process used above to find the corresponding y-values and place the results in the table of values.

x	y
0	0
4	5
−4	−5

Graph the line.

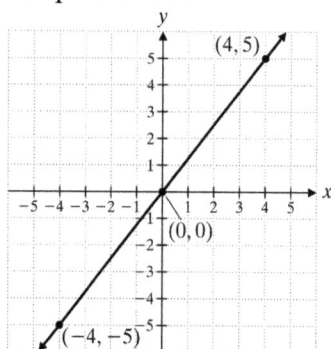

4. Graph $4x - 2y + 7 = 7$.

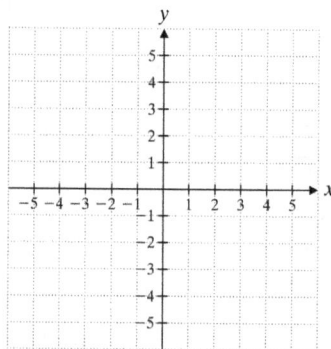

102

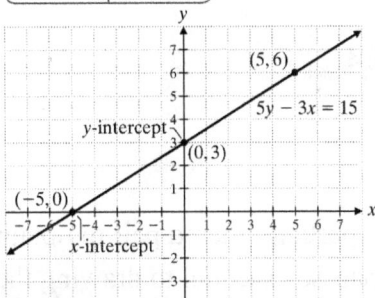

Example	Student Practice
5. Complete **(a)** and **(b)** for the equation $5y - 3x = 15$.	**6.** Complete **(a)** and **(b)** for the equation $y + 2x = -2$.

5. Complete **(a)** and **(b)** for the equation $5y - 3x = 15$.

(a) State the x- and y-intercepts.

Let $y = 0$.

$$5(0) - 3x = 15$$
$$-3x = 15$$
$$x = -5$$

$(-5, 0)$ is the x-intercept. Now let $x = 0$.

$$5y - 3(0) = 15$$
$$5y = 15$$
$$y = 3$$

$(0, 3)$ is the y-intercept.

(b) Use the intercept method to graph.

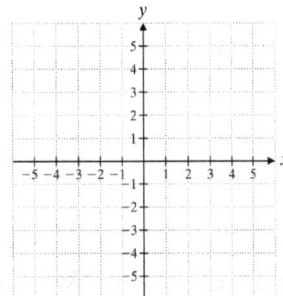

Find a third point and then graph. If we let $y = 6$, $x = 5$. The ordered pair is $(5, 6)$.

x	y
−5	0
0	3
5	6

6. Complete **(a)** and **(b)** for the equation $y + 2x = -2$.

(a) State the x- and y-intercepts.

(b) Use the intercept method to graph.

103

Example	Student Practice
7. Graph $2x+1=11$.	**8.** Graph $y+4=7$.

Notice that there is only one variable, x, in the equation. Simplifying the equation yields $x=5$. Since the x-coordinate of every point on this line is 5, we can see that the vertical line will be 5 units to the right of the y-axis.

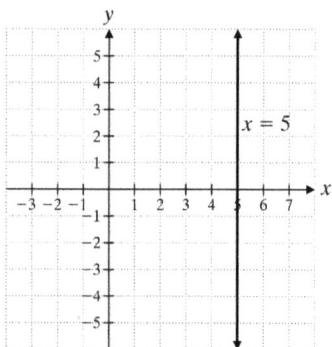

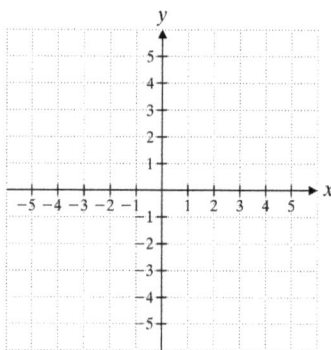

Extra Practice

1. Graph $y=2x-5$ by plotting three points and connecting them.

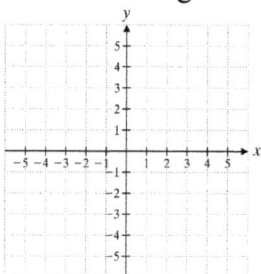

2. Graph $5x+2y=10$ by plotting three points and connecting them.

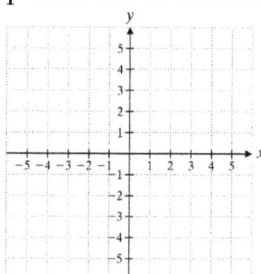

3. Graph $y=5-x$ by plotting intercepts and one other point.

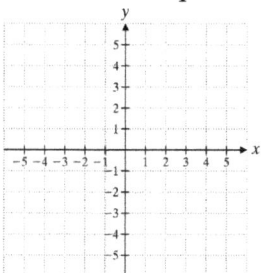

4. Graph $2x-5=y$ by plotting intercepts and one other point.

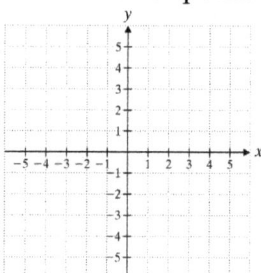

Concept Check

In graphing the equation $3y-7x=0$, what is the most important ordered pair to obtain before drawing a graph of the line? Why is that ordered pair so essential to drawing the graph?

Chapter 3 Graphing and Functions
3.3 The Slope of a Line

Vocabulary

slope • positive slope • negative slope • zero slope
undefined slope • slope-intercept form

1. In a coordinate plane, the _____ of a straight line is defined by the change in y divided by the change in x.

2. A vertical line is said to have _____.

3. If we know the slope and the y-intercept, we can write the equation of the line in _____.

Example	Student Practice
1. Find the slope of the line that passes through $(2,0)$ and $(4,2)$.	**2.** Find the slope of the line that passes through $(0,-4)$ and $(-3,-6)$.
Let $(2,0)$ be the first point (x_1, y_1) and $(4,2)$ be the second point (x_2, y_2).	
$$\text{slope} = m = \frac{y_2 - y_1}{x_2 - x_1} = \frac{2-0}{4-2} = \frac{2}{2} = 1$$	
Note that the slope of the line will be the same if we let $(4,2)$ be the first point (x_1, y_1) and $(2,0)$ be the second point (x_2, y_2).	
$$m = \frac{y_2 - y_1}{x_2 - x_1} = \frac{0-2}{2-4} = \frac{-2}{-2} = 1$$	
Thus, it does not matter which point you call (x_1, y_1) and which you call (x_2, y_2).	

Example	Student Practice
3. Find the slope of the line that passes through the given points. **(a)** $(0,2)$ and $(5,2)$ Calculate the slope. $$m = \frac{y_2 - y_1}{x_2 - x_1} = \frac{2-2}{5-0} = \frac{0}{5} = 0$$ The slope of a horizontal line is 0. **(b)** $(-4,0)$ and $(-4,-4)$ Calculate the slope. $$m = \frac{y_2 - y_1}{x_2 - x_1} = \frac{-4-0}{-4-(-4)} = \frac{-4}{0}$$ Recall that division by 0 is undefined. The slope of a vertical line is undefined.	**4.** Find the slope of the line that passes through the given points. **(a)** $(7,3)$ and $(7,-4)$ **(b)** $(5,3)$ and $(-1,3)$
5. What is the slope and the y-intercept of the line $5x + 3y = 2$? We want to solve for y and get the equation in the form $y = mx + b$. $5x + 3y = 2$ $\quad 3y = -5x + 2$ $\quad\quad y = -\frac{5}{3}x + \frac{2}{3}$ $m = -\frac{5}{3}$ and $b = \frac{2}{3}$ The slope is $-\frac{5}{3}$. The y-intercept is $\left(0, \frac{2}{3}\right)$.	**6.** What is the slope and the y-intercept of the line $9x + 3y = 12$?

Example	Student Practice
7. Find an equation of the line with slope $\frac{2}{5}$ and y-interept $(0,-3)$.	**8.** Find an equation of the line with slope $\frac{3}{4}$ and y-interept $(0,-7)$.
(a) Write the equation in slope-intercept form, $y = mx + b$.	**(a)** Write the equation in slope-intercept form, $y = mx + b$.
We are given that $m = \frac{2}{5}$ and $b = -3$. Thus we have the following.	
$y = mx + b$	
$y = \frac{2}{5}x + (-3)$	
$y = \frac{2}{5}x - 3$	
(b) Write the equation in the form $Ax + By = C$.	**(b)** Write the equation in the form $Ax + By = C$.
Clear the equation of fractions so that A, B, and C are integers.	
$y = \frac{2}{5}x - 3$	
$5y = 5\left(\frac{2x}{5}\right) - 5(3)$	
$5y = 2x - 15$	
Subtract $2x$ from each side. Then, multiply each term by -1, because the form $Ax + By = C$ is usually written with A as a positive integer.	
$5y = 2x - 15$	
$-2x + 5y = -15$	
$2x - 5y = 15$	

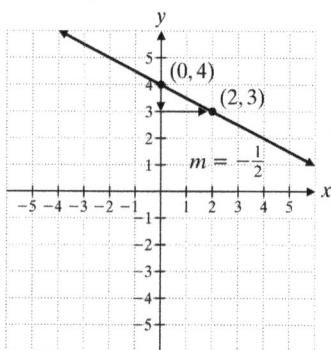

Example	Student Practice
9. Graph the equation $y = -\frac{1}{2}x + 4$.	**10.** Graph the equation $y = -\frac{3}{4}x + 1$.

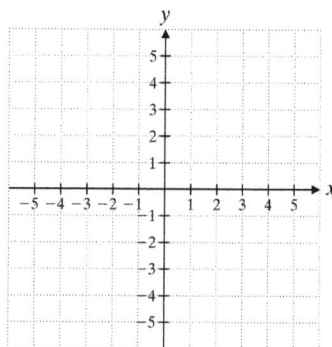

Begin with the y-intercept. Since $b = 4$, plot the point $(0,4)$. The slope, $-\frac{1}{2}$ can be written as $\frac{-1}{2}$. Begin at $(0,4)$ and go down 1 unit and to the right 2 units. This is the point $(2,3)$. Plot the point. Draw the line that connects the two points. This is the graph of the equation $y = -\frac{1}{2}x + 4$.

Extra Practice

1. Find the slope of a straight line that passes through the points $(-5,-2)$ and $(1,-4)$.

2. Find the slope and the y-intercept of the line $y = 5x$.

3. Write the equation of the line in slope-intercept form given $m = -4$ and the y-intercept is $\left(0, \frac{4}{5}\right)$.

4. Write the equation of the line in slope-intercept form given $m = 0$ and the y-intercept is $(0,-2)$.

Concept Check

Consider the formula for slope: $m = \frac{y_2 - y_1}{x_2 - x_1}$. Explain why we substitute the y coordinates in the numerator.

Name: _____   Date: _____

Instructor: _____   Section: _____

Chapter 3 Graphing and Functions
3.4 Writing the Equation of a Line

Vocabulary
slope • *y*-intercept • slope-intercept form • vertical units • horizontal units
parallel lines • perpendicular lines

1. Given the slope and a point on the line we can find the _____.

2. To find the slope of a line given the graph, we count the number of _____ and horizontal units from one point on the line to another.

3. _____ have slopes whose product is −1.

4. _____ have the same slope but different *y*-intercepts.

Example	**Student Practice**
1. Find an equation of the line that passes through $(-3, 6)$ with slope $-\dfrac{2}{3}$.	**2.** Find an equation of the line that passes through $(2, -5)$ with slope $-\dfrac{1}{2}$.

Example 1 (continued):

We are given the values $m = -\dfrac{2}{3}$, $x = -3$, and $y = 6$. Substitute the given values of x, y, and m into the equation $y = mx + b$. Solve for b.

$$y = mx + b$$
$$6 = \left(-\frac{2}{3}\right)(-3) + b$$
$$6 = 2 + b$$
$$4 = b$$

Use the values of b and m to write the equation in the form $y = mx + b$.

An equation of the line is $y = -\dfrac{2}{3}x + 4$.

Example	Student Practice
3. Find an equation of the line that passes through $(2,5)$ and $(6,3)$. We first find the slope of the line. Then proceed as in Example **1**. Substitute $(x_1, y_1) = (2,5)$ and $(x_2, y_2) = (6,3)$ into the formula. $$m = \frac{y_2 - y_1}{x_2 - x_1}$$ $$m = \frac{y_2 - y_1}{x_2 - x_1} = \frac{3-5}{6-2} = \frac{-2}{4} = -\frac{1}{2}$$ Choose either point, say $(2,5)$, to substitute into $y = mx + b$ as in Example **1**. Then solve for b. $$y = mx + b$$ $$5 = -\frac{1}{2}(2) + b$$ $$5 = -1 + b$$ $$6 = b$$ Use the values for b and m to write the equation. An equation of the line is $y = -\frac{1}{2}x + 6$. Note: We could have substituted the slope and the other point, $(6,3)$, into the slope-intercept form and arrived at the same answer. Try it.	**4.** Find an equation of the line that passes through $(5,4)$ and $(10,1)$.

Example	Student Practice
5. What is the equation of the line in the figure below?	**6.** What is the equation of the line in the figure below?

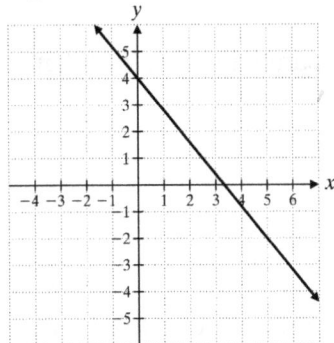

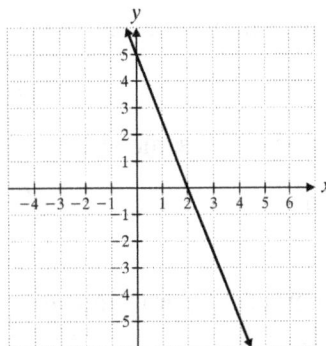

First, look for the y-intercept. The line crosses the y-axis at $(0,4)$. Thus $b = 4$.

Second, find the slope.

$$m = \frac{\text{change in } y}{\text{change in } x}$$

Look for another point on the line. We choose $(5,-2)$. Count the number of vertical units from 4 to -2 (rise). Count the number of horizontal units from 0 to 5 (run).

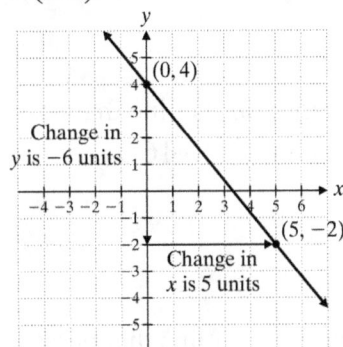

$$m = \frac{-6}{5}$$

Now, using $m = \frac{-6}{5}$ and $b = 4$, we can write an equation of the line.

$$y = mx + b$$

$$y = -\frac{6}{5}x + 4$$

Example	Student Practice
7. Line h has a slope of $-\dfrac{2}{3}$.	**8.** Line h has a slope of $\dfrac{5}{7}$.

7.

(a) If line f is parallel to line h, what is its slope?

Parallel lines have the same slope.

Line f has a slope of $-\dfrac{2}{3}$.

(b) If line g is perpendicular to line h, what is its slope?

Perpendicular lines have slopes whose product is -1.

$$m_1 m_2 = -1$$

$$-\frac{2}{3} m_2 = -1$$

$$m_2 = \frac{3}{2}$$

Line g has a slope of $\dfrac{3}{2}$.

8.

(a) If line f is parallel to line h, what is its slope?

(b) If line g is perpendicular to line h, what is its slope?

Extra Practice

1. Find the equation of the line that passes through $(3,1)$ and has slope $-\dfrac{1}{2}$.

2. Write an equation of the line that passes through $(3,4)$ and $(-1,-16)$.

3. Write an equation of the line that passes through $\left(1,\dfrac{1}{6}\right)$ and $\left(2,\dfrac{4}{3}\right)$.

4. Find the equation of a line that passes through $(3,-7)$ and is parallel to $y = -5x + 2$.

Concept Check

How would you find an equation of the line that passes through $(-2,-3)$ and has zero slope?

Chapter 3 Graphing and Functions
3.5 Graphing Linear Inequalities

Vocabulary
linear inequality • solution • solid line • dashed line • test point

1. The _____ of an inequality is the set of all possible ordered pairs that when substituted into the inequality will yield a true statement.

2. If the _____ is a solution of the inequality, we shade the region on the side of the line that includes the point.

3. We use a _____ to indicate that the points on the line are included in the solution of the inequality.

Example	**Student Practice**
1. Graph $5x + 3y > 15$.	**2.** Graph $4x + 3y > 12$.

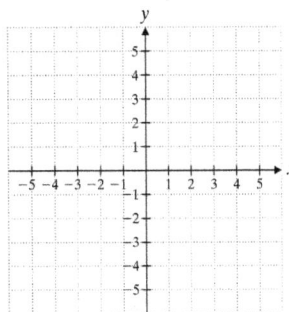

Begin by graphing the line $5x + 3y = 15$. Since there is no equals sign in the inequality, draw a dashed line to indicate that the line is not part of the solution set. The easiest test point to test is $(0,0)$. Substitute $(0,0)$ for (x,y).

$$5x + 3y > 15$$

$$5(0) + 3(0) > 15$$

$$0 > 15 \quad \text{false}$$

$(0,0)$ is not a solution. Shade the region on the side of the line that does not include $(0,0)$.

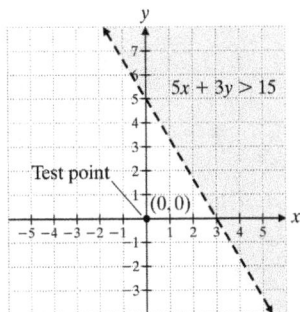

Example	Student Practice
3. Graph $2y \le -3x$.	**4.** Graph $4y \le -5x$.

First, graph $2y = -3x$. Since $\le$ is used, the line will be a solid line.

We see that the line passes through $(0,0)$. We choose another test point. We will choose $(-3,-3)$.

$$2y \le -3x$$
$$2(-3) \le -3(-3)$$
$$-6 \le 9 \quad \text{true}$$

Since $(-3,-3)$ is a solution to the inequality, shade the region that includes $(-3,-3)$, that is the region below the line.

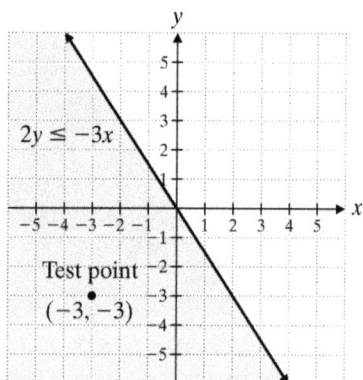

Example	Student Practice
5. Graph $x < -2$.	**6.** Graph $x > -3$.

First, graph $x = -2$. Since $<$ is used, the line will be dashed.

Second, test $(0,0)$ in the inequality.

$x < -2$

$0 < -2$ false

Since $(0,0)$ is not a solution to the inequality, shade the region that does not include $(0,0)$, that is the region to the left of the line $x = -2$.

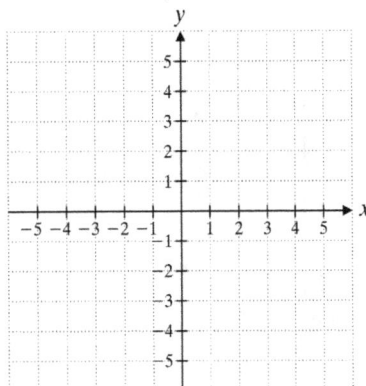

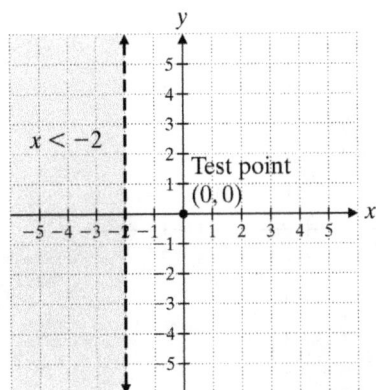

Observe that every point in the shaded region has an x-value that is less than -2.

Extra Practice

1. Graph $y < 2x + 1$.

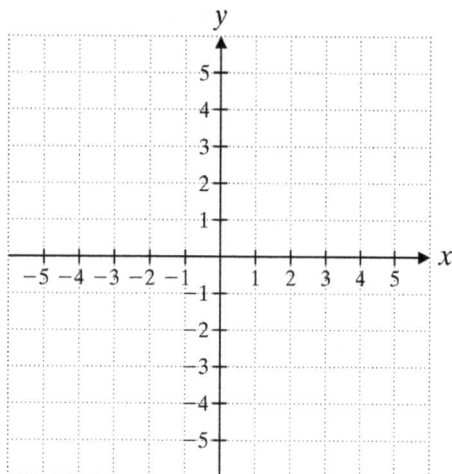

2. Graph $3x - 5y - 10 \geq 0$.

3. Graph $y \leq 4$.

4. Graph $3x + 6y - 9 < 0$.

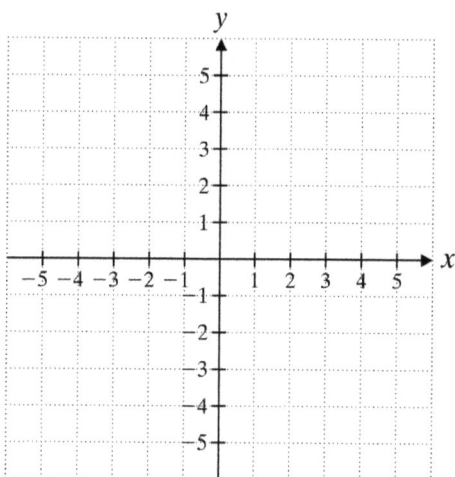

Concept Check

Explain how you would determine if you should shade the region above the line or below the line if you were to graph the inequality $y > -3x + 4$ using $(0, 0)$ as a test point.

Chapter 3 Graphing and Functions
3.6 Functions

Vocabulary

independent variable • dependent variable • relation • domain • range
function • absolute zero • vertical line test • function notation

1. A(n) _____ is any set of ordered pairs.

2. A(n) _____ is a relation in which no two different ordered pairs have the same first coordinate.

3. The _____ is used to determine whether a relation is a function.

4. All the first coordinates in all of the ordered pairs of the relation make up the _____ of the relation.

Example	**Student Practice**
1. State the domain and range of the relation. $\{(5,7),(9,11),(10,7),(12,14)\}$	**2.** State the domain and range of the relation. $\{(-3,6),(7,1),(-2,1),(4,6)\}$
The domain consists of all the first coordinates in the ordered pairs. The first coordinates are 5, 9, 10, and 12.	
The range consists of all the second coordinates in the ordered pairs. The second coordinates are 7, 11, 7, and 14.	
We usually list the values of a domain or range from smallest to largest.	
The domain is $\{5,9,10,12\}$.	
The range is $\{7,11,14\}$.	
Note that we list 7 only once.	

Example	Student Practice

3. Determine whether the relation is a function.

(a) $\{(3,9),(4,16),(5,9),(6,36)\}$

No two ordered pairs have the same first coordinate. Thus this set of ordered pairs defines a function.

(b) $\{(7,8),(9,10),(12,13),(7,14)\}$

Two different ordered pairs, $(7,8)$ and $(7,14)$ have the same first coordinate. Thus this relation is not a function.

4. Determine whether the relation is a function.

(a) $\{(9,3),(16,4),(9,5),(36,6)\}$

(b) $\{(-3,17),(2,1),(4,-3),(7,17)\}$

---

5. Graph $y = x^2$.

Begin by constructing a table of values. We select values for x and then determine by the equation the corresponding values of y. We then plot the ordered pairs and connect the points with a smooth curve.

x	$y = x^2$	y
-2	$y = (-2)^2 = 4$	4
-1	$y = (-1)^2 = 1$	1
0	$y = (0)^2 = 0$	0
1	$y = (1)^2 = 1$	1
2	$y = (2)^2 = 4$	4

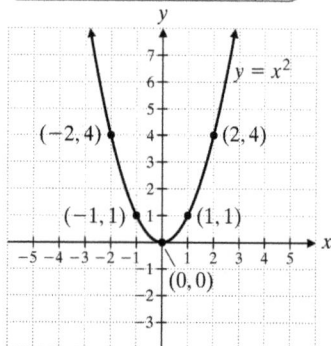

6. Graph $x = y^2$.

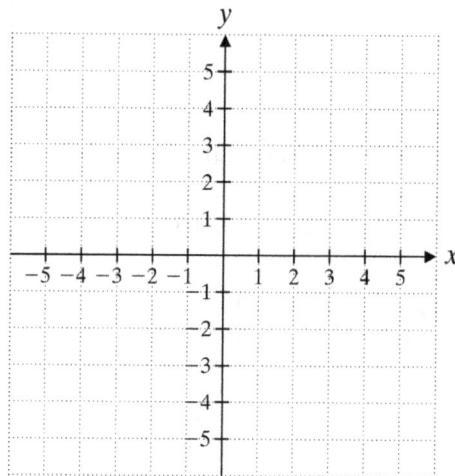

Example	Student Practice

7. Determine whether each of the following is the graph of a function.

(a)

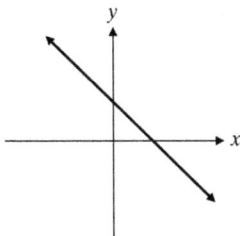

The graph of a straight line is a function. Any vertical line will cross this straight line in only one location.

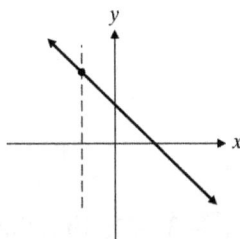

(b)

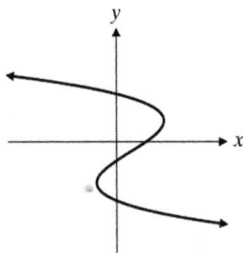

This graph is not the graph of a function. There exists a vertical line that will cross the curve in more than one place.

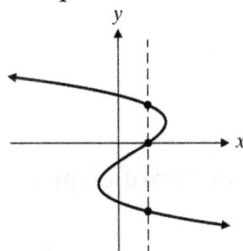

8. Determine whether each of the following is the graph of a function.

(a)

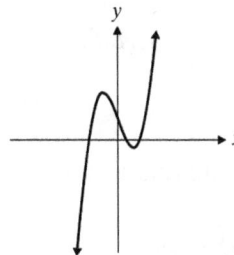

(b)

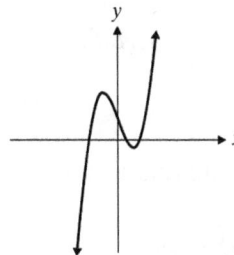

Copyright © 2017 Pearson Education, Inc.

Example	Student Practice

9. If $f(x) = 3x^2 - 4x + 5$, find each of the following.

(a) $f(-2)$

$$f(-2) = 3(-2)^2 - 4(-2) + 5$$
$$= 3(4) + -4(-2) + 5$$
$$= 12 + 8 + 5 = 25$$

(b) $f(4)$

$$f(4) = 3(4)^2 - 4(4) + 5$$
$$= 3(16) + -4(4) + 5$$
$$= 48 - 16 + 5 = 37$$

(c) $f(0)$

$$f(0) = 3(0)^2 - 4(0) + 5 = 5$$

10. If $f(x) = 4x^2 - 2x + 7$, find each of the following.

(a) $f(-3)$

(b) $f(2)$

(c) $f(0)$

Extra Practice

1. Find the domain and range of the relation. Determine whether the relation is a function.

$$\{(2.5, 3), (3.5, 0), (5.5, -2), (8.5, -6)\}$$

2. Given $f(x) = 2x^2 - 3x + 1$, find the indicated values.

(a) $f(0)$

(b) $f(-3)$

(c) $f(3)$

3. Determine whether the relation is a function.

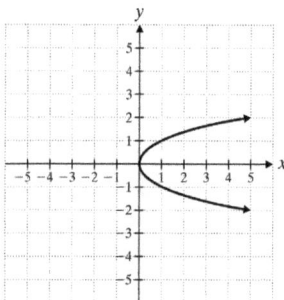

4. Graph $y = -3x^2$.

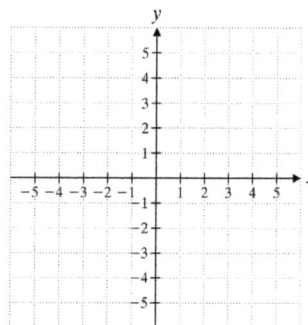

Concept Check

In the relation $\{(3, 4), (5, 6), (3, 8), (2, 9)\}$, why is there a different number of elements in the domain than the range?

MATH COACH

Mastering the skills you need to do well on the test.

Watch the **MATH COACH** videos in MyMathLab® or on YouTube™ while you work the problems below. These helpful hints will help you avoid making common errors on test problems.

Graphing a Linear Equation by Plotting Three Ordered Pairs—Problem 4   Graph $y = \dfrac{2}{3}x - 4$.

> **Helpful Hint:** Find three ordered pairs that are solutions to the equation. Plot those 3 points. Then draw a line through the points. When the equation is solved for y and there are fractional coefficients on x, it is sometimes a good idea to choose values for x that result in integer values for y. This will make graphing easier.

Choosing values for x that result in integer values for y means choosing 0 or multiples of 3 since 3 is the denominator of the fractional coefficient on x. When we multiply by these values, the result becomes an integer.

Did you choose values for x that result in values for y that are not fractions? Yes _____ No _____

If you answered No, try using $x = 0$, $x = 3$, and $x = 6$. Solve the equation for y in each case to find the y-coordinate. Remember that it will make the graphing process easier if you choose values for x that will clear the fraction from the equation.

Did you plot the three points and connect them with a line? Yes _____ No _____

If you answered No, go back and complete this step.

If you feel more comfortable using the slope-intercept method to solve this problem, simply identify the y-intercept from the equation in $y = mx + b$ form, and then use the slope m to find two other points.

If you answered Problem 4 incorrectly, go back and rework the problem using these suggestions.

Write the Equation of a Line Given Two Points—Problem 10
Find an equation for the line passing through $(5, -4)$ and $(-3, 8)$.

> **Helpful Hint:** When given two points (x_1, y_1) and (x_2, y_2), we can find the slope m using the slope formula $m = \dfrac{y_2 - y_1}{x_2 - x_1}$. Then you can substitute m into the equation $y = mx + b$ along with the coordinates of one of the points to find b, the y-intercept.

When you substituted the points into the slope formula to find m, did you obtain either $m = \dfrac{8 - (-4)}{-3 - 5}$ or

$m = \dfrac{-4 - 8}{5 - (-3)}$?

Yes _____ No _____

If you answered No, check your work to make sure you substituted the points correctly. Be careful to avoid any sign errors.

(continued on next page)

121

Did you use $m = -\dfrac{3}{2}$ and either of the points given when you substituted the values into the equation $y = mx + b$ to find the value of b? Yes ____ No ____

If you answered No, stop and make a careful substitution for $m = -\dfrac{3}{2}$ and either $x = 5$ and $y = -4$ or $x = -3$ and $y = 8$.
See if you can solve the resulting equation for b.

Now go back and rework the problem using these suggestions.

Graphing Linear Inequalities in Two Variables—Problem 12

Graph the region described by $-3x - 2y > 10$.

> **Helpful Hint:** First graph the equation $-3x - 2y = 10$. Determine if the line should be solid or dashed. Then pick a test point to see if it satisfies the inequality $-3x - 2y > 10$. If the test point satisfies the inequality, shade the side of the line on which the point lies. If the test point does not satisfy the inequality, shade the opposite side of the line.

Examine your work. Does the line $-3x - 2y = 10$ pass through the point $(0, -5)$? Yes ____ No ____

If you answered No, substitute $x = 0$ into the equation and solve for y. Check the calculations for each of the points you plotted to find the graph of this equation.

Did you draw a solid line? Yes ____ No ____

If you answered Yes, look at the inequality symbol. Remember that we only use a solid line with the symbols $\leq$ and $\geq$. A dashed line is used for $<$ and $>$.

Did you shade the area above the dashed line? Yes ____ No ____

If you answered Yes, stop now and use $(0, 0)$ as a test point and substitute it into the inequality $-3x - 2y > 10$. Then use the Helpful Hint to determine which side to shade.

If you answered Problem 12 incorrectly, go back and rework the problem using these suggestions.

Using Function Notation to Evaluate a Function—Problem 16(a) and 16(b)

For $f(x) = -x^2 - 2x - 3$: **(a)** find $f(0)$. **(b)** find $f(-2)$.

> **Helpful Hint:** Replace x with the number indicated. It is a good idea to place parentheses around the value to avoid any sign errors. Then use the order of operations to evaluate the function in each case.

(a) Did you replace x with 0 and write
$$f(0) = -(0)^2 - 2(0) - 3?$$
Yes ____ No ____

If you answered No, take time to go over your steps one more time, remembering that 0 times any number is 0.

(b) Did you replace x with -2 and write
$$f(0) = -(0)^2 - 2(0) - 3?$$
Yes ____ No ____

If you answered No, go over your steps again, remembering to place parentheses around -2.

Note that $(-2)^2 = 4$ and therefore
$$-(-2)^2 = -4.$$

Now go back and rework the problem again using these suggestions.

122

Name: _____   Date: _____

Instructor: _____   Section: _____

Chapter 4 Systems of Equations
4.1 Solving a System of Equations in Two Variables by Graphing

Vocabulary
system of equations • consistent • inconsistent • dependent • coordinate system
slope • linear equation • intersection

1. A linear system of equations that has one solution is said to be _____.

2. A(n) _____, is any set of equations in multiple variables that is considered at the same time.

3. A system of equations with infinite solutions is called _____.

4. A system of equations that has no solution is said to be _____.

Example	Student Practice
1. Solve by graphing. $\begin{array}{l}-2x+3y=6\\2x+3y=18\end{array}$	2. Solve by graphing. $\begin{array}{l}2x+2y=4\\-3x+y=-2\end{array}$

Solve each equation for y. Use a table to find and graph a series of ordered pairs.

$$y=\frac{2}{3}x+2$$

$$y=-\frac{2}{3}x+6$$

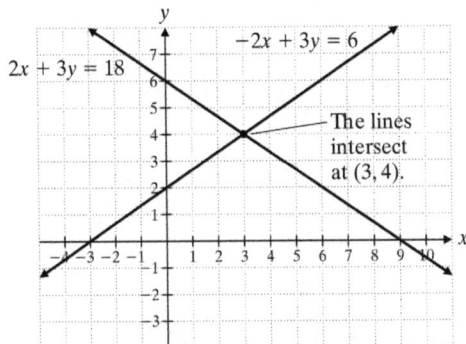

The unique solution is the point of intersection, $(3,4)$.

Example	Student Practice

Example

3. Solve by graphing. $\begin{array}{l} 3x - y = 1 \\ 3x - y = -7 \end{array}$

Graph both equations on the same coordinate system.

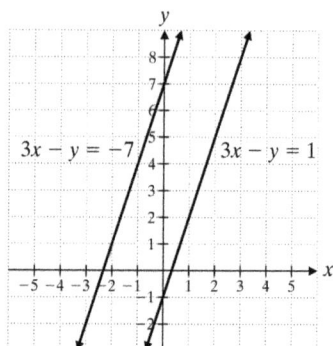

$3x - y = -7$ $3x - y = 1$

The two lines are parallel. A system of linear equations that does not intersect does not have a solutions and is called inconsistent.

5. Solve by graphing. $\begin{array}{l} x + y = 4 \\ 3x + 3y = 12 \end{array}$

Graph both equations on the same coordinate system.

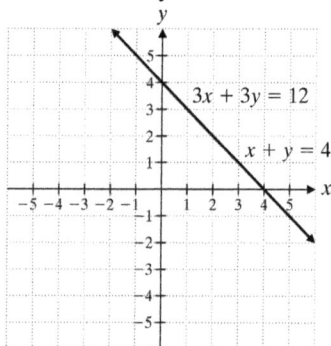

$3x + 3y = 12$

$x + y = 4$

The two equations represent the same line; they coincide. There is an infinite number of solutions to this system. Such equations are said to be dependent.

Student Practice

4. Solve by graphing. $\begin{array}{l} 2x + 4y = 0 \\ \dfrac{1}{2}x + y = 3 \end{array}$

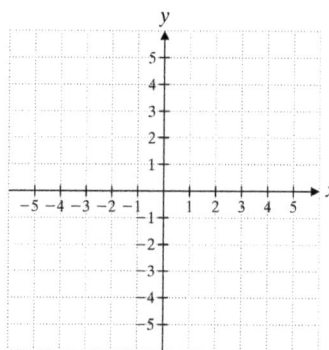

6. Solve by graphing. $\begin{array}{l} -3x + 6y = 24 \\ -x + 2y = 8 \end{array}$

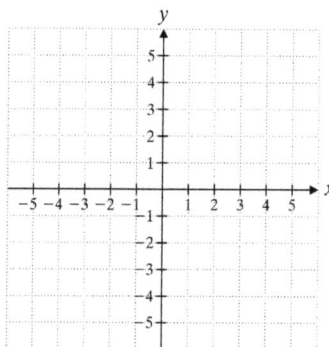

Example	Student Practice

7. Roberts Plumbing and Heating charges $40 for a house call and then $35 per hour for labor. Instant Plumbing Repairs charges $70 for a house call and then $25 per hour for labor.

(a) Create a cost equation for each company, where y is the total cost of plumbing repairs and x is the number of hours per labor. Write a system of equations.

For each company obtain a cost equation.

$$\begin{array}{cccc} \textit{Cost of} & \textit{house} & \textit{per} & \textit{labor} \\ \textit{plumbing} = & \textit{call} & + \textit{hour} & \times \textit{hours} \end{array}$$

$$y = 40 + 35 \times x$$
$$y = 70 + 25 \times x$$

(b) Graph the two equations and determine from your graph how many hours of plumbing repairs would be required for the two companies to charge the same.

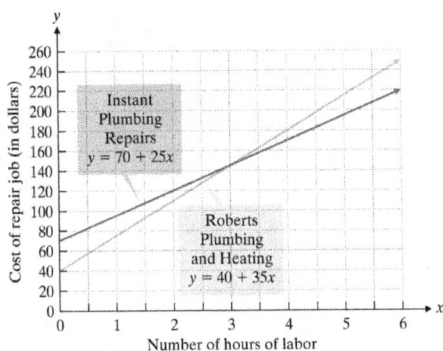

We see that the graphs of the two lines intersect at $(3,145)$. Thus the two companies will charge the same if 3 hours of plumbing repairs are required.

8. Suppose Walter and Barbara try two more plumbers. Plumbers A charges $50 for a house call and then $40 per hour for labor. Plumbers B charges $90 for a house call and then $30 per hour for labor.

(a) Create a cost equation for each company, where y is the total cost of plumbing repairs and x is the number of hours per labor. Write a system of equations.

(b) Graph the two equations.

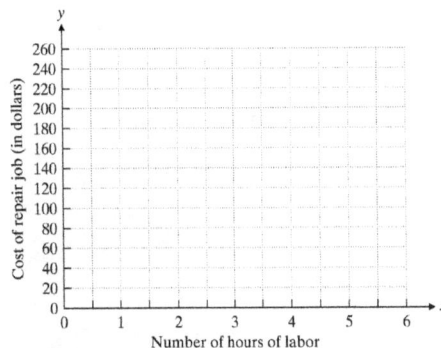

(c) Determine from your graph how many hours of plumbing repairs would be required for the two companies to charge the same.

(d) Determine from your graph which company charges less if the estimated amount of time to complete the plumbing repairs is 5 hours.

Extra Practice

1. Solve by graphing. If there isn't a unique solution to the system, state the reason.

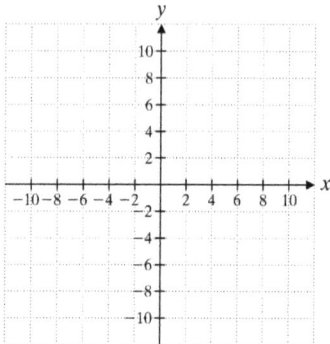

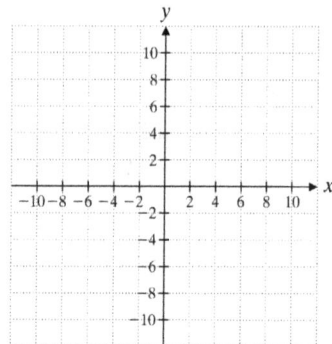

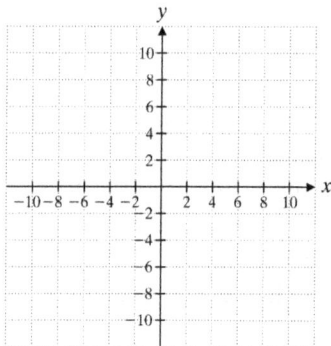

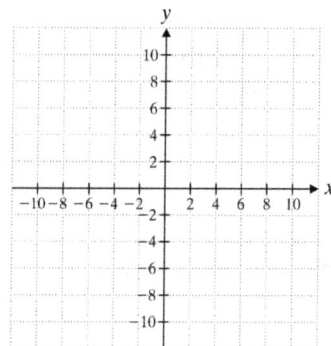

   $3x - y = 0$

   $2x + y = -10$

2. Solve by graphing. If there isn't a unique solution to the system, state the reason.

   $-6x + 2y = 5$

   $3x - y = 3$

3. Solve by graphing. If there isn't a unique solution to the system, state the reason.

   $2x + 3y = 18$

   $3x - 2y = 14$

4. Solve by graphing. If there isn't a unique solution to the system, state the reason.

   $4x - 8y = 20$

   $-2x + 4y = -10$

Concept Check

You are attempting to find the solution to this system of equations by graphing.

$-3x + 4y = -16$

$0 = 6x - 8y + 16$

What will you discover?

Name: _____   Date: _____

Instructor: _____   Section: _____

Chapter 4 Systems of Equations
4.2 Solving a System of Equations in Two Variables by the Substitution Method

Vocabulary
substitution method • system of equations • consistent • inconsistent
linear equation • dependent

1. An equation of the form $Ax + By = C$ is called a(n) _____.

2. The _____ is a strategy for solving a system of two linear equations that involves reducing the system to one equation in one variable which can be solved for.

Example	**Student Practice**
1. Find the solution. $x - 2y = 7$ (1) $-5x + 4y = -5$ (2)	**2.** Find the solution. $3x + y = -3$ $-5x - 2y = 4$

Solve equation (1) for x.

$x = 7 + 2y$

Substitute this expression into (2).

$-5x + 4y = -5$
$-5(7 + 2y) + 4y = -5$

Solve this equation for y.

$-35 - 10y + 4y = -5$
$-6y = 30$
$y = -5$

Use the value for y to find the value for x.

$x = 7 + 2y$
$x = 7 + 2(-5)$
$x = -3$
The solution is $(-3, -5)$.

Example	Student Practice
3. Find the solution. $\frac{3}{2}x + y = \frac{5}{2}$ (1) $-y + 2x = -1$ (2)	**4.** Find the solution. $\frac{4}{3}x - \frac{2}{3}y = \frac{1}{12}$ $-y - x = -1$

Clear the first equation of fractions.

$$2\left(\frac{3}{2}x\right) + 2(y) = 2\left(\frac{5}{2}\right)$$

$$3x + 2y = 5$$

The new system is as follows.

$$3x + 2y = 5 \quad (3)$$
$$-y + 2x = -1 \quad (2)$$

Solve for y in equation (2).

$$y = 1 + 2x$$

Substitute this expression into (3) and solve for x.

$$3x + 2(1 + 2x) = 5$$
$$7x + 2 = 5$$
$$x = \frac{3}{7}$$

Use the value for x to find the value for y.

$$y = 1 + 2x$$
$$y = 1 + 2\left(\frac{3}{7}\right)$$
$$y = \frac{13}{7}$$

The solution is $\left(\frac{3}{7}, \frac{13}{7}\right)$.

Extra Practice

1. Find the solution to the system of equations by the substitution method. Write your solution in the form (a, b).

    $5a + 7b = -21$

    $a + 18 = -6b$

2. Find the solution to the system of equations by the substitution method. Write your solution in the form (x, y).

    $x = 6 + 4y$

    $-3(x + y) = -36$

3. Find the solution to the system of equations by the substitution method. Write your solution in the form (x, y).

    $8x - 5y = -78$

    $6x - 3y = -54$

4. Find the solution to the system of equations by the substitution method. Write your solution in the form (x, y).

    $5x + 8(y - 3) = 15$

    $2(x + 2y) = 42$

Concept Check

Explain how you would solve the following system by the substitution method.

$3x - y = -4$

$-2x + 4y = 36$

Chapter 4 Systems of Equations
4.3 Solving a System of Equations in Two Variables by the Addition Method

Vocabulary
substitution method • system of equations • addition method • elimination method

1. The addition method is often referred to as the _____.

2. The _____ is a strategy for solving a system of two linear equations that works well for integer, fractional, and decimal coefficients.

Example

1. Solve by addition. $5x + 2y = 7$ (1)

$\qquad\qquad\qquad\quad 3x - y = 13$ (2)

Make the coefficients of the y-terms opposites by multiplying each term of equation (2) by 2.

$$2(3x) - 2(y) = 2(13)$$

Multiply, then add the equations.

$5x + 2y = 7$
$\underline{6x - 2y = 26}$
$11x \quad\;\; = 33$
$\qquad\;\; x = 3$

Substitute $x = 3$ into (1) and solve for y.

$5(3) + 2y = 7$
$\;\; 15 + 2y = 7$
$\qquad\;\; 2y = -8$
$\qquad\;\;\; y = -4$

The solution is $(3, -4)$.

Student Practice

2. Solve by addition. $5x - y = 21$

$\qquad\qquad\qquad\qquad 3x + 3y = 9$

Example	Student Practice
3. Solve. $x - \dfrac{5}{2}y = \dfrac{5}{2}$ (1) $\dfrac{4}{3}x + y = \dfrac{23}{3}$ (2)	**4.** Solve. $\dfrac{3}{4}x - \dfrac{5}{4}y = -\dfrac{9}{4}$ $\dfrac{1}{3}x + y = \dfrac{11}{3}$

Multiply each term of (1) by 2 and each term of (2) by 3 to clear the fractions.

$$2(x) - 2\,\frac{5}{2}y = 2\,\frac{5}{2}$$

$$3\,\frac{4}{3}x + 3(y) = 3\,\frac{23}{3}$$

$$2x - 5y = 5 \qquad (3)$$
$$4x + 3y = 23 \qquad (4)$$

Make the coefficients of the x-terms opposites by multiplying each term of equation (3) by -2. Add and solve for y.

$$-4x + 10y = -10$$
$$\underline{4x + 3y = 23}$$
$$13y = 13$$
$$y = 1$$

Substitute $y = 1$ into (4) and solve for x.

$$4x + 3(1) = 23$$
$$4x = 20$$
$$x = 5$$

The solution is $(5, 1)$.

The check is left to the student.

Example	Student Practice

5. Solve. $0.12x + 0.05y = -0.02$ (1)
      $0.08x - 0.03y = -0.14$ (2)
	6. Solve. $0.12x - 0.67y = -1.03$
	            $0.50x + 0.75y = -0.75$

Since the decimals are hundredths, multiply each term of both equations by 100.

$$100(0.12x) + 100(0.05y) = 100(-0.02)$$
$$100(0.08x) - 100(0.03y) = 100(-0.14)$$

$$12x + 5y = -2 \qquad (3)$$
$$8x - 3y = -14 \qquad (4)$$

Make the coefficients of the y-terms opposites by multiplying each term of (3) by 3 and each term of (4) by 5.

$$36x + 15y = -6$$
$$\underline{40x - 15y = -70}$$
$$76x = -76$$
$$x = -1$$

Substitute x into (3) and solve for y.

$$12(-1) + 5y = -2$$
$$y = 2$$

The solution is $(-1, 2)$.

To check, substitute the solution into the original equations.

$$0.12(-1) + 0.05(2) = -0.02$$
$$-0.02 = -0.02 \quad \text{Checks}$$
$$0.08(-1) - 0.03(2) = -0.14$$
$$-0.14 = -0.14 \quad \text{Checks}$$

Extra Practice

1. Find the solution by the addition method. Check your answers.

$$x - y = 6$$
$$7x - 9y = -4$$

2. Find the solution by the addition method. Check your answers.

$$7x + 13 = 9y$$
$$3x + 5y = -41$$

3. Find the solution by the addition method. Check your answers.

$$x + \frac{3}{2}y = 3$$
$$7x - 9y = -5$$

4. Find the solution by the addition method. Check your answers.

$$6x - y = 5$$
$$6x + 3y = -3$$

Concept Check

Explain how you would obtain an equivalent system that does not have decimals if you wanted to solve the following system.

$$0.5x + 0.4y = -3.4$$
$$0.02x + 0.03y = -0.08$$

Chapter 4 Systems of Equations
4.4 Review of Methods for Solving Systems of Equations

Vocabulary
substitution method • addition method • inconsistent • identity • dependent

1. The equation $7 = 7$ is always true; it is an example of a(n) _____.

2. The _____ works well if one or more variables have a coefficient of 1 or -1.

3. A(n) _____ system of linear equations is a system of parallel lines that are not equal.

4. _____ equations produce a system with an unlimited number of solutions.

Example	**Student Practice**
1. Select a method and solve the system of equations.	**2.** Select a method and solve the system of equations.

Example:

$$x + y = 3080$$
$$2x + 3y = 8740$$

Student Practice:

$$3x - 4y = 13$$
$$9x - 2y = 119$$

Use substitution since there are x- and y-values that have coefficients of 1. Solve the first equation for y and substitute it into the other to find x.

$$y = 3080 - x$$
$$2x + 3(3080 - x) = 8740$$
$$2x + 9240 - 3x = 8740$$
$$x = 500$$

Substitute 500 for x and solve for y.

$$y = 3080 - 500$$
$$y = 2580$$

The solution is $(500, 2580)$.

Example	Student Practice

3. Solve algebraically. $\quad 3x - y = -1$
$\quad 3x - y = -7$

The addition method would be very convenient in this case.

Multiply each term in the second equation by -1 and then add the two equations.

$\quad 3x - y = -1$
$\underline{-3x + y = +7}$
$\qquad 0 = 6$

The statement $0 = 6$ is not true, thus there is no solution to this system of equations. The lines are parallel and the system is inconsistent.

4. Solve algebraically. $\quad -3x - y = -5$
$\quad 6x + 2y = -1$

---

5. Solve algebraically. $\quad x + y = 4$
$\quad 3x + 3y = 12$

Let us use the substitution method. Solve the first equation for y and substitute it into the second equation.

$\qquad y = 4 - x$
$3x + 3(4 - x) = 12$
$\quad 3x + 12 - 3x = 12$
$\qquad\qquad 12 = 12$

$12 = 12$ is always true; it is an identity. This means all the solutions of one equation of the system are also solutions of the other equation. Therefore, the lines coincide and there are an infinite number of solutions to this system of equations.

6. Solve algebraically. $\quad 4x - y = 3$
$\quad 12x - 3y = 9$

Extra Practice

1. If possible, solve by an algebraic method. Otherwise, state that the problem has no solution or an infinite number of solutions.

$$9x + 1 = 7y$$
$$-2x - 4y = 28$$

2. If possible, solve by an algebraic method. Otherwise, state that the problem has no solution or an infinite number of solutions.

$$9x - 3y = 10$$
$$45x - 15y = 20$$

3. If possible, solve by an algebraic method. Otherwise, state that the problem has no solution or an infinite number of solutions.

$$\frac{3}{2}x + \frac{1}{2}y - 7 = 0$$
$$\frac{2}{5}x - \frac{3}{5}y + \frac{9}{5} = 0$$

4. If possible, solve by an algebraic method. Otherwise, state that the problem has no solution or an infinite number of solutions.

$$3x + y = -14$$
$$y = 3x + 4$$

Concept Check

Explain how you would remove the fractions in order to solve the following system.

$$\frac{3}{10}x + \frac{2}{5}y = \frac{1}{2}$$
$$\frac{1}{8}x - \frac{3}{16}y = -\frac{1}{2}$$

Name: _____ Date: _____
Instructor: _____ Section: _____

Chapter 4 Systems of Equations
4.5 Solving Word Problems Using Systems of Equations

Vocabulary
substitution method • addition method • elimination method • system of equations

1. When solving a word problem with two equations and two unknown it is often helpful to set up a(n) _____.

2. If one or more variables have a coefficient of 1 or −1, the _____ works well.

Example	**Student Practice**
1. A worker in a large post office is trying to verify the rate at which two electronic card-sorting machines operate. Yesterday the first machine sorted for 3 minutes and the second machine sorted for 4 minutes. The total workload both machines processed during that time was 10,300 cards. Two days ago the first machine sorted for 2 minutes and the second machine sorted for 3 minutes. The total workload both machines processed during that time period was 7400 cards. Can you determine the number of cards per minute sorted by each machine?	**2.** A landscape company is trying to verify the rate at which two employees can plant bulbs in a large park. Yesterday the first employee planted for 5 hours and the second employee planted for 2 hours. The total number of bulbs planted during that time was 115. Last week the first employee planted for 3 hours and the second employee planted for 4 hours. The total number of bulbs planted during that time was 125. Can you determine the number of bulbs planted per hour by each employee?
Let $x =$ the number of cards per minute sorted by the first machine and $y =$ the number of cards per minute sorted by the second machine and write a system of two equations with two unknowns using the information from the two days. $3x + 4y = 10,300$ Yesterday $2x + 3y = 7400$ Two days ago	
Solve the system using the addition method. The first machine sorts 1300 cards per minute and the second machine sorts 1600 cards per minute.	

Example	Student Practice

3. A lab technician is required to prepare 200 liters of a solution. The prepared solution must contain 42% fungicide. The technician wishes to combine a solution that contains 30% fungicide with a solution that contains 50% fungicide. How much of each solution should he use?

We need to find the amount of each solution that will give us 200 liters of a 42% solution. Let x = the number of liters of the 30% solution needed and y = the number of liters of the 50% solution needed.

Write an equation for combining x and y to get 200 liters.

$$x + y = 200$$

Write an equation for the percent of fungicide in each solution (30% of x liters, 50% of y liters, and 42% of 200 liters).

$$0.3x + 0.5y = 0.42(200) = 84$$

Solve the following system of equations using the addition method.

$$x + y = 200$$
$$0.3x + 0.5y = 84$$

The technician should use 80 liters of the 30% fungicide and 120 liters of the 50% fungicide to obtain the required solution.

The check is left to the student.

4. A chemistry student is required to prepare 150 milliliters of a solution. The prepared solution must contain 15% alcohol. The student wishes to combine a solution that contains 25% alcohol with a solution that contains 10% alcohol. How much of each solution should he use?

Example	Student Practice
5. Mike recently rode his boat on Lazy River. He took a 48-mile trip up the river against the current in exactly 3 hours. He refueled and made the return trip in exactly 2 hours. What was the speed of his boat in still water and the speed of the current in the river?	**6.** Gary flew his small airplane 150 nautical miles to visit a friend. On the way there, he flew against a strong wind for 2.5 hours. On the way home, he flew in the same direction as the wind and made the trip in exactly 1.5 hours. What was the speed of his plane in still air and the speed of the wind that day?

5. Mike recently rode his boat on Lazy River. He took a 48-mile trip up the river against the current in exactly 3 hours. He refueled and made the return trip in exactly 2 hours. What was the speed of his boat in still water and the speed of the current in the river?

Let b = the speed of the boat in still water in miles/hour and let c = the speed of the river current in miles/hour.

When we travel against the current, the current is slowing us down. Since the current's speed opposes the boat's speed in still water, we must subtract: $b - c$.

When we travel with the current, the current is helping us travel forward. The current's speed is added to the boat's speed in still water, we must add: $b + c$.

Use the distance formula,

$$\text{distance} = \text{rate} \times \text{time}$$

to write a system of equations.

$48 = (b - c) \cdot 3$ Against the current
$48 = (b + c) \cdot 2$ With the current

Remove the parenthesis to get the following system of equations.

$48 = 3b - 3c$ Against the current
$48 = 2b + 2c$ With the current

Solve the system of equations using the addition method to find that the speed of Mike's boat in still water was 20 miles/hour and the speed of the current in Lazy River was 4 miles/hour.

Extra Practice

1. Elias and Allison found 26 coins. The coins were either quarters or nickels. The total value they had was $4.70. How many quarters and nickels did they find?

2. A bulk food store wants to make a 10-pound box of snack bars that sells for $50. The box will contain large snack bars that sell for $6.00 per pound and small snack bars that sell for $4.00 per pound. How many pounds of each snack bar should they include in the box to obtain the desired mixture?

3. Ronaldo and Lee both work at the same construction site. Ronaldo worked for 5 hours and Lee worked for 6 hours. Together they hammered 875 nails. The next day, Ronaldo worked for 8 hours and Lee worked for 5 hours. Together, they hammered 1055 nails. How many nails does Ronaldo hammer per hour? How many nails does Lee hammer per hour?

4. The present population of Anytown is 45,600. The town is growing at the rate of 400 people per year. The population of the city of Anyplace is 53,100 and is increasing at the rate of 100 people per year. How many years will it be until the populations of Anytown and Anyplace are the same? What will each population be?

Concept Check

A new company is making a drink that is 45% pure fruit juice. They make a test batch of 500 gallons of the new drink. They are using some juice that is 50% pure fruit juice and some juice that is 30% pure fruit juice. They want to find out how many gallons of each of these two types they will need. Explain how you would set up a system of two equations using the variables x and y to solve this problem.

MATH COACH

Mastering the skills you need to do well on the test.

Watch the **MATH COACH** videos in MyMathLab° or on You Tube™
while you work the problems below. These helpful hints will
help you avoid making common errors on test problems.

**Solving a System of Equations by the Substitution
Method—Problem 1** Solve by the substitution method.

$$3x - y = -5$$
$$-2x + 5y = -14$$

> **Helpful Hint:** If one equation contains $-y$, it is easier to solve for y by
> adding $+y$ to each side of the equation. Use this result to substitute for y
> in the second equation.

Did you add y to each side of the first equation and then add
5 to each side to obtain $3x + 5 = y$? Yes _____ No _____

If you answered No, consider why solving for y in the first
equation is the most logical first step when using the
substitution method to solve this system.

Did you substitute $(3x + 5)$ for y in the second equation to
obtain $-2x + 5(3x + 5) = -14$ and then simplify and solve for
x? Yes _____ No _____

If you answered No, stop and perform these
steps. Remember to substitute your final value
of x into one of the original equations to find
y.

If you answered Problem 1 incorrectly, go
back and rework the problem using these
suggestions.

Solving a system of Equations with Fractional Coefficients—Problem 7

Solve by any method.
$$\frac{2}{3}x - \frac{1}{5}y = 2$$
$$\frac{4}{3}x + 4y = 4$$

> **Helpful Hint:** Find the LCD of the fractions in the top equation. Multiply each term of that equation by this
> LCD. Find the LCD of the fractions in the bottom equation. Multiply each term of that equation by this LCD.
> Now use these two new equations to solve the system.

Did you find that 15 is the LCD of the top equation? Did you
multiply all three terms of the top equation by 15 to obtain
$10x - 3y = 30$? Yes _____ No _____

If you answered No to theses questions, consider why the
LCD is 15 and then carefully multiply each term of the top
equation by 15 to find the correct result.

Did you find that 3 is the LCD of the bottom equation? Did
you multiply all three terms of the bottom equation by 3 to
obtain $4x + 12y = 12$? Yes _____ No _____

If you answered No to these questions,
consider why the LCD is 3 and then carefully
multiply each term of the bottom equation by
3 to find the correct result. Then use these two
new equations to solve the system.

Now go back and rework the problem again
using these suggestions.

Solving an Inconsistent or Dependent System of Equations—Problem 9

Solve by any method. If there is not one solution to a system, state why.

$$5x - 2 = y$$
$$10x = 4 + 2y$$

> **Helpful Hint:** When you try to solve a system and get a false statement such as $3 = 0,$ then the system has no solution and is inconsistent. When you try to solve a system and get a statement that is always true such as $4 = 4,$ then the system has an infinite number of solutions and is dependent.

Since the top equation is already solved for $y,$ it makes sense to use the substitution method.

Did you substitute $(5x - 2)$ for y in the bottom equation to obtain $10x = 4 + 2(5x - 2)$? Yes _____ No _____

If you answered No, go back and make this substitution.

Did you simplify the equation to obtain $0 = 0$?
Yes _____ No _____

If you used the addition method, you should still obtain $0 = 0.$ Did you determine that this system has an infinite number of solutions?

Yes _____ No _____

If you answered No to these questions, examine your work for any errors and review the definition of inconsistent and dependent systems. In your final description, state whether the system is inconsistent or dependent and also state whether the system has no solution or an infinite number of solutions.

If you answered Problem 9 incorrectly, go back and rework the problem using these suggestions.

Solving a System of Equations with Decimal Coefficients—Problem 10

Solve by any method.
$$0.3x + 0.2y = 0$$
$$1.0x + 0.5y = -0.5$$

> **Helpful Hint:** Since the decimals are tenths, multiply each term of each equation by 10 to find an equivalent system of equations without decimal coefficients. Then solve the system.

Did you multiply both equation by 10 to obtain the system
$$3x + 2y = 0$$
$$10 + 5y = -5?$$

Yes _____ No _____

If you answered No, carefully go through each equation and move the decimal point one place to the right for each number. This represents multiplying by 10. See if you get the above result.

With the new system, a good approach is to multiply the top equation by 5 and the bottom equation by −2. If you follow these steps, do you get the system
$$15x + 10y = 0$$
$$-20x - 10y = 10?$$

Yes _____ No _____

If you answered No, stop and carefully multiply every term of the top equation by 5. Remember that $5 \times 0 = 0.$

Next, multiply every term of the bottom equation by −2 and complete the problem. Watch out for + and − signs. Remember to substitute your final value for x into either of the original equations to find $y.$

Now go back and rework the problem using these suggestions.

Chapter 5 Exponents and Polynomials
5.1 The Rules of Exponents

Vocabulary
exponent   •   base   •   exponential expression   •   the product rule
numerical coefficient   •   the quotient rule

1. In the exponential expression x^a, x is called the _____.

2. In the exponential expression x^a, a is called the _____.

3. Simplifying the exponential expression $\dfrac{x^a}{x^b}$ requires using the _____.

4. A(n) _____ is a number that is multiplied by a variable, such as the 4 in $4x^2$.

Example	Student Practice
1. Multiply.	**2.** Multiply.
(a) $x^3 \cdot x^6$	**(a)** $z^4 \cdot z$
The expressions have the same base so we can add the exponents.	
$x^3 \cdot x^6 = x^{3+6} = x^9$	
(b) $x \cdot x^5$	**(b)** $2^2 \cdot 2^6$
Every variable that does not have a written exponent is understood to have an exponent of 1. Thus, $x = x^1$.	
$x \cdot x^5 = x^{1+5} = x^6$	

Example	Student Practice
3. Multiply. $(5ab)\left(-\dfrac{1}{3}a\right)(9b^2)$	**4.** Multiply. $(-4x^3)(xy^2)(2x^2y)$

Multiply the numerical coefficients and group like bases.

$$(5ab)\left(-\dfrac{1}{3}a\right)(9b^2)$$

$$= (5)\left(-\dfrac{1}{3}\right)(9)(a \cdot a)(b \cdot b^2)$$

Use the rule for multiplying expression with exponents. Add the exponents.

$$(5)\left(-\dfrac{1}{3}\right)(9)(a \cdot a)(b \cdot b^2) = -15a^2b^3$$

Example	Student Practice
5. Divide. $\dfrac{2^{16}}{2^{11}}$	**6.** Divide. $\dfrac{x^9}{x^5}$

The expressions have the same base so we can subtract the exponents.

$$\dfrac{2^{16}}{2^{11}} = 2^{16-11} = 2^5$$

Example	Student Practice
7. Divide. $\dfrac{12^{17}}{12^{20}}$	**8.** Divide. $\dfrac{n^6}{n^{10}}$

The expressions have the same base so we can subtract the exponents. Notice that the larger exponent is in the denominator.

$$\dfrac{12^{17}}{12^{20}} = \dfrac{1}{12^{20-17}} = \dfrac{1}{12^3}$$

Example	Student Practice
9. Simplify. $\dfrac{4x^0y^2}{8^0y^5z^3}$ Any number (except 0) to the 0 power equals 1. $\dfrac{4x^0y^2}{8^0y^5z^3} = \dfrac{4(1)y^2}{(1)y^5z^3}$ $= \dfrac{4y^2}{y^5z^3}$ $= \dfrac{4}{y^3z^3}$	**10.** Simplify. $\dfrac{(2y^4)(3x^2y)}{12x^0y^{10}}$
11. Simplify. **(a)** $\left(x^3\right)^5$ We are raising a power to a power, so multiply exponents. $\left(x^3\right)^5 = x^{3\cdot5} = x^{15}$ **(b)** $(-1)^8$ Since n is even, a positive number results. $\left(-1^8\right) = +1$	**12.** Simplify. **(a)** $\left(y^8\right)^4$ **(b)** $(-1)^{13}$
13. Simplify. $\left(\dfrac{x}{y}\right)^5$ The fraction within the parentheses is raised to a power. Raise both the numerator and the denominator to that power. $\left(\dfrac{x}{y}\right)^5 = \dfrac{x^5}{y^5}$	**14.** Simplify. $\left(\dfrac{y}{z^2}\right)^3$

Example	Student Practice
15. Simplify. $\left(\dfrac{-3x^2z^0}{y^3}\right)^4$	**16.** Simplify. $\left(\dfrac{4x^5y}{-16x^0y^3}\right)^3$

Simplify inside the parentheses first. Apply the rules for raising a power to a power and simplify.

$$\left(\dfrac{-3x^2z^0}{y^3}\right)^4 = \left(\dfrac{-3x^2}{y^3}\right)^4$$

$$= \dfrac{(-3)^4 x^8}{y^{12}} = \dfrac{81x^8}{y^{12}}$$

Extra Practice

1. Multiply. Leave your answer in exponent form. $\left(12x^4y^2\right)\left(2x^2y^3\right)$

2. Simplify. Leave your answer in exponent form. Assume that all variables in any denominator are nonzero. $\dfrac{9a^3}{a^3}$

3. Simplify. Leave your answer in exponent form. Assume that all variables in any denominator are nonzero. $\left(\dfrac{b}{b^3}\right)^2$

4. Simplify. Leave your answer in exponent form. Assume that all variables in any denominator are nonzero. $\left(\dfrac{9x^2y^0}{14x^5}\right)^2$

Concept Check

Explain the steps you would need to follow to simplify the expression. $\dfrac{\left(4x^3\right)^2}{\left(2x^4\right)^3}$

Chapter 5 Exponents and Polynomials
5.2 Negative Exponents and Scientific Notation

Vocabulary

exponent • base • negative exponent • scientific notation • significant digits

1. All digits in a number, excluding the zeros that allow the decimal point to be properly located, are considered _____.

2. In the number 1.23×10^5, "10" is referred to as the _____.

3. A useful way to express very large or very small numbers is to use _____.

4. When using scientific notation to express a number less than zero, it is necessary to use a _____ on the base of ten.

Example	Student Practice
1. Evaluate. 3^{-2} First write the expression with a positive exponent. Then evaluate. $3^{-2} = \dfrac{1}{3^2} = \dfrac{1}{9}$	**2.** Evaluate. 7^{-4}
3. Simplify. Write the expression with no negative exponents. $\left(3x^{-4}y^2\right)^{-3}$ Use the power to power rule. $\left(3x^{-4}y^2\right)^{-3} = 3^{-3}x^{12}y^{-6}$ Now write the expression with positive exponents only and simplify. $3^{-3}x^{12}y^{-6} = \dfrac{x^{12}}{3^3 y^6} = \dfrac{x^{12}}{27y^6}$	**4.** Simplify. Write the expression with no negative exponents. $\left(2xy^{-2}z^3\right)^{-4}$

Example	Student Practice
5. Write in scientific notation. $157{,}000{,}000$	**6.** Write in scientific notation. 2564

5. Move the decimal point 8 places to the left and multiply by $100{,}000{,}000$.

$$157{,}000{,}000 = 1.\underbrace{57000000}_{\text{8 places}} \times 1\underbrace{00000000}_{\text{8 zeros}}$$

$$= 1.57 \times 10^{8}$$

---

7. Write in scientific notation.	**8.** Write in scientific notation.

7. (a) 0.061

Move the decimal point 2 places to the right and multiply by 10^{-2}.

$$0.061 = 6.1 \times 10^{-2}$$

(b) 0.000052

$$0.000052 = 5.2 \times 10^{-5}$$

8. (a) 0.0013

(b) 0.000001

---

9. Write in decimal notation.	**10.** Write in decimal notation.

9. (a) 1.568×10^{2}

The exponent 2 tells us to move the decimal point 2 places to the right.

$$1.568 \times 10^{2} = 156.8$$

(b) 7.432×10^{-3}

The exponent -3 tells us to move the decimal point 3 places to the left.

$$7.432 \times 10^{-3} = 7.432 \times \frac{1}{1000}$$

$$= 0.007432$$

10. (a) 5.28×10^{-4}

(b) 3.2214×10^{6}

Example	Student Practice
11. The approximate distance from Earth to the star Polaris is 208 parsecs (pc). A parsec is a distance of approximately 3.09×10^{13} km. How long would it take a space probe traveling at 40,000 km/hr to reach the star? Round to three significant digits.	**12.** Use the information in Example **11** to answer the following. How long would it take the space probe to reach a star that is 500 parsecs from Earth? Round to three significant digits.

Understand the problem. We need to change the distance from parsecs to kilometers using the given relationship.

$$208 \text{ pc} = \frac{(208 \text{ pc})(3.09 \times 10^{13} \text{ km})}{1 \text{ pc}}$$

$$= 642.72 \times 10^{13}$$

Write an equation using the distance formula, $\text{distance} = \text{rate} \times \text{time}$, or $d = r \times t$.

Substitute known values and change all given values to scientific notation.

$$6.4272 \times 10^{15} \text{ km} = \frac{4 \times 10^4 \text{ km}}{1 \text{ hr}} \times t$$

Multiply both sides by the reciprocal of $\frac{4 \times 10^4 \text{ km}}{1 \text{ hr}}$ and simplify.

$$6.4272 \times 10^{15} \text{ km} \times \frac{1 \text{ hr}}{4 \times 10^4 \text{ km}} = t$$

$$\frac{(6.4272 \times 10^{15} \text{ km})(1 \text{ hr})}{4 \times 10^4 \text{ km}} = t$$

$$1.6068 \times 10^{11} \text{ hr} = t$$

Round to three significant figures.
$$1.6068 \times 10^{11} \text{ hr} \approx 1.61 \times 10^{11} \text{hr}$$

The check is left to the student.

Extra Practice

1. Simplify. Express your answer with positive exponents. Assume that all variables are nonzero. 7^{-3}

2. Simplify. Express your answer with positive exponents. Assume that all variables are nonzero. $\left(\dfrac{3a^3b^{-2}}{c^3}\right)^{-4}$

3. Write in decimal notation. 6.34×10^3

4. Evaluate by using scientific notation and the laws of exponents. Leave your answer in scientific notation. $\dfrac{0.0046}{0.023}$

Concept Check

Explain how you would simplify a problem like the following so that your answer has only positive exponents. $\left(4x^{-3}y^4\right)^{-3}$

Chapter 5 Exponents and Polynomials
5.3 Fundamental Polynomial Operations

Vocabulary
polynomial • multivariable polynomial • degree of a term • degree of a polynomial
monomial • binomial • trinomial • decreasing order • evaluate

1. A(n) _____ is a polynomial with two terms.

2. In a term, the sum of the exponents of all of the variables is called the _____.

3. When a polynomial is written in _____, the value of each exponent decreases as we move from left to right.

4. The highest degree of all of the terms in a polynomial is called the _____.

Example	Student Practice
1. State the degree of the polynomial, and whether it is a monomial, a binomial, or a trinomial.	**2.** State the degree of the polynomial, and whether it is a monomial, a binomial, or a trinomial.
(a) $5xy + 3x^3$	**(a)** $7y^3z^4$
This polynomial is of degree 3. It has two terms, so it is a binomial.	
(b) $-7a^5b^2$	**(b)** $x^2 + 5x - 4$
The sum of the exponents is $5 + 2 = 7$. Therefore, this polynomial is of degree 7. It has one term, so it is a monomial.	
(c) $8x^4 - 9x - 15$	**(c)** $2xy^6 + 7x^2$
This polynomial is of degree 4. It has three terms, so it is a trinomial.	

Example	Student Practice
3. Add. $\left(5x^2 - 6x - 12\right) + \left(-3x^2 - 9x + 5\right)$	**4.** Add. $\left(-x^2 + x - 5\right) + \left(5x^2 - 9x + 10\right)$

Group like terms.

$$\left(5x^2 - 6x - 12\right) + \left(-3x^2 - 9x + 5\right)$$
$$= \left[5x^2 + \left(-3x^2\right)\right] + \left[-6x + \left(-9x\right)\right] + \left[-12 + 5\right]$$

Add like terms.

$$\left[\left(5-3\right)x^2\right] + \left[\left(-6-9\right)x\right] + \left[-12+5\right]$$
$$= 2x^2 + \left(-15x\right) + \left(-7\right)$$
$$= 2x^2 - 15x - 7$$

5. Add. $\left(\dfrac{1}{2}x^2 - 6x + \dfrac{1}{3}\right) + \left(\dfrac{1}{5}x^2 - 2x - \dfrac{1}{2}\right)$	**6.** Add. $\left(\dfrac{3}{4}x^2 - 2x + \dfrac{1}{8}\right) + \left(x^2 + \dfrac{2}{5}x + \dfrac{1}{2}\right)$

The numerical coefficients of polynomials may be any real number. Thus, polynomials may have numerical coefficients that are decimals or fractions.

To add, first group like terms.

$$\left(\frac{1}{2}x^2 - 6x + \frac{1}{3}\right) + \left(\frac{1}{5}x^2 - 2x - \frac{1}{2}\right)$$
$$= \left[\frac{1}{2}x^2 + \frac{1}{5}x^2\right] + \left[-6x + \left(-2x\right)\right] + \left[\frac{1}{3} + \left(-\frac{1}{2}\right)\right]$$

Add like terms.

$$\left[\left(\frac{1}{2} + \frac{1}{5}\right)x^2\right] + \left[\left(-6 - 2\right)x\right] + \left[\frac{1}{3} + \left(-\frac{1}{2}\right)\right]$$
$$= \left[\left(\frac{5}{10} + \frac{2}{10}\right)x^2\right] + \left(-8x\right) + \left[\frac{2}{6} - \frac{3}{6}\right]$$
$$= \frac{7}{10}x^2 - 8x - \frac{1}{6}$$

Example	Student Practice
7. Subtract. $\left(7x^2 - 6x + 3\right) - \left(5x^2 - 8x - 12\right)$	**8.** Subtract. $$\left(-5x^2 - 3x + 1\right) - \left(-2x^2 + 7x - 4\right)$$

7. Subtract. $\left(7x^2 - 6x + 3\right) - \left(5x^2 - 8x - 12\right)$

We change the sign of each term in the second polynomial and then add.

$$\left(7x^2 - 6x + 3\right) - \left(5x^2 - 8x - 12\right)$$

$$= \left(7x^2 - 6x + 3\right) + \left(-5x^2 + 8x + 12\right)$$

$$= \left(7 - 5\right)x^2 + \left(-6 + 8\right)x + \left(3 + 12\right)$$

$$= 2x^2 + 2x + 15$$

---

9. Automobiles sold in the United States have become more fuel efficient over the years due to regulations from Congress. The number of miles per gallon obtained by the average automobile in the United States can be described by the polynomial $0.3x + 12.9$, where x is the number of years since 1970. (*Source:* U.S. Federal Highway Administration.) Use this polynomial to estimate the number of miles per gallon obtained by the average automobile in 1972.

The year 1972 is two years later than 1970, so $x = 2$.

Thus, the number of miles per gallon obtained by the average automobile in 1972 can be estimated by evaluating $0.3x + 12.9$ when $x = 2$.

$$0.3(2) + 12.9 = 0.6 + 12.9$$

$$= 13.5$$

We estimate that the average car in 1972 obtained 13.5 miles per gallon.

10. Use the information in example **9** to answer the following. Estimate the number of miles per gallon that will be obtained by the average automobile in the United States in 2015.

Extra Practice

1. State the degree of the polynomial and whether it is a monomial, a binomial, or a trinomial. $23x^6 - 14x^3 + 5$

2. Add. $(6.4x - 3) + (4.4x - 11)$

3. Subtract.

   $$(2r^4 - 3r^2 + 14) - (-3r^4 - 2r^2 + 6)$$

4. Buses sold in the U.S. have become more fuel efficient over the years. The number of miles per gallon obtained by a certain type of bus can be described by the polynomial $0.37x + 5.31$, where x is the number of years since 1970. Use this polynomial to estimate the number of miles per gallon obtained by this type of bus in 1980.

Concept Check

Explain how you would determine the degree of the following polynomial and how you would decide if it is a monomial, a binomial, or a trinomial. $2xy^2 - 5x^3y^4$

Chapter 5 Exponents and Polynomials
5.4 Multiplying Polynomials

Vocabulary
polynomial • monomial • binomial • distributive property • FOIL

1. A _____ is a polynomial with only one term.

2. The process of using the distributive property to multiply two binomials is often referred to as _____.

3. The _____ states that for all real numbers a, b, and c, $a(b+c) = ab + ac$.

Example	Student Practice
1. Multiply. $3x^2(5x-2)$	**2.** Multiply. $5y^2(-y+4)$
Use the distributive property and multiply each term by $3x^2$.	
$3x^2(5x-2) = 3x^2(5x) + 3x^2(-2)$	
Simplify.	
$3x^2(5x) + 3x^2(-2)$ $= (3 \cdot 5)(x^2 \cdot x) + (3)(-2)x^2$ $= 15x^3 - 6x^2$	
3. Multiply. $(x^2 - 2x + 6)(-2xy)$	**4.** Multiply. $(x^2 - 7x - 10)(2x^2 y)$
Use the distributive property and multiply each term by $-2xy$.	
$(x^2 - 2x + 6)(-2xy)$ $= -2x^3 y + 4x^2 y - 12xy$	

Example	Student Practice
5. Multiply. $(2x-1)(3x+2)$	**6.** Multiply. $(3x-1)(5x+3)$

Multiply the First terms, $2x$ and $3x$.
$(2x)(3x)=6x^2$

Multiply the Outer terms, $2x$ and 2.
$(2x)(2)=4x$

Multiply the Inner terms, -1 and $3x$.
$(-1)(3x)=-3x$

Multiply the Last terms, -1 and 2.
$(-1)(2)=-2$

Add the results and combine like terms.
  First $+$ Outer $+$ Inner $+$ Last

$= 6x^2 \ + \ 4x \ - \ 3x \ - \ 2$

$= 6x^2 + x - 2$

7. Multiply. $(3x+2y)(5x-3z)$	**8.** Multiply. $(7x-3y)(x+2z)$

Multiply the First terms, $3x$ and $5x$.
$(3x)(5x)=15x^2$

Multiply the Outer terms, $3x$ and $-3z$.
$(3x)(-3z)=-9xz$

Multiply the Inner terms, $2y$ and $5x$.
$(2y)(5x)=10xy$

Multiply the Last terms, $2y$ and $-3z$.
$(2y)(-3z)=-6yz$

Add the results and combine like terms.
$(3x+2y)(5x-3z)$

$=15x^2-9xz+10xy-6yz$

Example	Student Practice

9. Multiply. $(7x-2y)^2$

When we square a binomial, it is the same as multiplying the binomial by itself.

$(7x-2y)(7x-2y)$

Multiply the First terms, $7x$ and $7x$.
$(7x)(7x)=49x^2$

Multiply the Outer terms, $7x$ and $-2y$.
$(7x)(-2y)=-14xy$

Multiply the Inner terms, $-2y$ and $7x$.
$(-2y)(7x)=-14xy$

Multiply the Last terms, $-2y$ and $-2y$.
$(-2y)(-2y)=4y^2$

Add the results and combine like terms.

$(7x-2y)^2 = 49x^2 - 14xy - 14xy + 4y^2$
$\quad = 49x^2 - 28xy + 4y^2$

10. Multiply. $(4x+3y)^2$

---

11. Multiply. $(3x^2+4y^3)(2x^2+5y^3)$

Use the FOIL method and the rules for multiplying expressions with exponents.

$(3x^2+4y^3)(2x^2+5y^3)$
$= 6x^4 + 15x^2y^3 + 8x^2y^3 + 20y^6$
$= 6x^4 + 23x^2y^3 + 20y^6$

12. Multiply. $(10x^2+5y^4)(2x^2-3y^4)$

Example	Student Practice

13. The width of a living room is $(x+4)$ feet. The length of the room is $(3x+5)$ feet. What is the area of the room in square feet?

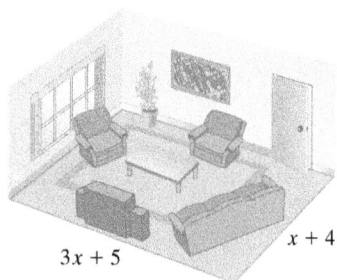

$3x + 5$ $x + 4$

Use the area formula and solve.

$A = (\text{length})(\text{width})$

$A = (3x+5)(x+4)$

$A = 3x^2 + 12x + 5x + 20$

$A = 3x^2 + 17x + 20$

There are $\left(3x^2 + 17x + 20\right)$ square feet in the room.

14. The width of a brick patio is $(2x+1)$ feet. The length of the patio is $(4x-2)$ feet. What is the area of the patio in square feet?

Extra Practice

1. Multiply. $5x\left(-2x^3 + 3x\right)$

2. Multiply. $(x-3)(x-11)$

3. Multiply. $(3x-7y)(5x+8y)$

4. Multiply. $\left(2x^2 - 3y^3\right)\left(4x^2 + 5y^3\right)$

Concept Check

Explain how you would multiply $(7x-3)^2$.

Chapter 5 Exponents and Polynomials
5.5 Multiplication: Special Cases

Vocabulary
polynomial   •   term   •   binomial   •   FOIL   •   square of a sum
square of a difference   •   vertical multiplication   •   horizontal multiplication

1. One method of multiplying polynomials that uses a method similar to that used in arithmetic for multiplying whole numbers is called _____.

2. A binomial of the form $(a-b)^2$ is referred to as the _____.

3. One way to multiply polynomials with more than two terms is to use _____ horizontal multiplication.

Example	Student Practice
1. Multiply. $(7x+2)(7x-2)$ Use the rule for multiplying a sum and a difference. $(a+b)(a-b)=a^2-b^2$ Here, $a=7x$ and $b=2$. $(7x+2)(7x-2)=(7x)^2-(2)^2$ $\qquad\qquad\qquad=49x^2-4$ The check is left to the student.	**2.** Multiply. $(8y+4)(8y-4)$
3. Multiply. $(5x-8y)(5x+8y)$ Use the rule for multiplying a sum and a difference. Here, $a=5x$ and $b=8y$. $(5x-8y)(5x+8y)=(5x)^2-(8y)^2$ $\qquad\qquad\qquad\qquad=25x^2-64y^2$	**4.** Multiply. $(2x-9y)(2x+9y)$

Example	Student Practice
5. Multiply. $(5y-2)^2$	**6.** Multiply. $(7x+y)^2$

5. Multiply. $(5y-2)^2$

Use a rule for a binomial squared.

$(a+b)^2 = a^2 + 2ab + b^2$

$(a-b)^2 = a^2 - 2ab + b^2$

Here, $a=5y$ and $b=2$.

$(5y-2)^2$

$=(5y)^2 - (2)(5y)(2) + (2)^2$

$= 25y^2 - 20y + 4$

---

6. Multiply. $(7x+y)^2$

---

7. Multiply vertically.

$(3x^3 + 2x^2 + x)(x^2 - 2x - 4)$

Place one polynomial over the other. Find the partial products and line them up under the original polynomials. Note that the answers for each partial product are placed so that like terms are underneath each other.

$$
\begin{array}{r}
3x^3 + 2x^2 + x \\
x^2 - 2x - 4 \\
\hline
-12x^3 - 8x^2 - 4x \\
-6x^4 - 4x^3 - 2x^2 \\
3x^5 + 2x^4 + x^3 \\
\end{array}
$$

Find the sum of the three partial products.

$$
\begin{array}{r}
3x^3 + 2x^2 + x \\
x^2 - 2x - 4 \\
\hline
-12x^3 - 8x^2 - 4x \\
-6x^4 - 4x^3 - 2x^2 \\
3x^5 + 2x^4 + x^3 \\
\hline
3x^5 - 4x^4 - 15x^3 - 10x^2 - 4x
\end{array}
$$

8. Multiply vertically.

$(x^2 + 3x - 6)(2x^2 - 4x - 5)$

Example	Student Practice
9. Multiply horizontally. $$\left(x^2+3x+5\right)\left(x^2-2x-6\right)$$ Use the distributive property repeatedly. $$\left(x^2+3x+5\right)\left(x^2-2x-6\right)$$ $$=x^2\left(x^2-2x-6\right)+3x\left(x^2-2x-6\right)$$ $$+5\left(x^2-2x-6\right)$$ $$=x^4-2x^3-6x^2+3x^3-6x^2-18x$$ $$+5x^2-10x-30$$ $$=x^4+x^3-7x^2-28x-30$$	**10.** Multiply horizontally. $$\left(x^2-8x+10\right)\left(x^2+x+4\right)$$
11. Multiply. $(2x-3)(x+2)(x+1)$ Multiply the first pair of binomials. Note that it does not matter which two binomials are multiplied first. $$(2x-3)(x+2)=2x^2+4x-3x-6$$ $$=2x^2+x-6$$ Replace the first two factors with their resulting product. $$\left(2x^2+x-6\right)(x+1)$$ Multiply again and combine like terms. $$\left(2x^2+x-6\right)(x+1)$$ $$=\left(2x^2+x-6\right)x+\left(2x^2+x-6\right)1$$ $$=2x^3+x^2-6x+2x^2+x-6$$ $$=2x^3+3x^2-5x-6$$	**12.** Multiply. $(x-2)(x+3)(2x+4)$

Extra Practice

1. Multiply. Use the special formula that applies. $(x+10)(x-10)$

2. Multiply. Use the special formula that applies. $(5a+3b)(5a-3b)$

3. Multiply. $(8x^2-2x+3)(3x+1)$

4. Multiply. $(x+2)(x-1)(x-5)$

Concept Check

Using the formula $(a+b)^2 = a^2+2ab+b^2$, explain how to multiply $(6x-9y)^2$.

Name: _____   Date: _____

Instructor: _____   Section: _____

Chapter 5 Exponents and Polynomials
5.6 Dividing Polynomials

Vocabulary
polynomial • subtraction • long division • descending order • binomial • monomial

1. To divide a polynomial by a _____, divide each term of the numerator by the denominator, then write the sum of the results.

2. _____ is a process used to divide polynomials when the divisor has two or more terms.

3. When dividing a polynomial by a binomial, you must first place the terms of each in _____, inserting a 0 for any missing terms.

4. When performing long division with polynomials, take great care on the_____ step when negative numbers are involved.

Example	**Student Practice**
1. Divide. $\dfrac{8y^6 - 8y^4 + 24y^2}{8y^2}$	2. Divide. $\dfrac{21x^5 + 7x^3 - 28x^2}{7x^2}$

Divide each term of the polynomial by the monomial.

$$\frac{8y^6 - 8y^4 + 24y^2}{8y^2} = \frac{8y^6}{8y^2} - \frac{8y^4}{8y^2} + \frac{24y^2}{8y^2}$$

Use the property $\dfrac{x^a}{x^b} = x^{a-b}$ to divide each term.

$$\frac{8y^6}{8y^2} - \frac{8y^4}{8y^2} + \frac{24y^2}{8y^2} = y^4 - y^2 + 3$$

Example	Student Practice

3. Divide. $\left(x^3 + 5x^2 + 11x + 4\right) \div \left(x + 2\right)$

Notice the terms are already in descending order with no missing terms. Divide the first term of the polynomial by the first term of the binomial.

$$x + 2 \overline{)\begin{array}{l} x^2 \\ x^3 + 5x^2 + 11x + 4 \end{array}}$$

Multiply x^2 by $x + 2$ and subtract the result from the first two terms.

$$x + 2 \overline{)\begin{array}{l} x^2 + 3x \\ x^3 + 5x^2 + 11x + 4 \\ \underline{x^3 + 2x^2} \\ 3x^2 + 11x \end{array}}$$

Continue this process until the degree of the remainder is less than the degree of the divisor.

$$x + 2 \overline{)\begin{array}{l} x^2 + 3x + 5 \\ x^3 + 5x^2 + 11x + 4 \\ \underline{x^3 + 2x^2} \\ 3x^2 + 11x \\ \underline{3x^2 + 6x} \\ 5x + 4 \\ \underline{5x + 10} \\ -6 \end{array}}$$

The remainder is written as the numerator of a fraction that has the binomial divisor as its denominator.

$$x^2 + 3x + 5 + \frac{-6}{x + 2}$$

The check is left to the student.

4. Divide. $\left(2x^3 + 4x^2 + 3x + 6\right) \div \left(x + 1\right)$

166

Example	Student Practice
5. Divide. $\left(5x^3 - 24x^2 + 9\right) \div \left(5x + 1\right)$	**6.** Divide. $\left(8x^3 - 8x + 5\right) \div \left(2x + 1\right)$

Insert $0x$ into the polynomial to represent the missing x-term.

$$\left(5x^3 - 24x^2 + 0x + 9\right) \div \left(5x + 1\right)$$

Divide the first term of the polynomial, $5x^3$, by the first term of the binomial, $5x$.

$$
\begin{array}{r}
x^2 \\
5x+1{\overline{\smash{\big)}\,5x^3 - 24x^2 + 0x + 9}} \\
\underline{5x^3 + x^2} \\
-25x^2
\end{array}
$$

Divide $-25x^2$ by $5x$ and then divide $5x$ by $5x$. Be cautious with the negative signs.

$$
\begin{array}{r}
x^2 - 5x + 1 \\
5x+1{\overline{\smash{\big)}\,5x^3 - 24x^2 + 0x + 9}} \\
\underline{5x^3 + x^2} \\
-25x^2 + 0x \\
\underline{-25x^2 - 5x} \\
5x + 9 \\
\underline{5x + 1} \\
8
\end{array}
$$

Write the remainder as the numerator of a fraction that has the binomial divisor as its denominator.

The answer is $x^2 - 5x + 1 + \dfrac{8}{5x + 1}$.

The check is left to the student.

167

Example	Student Practice
7. Divide and check.	**8.** Divide and check.
$(12x^3 - 11x^2 + 8x - 4) \div (3x - 2)$	$(8x^3 - 4x^2 + 6) \div (2x - 3)$

$$\begin{array}{r} 4x^2 - x + 2 \\ 3x-2\overline{)12x^3 - 11x^2 + 8x - 4} \\ \underline{12x^3 - 8x^2} \\ -3x^2 + 8x \\ \underline{-3x^2 + 2x} \\ 6x - 4 \\ \underline{6x - 4} \\ 0 \end{array}$$

Check the answer.

$$(3x - 2)(4x^2 - x + 2) = 12x^3 - 11x^2 + 8x - 4$$

Extra Practice

1. Divide. $\dfrac{12a^7 - 4a^5 + 8a^3 - 2a^2}{2a^2}$

2. Divide. $(12y^4 - 18y^3 + 27y^2) \div 3y^2$

3. Divide and check. $\dfrac{3x^3 - 5x^2 + 7x - 5}{x - 1}$

4. Divide and check. $\dfrac{y^3 - 2y - 4}{y - 1}$

Concept Check

Explain how you would check if $x^2 + 2x + 8 + \dfrac{13}{x - 2}$ is the correct answer to the problem $(x^3 + 4x - 3) \div (x - 2)$. Perform the check. Does the answer check?

MATH COACH

Mastering the skills you need to do well on the test.

Watch the **MATH COACH** videos in MyMathLab® or on YouTube™ while you work the problems below. These helpful hints will help you avoid making common errors on test problems.

Raising Monomials to a Power—Problem 8 Simplify $\dfrac{\left(3x^2\right)^3}{\left(6x\right)^2}$.

Helpful Hint: Do the problem in three stages. First, use the power to a power rule to raise the numerator to the third power. Second, raise the denominator to the second power. Then divide the monomials using the rules of exponents. Be sure to simplify any fractions.

Did you use the power to a power rule to raise both 3^1 and x^2 to the third power in the numerator and both 6^1 and x^1 to the second power in the denominator?
Yes _____   No _____

If you answered No, stop and review the power to a power rule before completing these steps again.

Did you remember to simplify the fraction $\dfrac{27}{36}$?

Yes _____   No _____

Finally, did you remember to use the quotient rule to subtract the exponents in the x terms?
Yes _____   No _____

If you answered No to either of these questions, go back and examine your work carefully before completing these steps again.

If you answered Problem 8 incorrectly, go back and rework the problem using these suggestions.

Simplifying Monomials Involving Negative Exponents—Problem 11

Simplify and write with only positive exponents. $\dfrac{3x^{-3}y^2}{x^{-4}y^{-5}}$

Helpful Hint: First, use the definition of a negative exponent to rewrite the expression using only positive exponents. Then use the rules for exponents to simplify the resulting expression.

Did you remove the negative exponents by rewriting the expression as $\dfrac{3x^4y^2y^5}{x^3}$? Yes _____   No _____

If you answered No, review the definition of negative exponents in Section 4.2 and complete this step again.

Did you use the quotient rule to simplify the x terms and the product rule to simplify the y terms? Yes _____   No _____

If you answered No, review the rules for exponents in Sections 4.1 and 4.2 and simplify the expression again.

Now go back and rework the problem using these suggestions.

Multiplying Three Binomials—Problem 20   Multiply $(3x+2)(2x+1)(x-3)$.

Helpful Hint: A good approach is to start by multiplying the first two binomials. Then multiply that result by the third binomial. Be careful to avoid sign errors when multiplying, and be careful to write down the correct exponent for each term.

Did you use the FOIL method to multiply the first two binomials and obtain $6x^2+7x+2$? Yes _____ No _____

If you answered No, stop and complete this step.

Did you multiply the result above by $(x-3)$?
Yes _____ No _____

Did you multiply each term of $6x^2+7x+2$ by x?
Yes _____ No _____

Did you multiply each term of $6x^2+7x+2$ by -3?
Yes _____ No _____

If you answered No to any of these questions, go back and examine each step of the multiplication carefully.

Be sure to write the correct exponent each time that you multiply. Be careful to avoid sign errors when multiplying by -3. Then combine like terms before writing your final answer.

If you answered Problem 20 incorrectly, go back and rework the problem using these suggestions.

Dividing a Polynomial by a Binomial—Problem 27   Divide $\left(2x^3-6x-36\right)\div(x-3)$.

Helpful Hint: Review the procedure for dividing a polynomial by a binomial in Section 4.6. Make sure you understand each step. Be sure you understand where the expression $0x^2$ came from in the dividend. Be careful with subtraction. Write out the subtraction steps to avoid sign errors.

Did you write the division problem in the form

$x-3\overline{)2x^3+0x^2-6x-36}$? Yes _____ No _____

If you answered No, remember that every power must be represented. We must use $0x^2$ as a placeholder so that we can perform our division.

When you carried out the first step of division, did you obtain $2x^2$ as the first part of your answer?
Yes _____ No _____

When you multiplied $x-3$ by $2x^2$, and then subtracted, did you get the result $6x^2$? Yes _____ No _____

If you answered No to these questions, stop and examine your first division step carefully. Make sure that you subtracted carefully too. Remember to write out the subtraction steps: $0x^2-\left(-6x^2\right)=6x^2$.

Next, did you bring down $-6x$ from the dividend to obtain $6x^2-6x$?
Yes _____ No _____

If you answered No, go back and look at the dividend again and see how to obtain this result.

Now go back and rework the problem using these suggestions.

Chapter 6 Factoring
6.1 Removing a Common Factor

Vocabulary
factor   •   to factor   •   common factor   •   greatest common factor

1. When two or more numbers, variables, or algebraic expressions are multiplied, each is called a _____.

2. A _____ is a factor that both terms have in common.

3. When you are asked _____ a number or an algebraic expression, you are being asked, "What factors, when multiplied, will give that number or expression?"

4. When we factor, we begin by looking for the _____.

Example	Student Practice
1. Factor. (a) $3x - 6y$ Begin by looking for a common factor, a factor that both terms have in common. Then rewrite the expression as a product. $3x - 6y = 3(x - 2y)$ This is true because $3(x - 2y) = 3x - 6y.$ (b) $9x + 2xy$ $9x + 2xy = x(9 + 2y)$ This is true because $x(9 + 2y) = 9x + 2xy.$ Notice that factoring is using the distributive property in reverse.	**2.** Factor. (a) $5y + 15z$ (b) $12y - 5yz$

Example	Student Practice
3. Factor $24xy + 12x^2 + 36x^3$. Remember to remove the greatest common factor.	**4.** Factor $44y^3 + 55y^2 - 11xy$. Remember to remove the greatest common factor.

3. Factor $24xy + 12x^2 + 36x^3$. Remember to remove the greatest common factor.

We start by finding the greatest common factor of 24, 12, and 36.

You may want to factor each number, or you may notice that 12 is a common factor. 12 is the greatest numerical common factor.

Notice also that x is a factor of each term. Thus, $12x$ is the greatest common factor.

$$24xy + 12x^2 + 36x^3 = 12x\left(2y + x + 3x^2\right)$$

4. Factor $44y^3 + 55y^2 - 11xy$. Remember to remove the greatest common factor.

---

5. Factor.

(a) $12x^2 - 18y^2$

Note that the largest integer that is common to both terms is 6 (not 3 or 2).

$$12x^2 - 18y^2 = 6\left(2x^2 - 3y^2\right)$$

(b) $x^2y^2 + 3xy^2 + y^3$

Although y is common to all of the terms, we factor out y^2 since 2 is the largest exponent of y that is common to all terms.

We do not factor out x, since x is not common to all of the terms.

$$x^2y^2 + 3xy^2 + y^3 = y^2\left(x^2 + 3x + y\right)$$

6. Factor.

(a) $21m^2 - 28n^2$

(b) $m^3n^2 + 9m^2n^2 + 3m^4$

Example	Student Practice
7. Factor. $8x^3y + 16x^2y^2 - 24x^3y^3$	**8.** Factor. $27xy^3 - 36x^2y^2 - 9x^3y^3$

7. Factor. $8x^3y + 16x^2y^2 - 24x^3y^3$

We see that 8 is the largest integer that will divide evenly into the three numerical coefficients. We can factor x^2 out of each term. We can also factor y out of each term.

$8x^3y + 16x^2y^2 - 24x^3y^3$
$= 8x^2y\left(x + 2y - 3xy^2\right)$

Check.

$8x^2y\left(x + 2y - 3xy^2\right)$
$= 8x^3y + 16x^2y^2 - 24x^3y^3$

---

9. Factor. $9a^3b^2 + 9a^2b^2$

We observe that both terms contain a common factor of 9. We can also factor a^2 and b^2 out of each term. Thus, the greatest common factor is $9a^2b^2$.

$9a^3b^2 + 9a^2b^2 = 9a^2b^2\left(a + 1\right)$

10. Factor. $11m^4n^2 + 11m^3n^2$

---

11. Factor. $7x^2\left(2x - 3y\right) - \left(2x - 3y\right)$

The common factor of the terms is $\left(2x - 3y\right)$. What happens when we factor out $\left(2x - 3y\right)$? What are we left with in the second term?
Recall that $\left(2x - 3y\right) = 1\left(2x - 3y\right)$.

$7x^2\left(2x - 3y\right) - \left(2x - 3y\right)$
$= 7x^2\left(2x - 3y\right) - \mathbf{1}\left(2x - 3y\right)$
$= \left(2x - 3y\right)\left(7x^2 - 1\right)$

12. Factor. $5y^2\left(3x + 4y\right) - \left(3x + 4y\right)$

Example	Student Practice
13. A computer programmer is writing a program to find the total area of 4 circles. She uses the formula $A = \pi r^2$. The radii of the circles are a, b, c, and d, respectively. She wants the final answer to be in factored form with the value of π occurring only once, in order to minimize the rounding error. Write the total area of the 4 circles with a formula that has π occurring only once.	**14.** Find the total area of 3 circles using the formula $A = \pi r^2$. The radii of the circles are m, $2n$, and $3z$, respectively. Write the total area of the 3 circles with a formula that has π occurring only once.

For each circle, $A = \pi r^2$, where $r = a$, b, c, or d. We add the area of each of the 4 circles.

The total area is $\pi a^2 + \pi b^2 + \pi c^2 + \pi d^2$.

In factored form the total area = $\pi\left(a^2 + b^2 + c^2 + d^2\right)$.

Extra Practice

1. Remove the largest possible common factor. Check your answer by multiplication. $3x^3 + 12x^2 - 21x$

2. Remove the largest possible common factor. Check your answer by multiplication. $60x^2 y + 18xy - 24x$

3. Remove the largest possible common factor. Check your answer by multiplication. $8a(x+3y) - b(x+3y)$

4. Remove the largest possible common factor. Check your answer by multiplication. $5a(4x-3) - (4x-3)$

Concept Check

Explain how you would remove the greatest common factor from the following polynomial.
$36a^3 b^2 - 72a^2 b^3$

Chapter 6 Factoring
6.2 Factoring by Grouping

Vocabulary

factoring by grouping • common factor • commutative property • FOIL

1. Sometimes you will need to factor out a negative _____ from the second two terms to obtain two terms that contain the same parenthetical expression.

2. A procedure used to factor a four-term polynomial is called _____.

3. Rearrange the order using the _____ of addition.

4. To check, we multiply the two binomials using the _____ procedure.

Example	Student Practice
1. Factor. $x(x-3)+2(x-3)$ Observe each term: $\underbrace{x(x-3)}_{\substack{\text{first} \\ \text{term}}}+\underbrace{2(x-3)}_{\substack{\text{second} \\ \text{term}}}$ The common factor of the first and second terms is the quantity $(x-3)$. $x(x-3)+2(x-3)=(x-3)(x+2)$	**2.** Factor. $z(2z+5)-4(2z+5)$
3. Factor. $2x^2+3x+6x+9$ Factor out a common factor of x from the first two terms. Factor out a common factor of 3 from the second two terms. $2x^2+3x+6x+9=x(2x+3)+3(2x+3)$ The expression in parentheses is now a common factor of the terms. $x(2x+3)+3(2x+3)=(2x+3)(x+3)$	**4.** Factor. $12x^2+4x+15x+5$

Example	Student Practice
5. Factor. $4x + 8y + ax + 2ay$	**6.** Factor. $7y + 21z + xy + 3xz$

5. Factor out a common factor of 4 from the first two terms. Factor out a common factor of a from the second two terms.

$4x + 8y + ax + 2ay$
$= 4(x + 2y) + a(x + 2y)$

The common factor of the terms is the expression in parentheses, $x + 2y$.

$4(x + 2y) + a(x + 2y) = (x + 2y)(4 + a)$

Example	Student Practice
7. Factor. $bx + 4y + 4b + xy$	**8.** Factor. $nz + 6y + 6z + ny$

7. Rearrange the terms so that the first two terms have a common factor.

$bx + 4y + 4b + xy = bx + 4b + xy + 4y$

Factor out the common factor b from the first two terms and the common factor y from the second two terms.

$bx + 4b + xy + 4y = b(x + 4) + y(x + 4)$
$= (x + 4)(b + y)$

Example	Student Practice
9. Factor. $2x^2 + 5x - 4x - 10$	**10.** Factor. $2x^2 + 3x - 6x - 9$

9. Factor out the common factor x from the first two terms and the common factor -2 from the second two terms.

$2x^2 + 5x - 4x - 10$
$= x(2x + 5) - 4x - 10$
$= x(2x + 5) - 2(2x + 5)$
$= (2x + 5)(x - 2)$

Example	Student Practice
11. Factor. $2ax - a - 2bx + b$	**12.** Factor. $5nz - n - 5mz + m$

11. Factor. $2ax - a - 2bx + b$

Factor out the common factor a from the first two terms and the common factor $-b$ from the second two terms.

$$2ax - a - 2bx + b = a(2x-1) - b(2x-1)$$

Since the two resulting terms contain the same parenthetical expression, we can complete the factoring.

$$a(2x-1) - b(2x-1) = (2x-1)(a-b)$$

12. Factor. $5nz - n - 5mz + m$

---

13. Factor and check your answer.
$$8ad + 21bc - 6bd - 28ac$$

Use the commutative property of addition to rearrange the order so the first two terms have a common factor.

$$8ad + 21bc - 6bd - 28ac$$
$$= 8ad - 6bd - 28ac + 21bc$$

Factor out the common factor $2d$ from the first two terms and the common factor $-7c$ from the last two terms.

$$= 2d(4a - 3b) - 7c(4a - 3b)$$

Factor out the common factor $(4a - 3b)$.

$$= (4a - 3b)(2d - 7c)$$

To check, we multiply the two binomials using the FOIL procedure.

$$(4a - 3b)(2d - 7c)$$
$$= 8ad - 28ac - 6bd + 21bc$$
$$= 8ad + 21bc - 6bd - 28ac$$

14. Factor and check your answer.
$$9wz + 35xy - 15xz - 21wy$$

Extra Practice

1. Factor by grouping. Check you answer.

   $x(x+1)-2(x+1)$

2. Factor by grouping. Check you answer.

   $x^2-2x+5x-10$

3. Factor by grouping. Check you answer.

   $6x^2+15x-4x-10$

4. Factor by grouping. Check you answer.

   $12ax+15ay+8xb+10by$

Concept Check

Explain how you would factor the following polynomial.

   $10ax+b^2+2bx+5ab$

Chapter 6 Factoring

6.3 Factoring Trinomials of the Form $x^2 + bx + c$

Vocabulary

first terms   •   outer and inner terms   •   second terms   •   last terms

1. When multiplying an expression of the form $(x+m)(x+n)$, the product of the

    _____ in the factors produces the first term of the polynomial.

2. When multiplying an expression of the form $(x+m)(x+n)$, the sum of the

    _____ in the factors gives the coefficient of the middle term of the polynomial.

3. When multiplying an expression of the form $(x+m)(x+n)$, the sum of the products of

    the _____ in the factors produces the middle term of the polynomial.

4. When multiplying an expression of the form $(x+m)(x+n)$, the product of the

    _____ of the factors gives the last term of the polynomial.

Example	Student Practice
1. Factor. $x^2 + 7x + 12$ The answer will be of the form $(x+m)(x+n)$. We want to find the two numbers, m and n, that you can multiply to get 12 and add to get 7. The numbers are 3 and 4. $x^2 + 7x + 12 = (x+3)(x+4)$	**2.** Factor. $x^2 + 9x + 18$
3. Factor. $x^2 + 12x + 20$ We want two numbers that have a product of 20 and a sum of 12. The numbers are 10 and 2. $x^2 + 12x + 20 = (x+10)(x+2)$	**4.** Factor. $x^2 + 14x + 33$

Example	Student Practice
5. Factor. $x^2 - 8x + 15$ We want two numbers that have a product of $+15$ and a sum of -8. They must be negative numbers since the sign of the middle term is negative and the sign of the last term is positive. The sum $-5 + (-3)$ is -8 and the product $(-5)(-3)$ is $+15$. $x^2 - 8x + 15 = (x - 5)(x - 3)$ Multiply using FOIL to check.	**6.** Factor. $x^2 - 11x + 28$
7. Factor. $x^2 - 3x - 10$ We want two numbers whose product is -10 and whose sum is -3. The two numbers are -5 and $+2$. $x^2 - 3x - 10 = (x - 5)(x + 2)$	**8.** Factor. $x^2 - 2x - 24$
9. Factor and check your answer. $y^2 + 10y - 24$ The two numbers whose product is -24 and whose sum is $+10$ are $+12$ and -2. $y^2 + 10y - 24 = (y + 12)(y - 2)$ It is very easy to make a sign error in these problems. Make sure that you mentally check your answer by FOIL to obtain the original expression. Check. $(y + 12)(y - 2) = y^2 - 2y + 12y - 24$ $\qquad\qquad\qquad = y^2 + 10y - 24$	**10.** Factor and check your answer. $z^2 + 12z - 28$

Example	Student Practice
11. Factor. $y^4 - 2y^2 - 35$	**12.** Factor. $x^4 - 3x^2 - 10$

11. Factor. $y^4 - 2y^2 - 35$

Notice that $y^4 = (y^2)(y^2)$. This will be the first term in each set of parentheses.

$$(y^2 \quad)(y^2 \quad)$$

The last term of the polynomial is negative. Thus, the signs of m and n will be different.

$$(y^2 + \quad)(y^2 - \quad)$$

Now think of factors of 35 whose difference is 2.

$$(y^2 + 5)(y^2 - 7)$$

Multiply using FOIL to check.

---

13. Factor. $3x^2 + 9x - 162$

First factor out the common factor 3 from each term of the polynomial.

$$3x^2 + 9x - 162 = 3(x^2 + 3x - 54)$$

Then factor the remaining polynomial.

$$3(x^2 + 3x - 54) = 3(x - 6)(x + 9)$$

The final answer is $3(x - 6)(x + 9)$. Be sure to include the 3.

Check.

$$3(x - 6)(x + 9) = 3(x^2 + 3x - 54)$$
$$= 3x^2 + 9x - 162$$

14. Factor. $4x^2 - 16x - 84$

Example	Student Practice
15. Find a polynomial in factored form for the shaded area in the figure.	**16.** Find a polynomial in factored form for the shaded area in the figure.

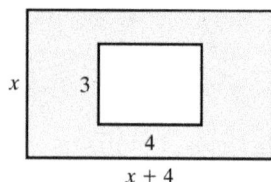

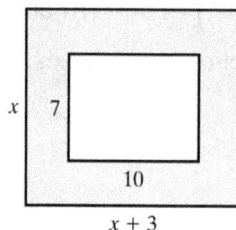

To obtain the shaded area, we find the area of the larger rectangle and subtract from it the area of the smaller rectangle. Thus, we have the following:

$$\text{shaded area} = x(x+4)-(4)(3)$$
$$= x^2 + 4x - 12$$

Now we factor this polynomial to obtain the shaded area $= (x+6)(x-2)$.

Extra Practice

1. Factor. $x^2 + 6x + 8$

2. Factor. $a^2 - 7a + 12$

3. Factor. $x^2 - x - 12$

4. Factor. $2x^2 + 18x + 28$

Concept Check

Explain how you would completely factor $4x^2 - 4x - 120$.

Chapter 6 Factoring
6.4 Factoring Trinomials of the Form $ax^2 + bx + c$

Vocabulary

trial-and-error method • grouping method • multiplying • greatest common factor

1. The _____ requires listing the possible factoring combinations and computing the middle terms by the FOIL method.

2. Always check by _____ the factors to see if the original trinomial is obtained.

3. First factor out the _____ if any, and then factor the trinomial.

4. The _____ for factoring trinomials of the form $ax^2 + bx + c$ requires writing the polynomial with four terms and then factoring the polynomial by grouping.

Example	**Student Practice**
1. Factor. $2x^2 + 5x + 3$	**2.** Factor. $2x^2 + 13x + 20$
In order for the coefficient of the x^2-term of the polynomial to be 2, the coefficients of the x-terms in the factors must be 2 and 1. Thus,	
$2x^2 + 5x + 3 = (2x \quad)(x \quad)$.	
In order for the last term of the polynomial to be 3, the constants in the factors must be 3 and 1. Since all signs in the polynomial are positive, each factor in parentheses will be positive. We have two possibilities. We check them by multiplying.	
$(2x+1)(x+3) = 2x^2 + 7x + 3$	
$(2x+3)(x+1) = 2x^2 + 5x + 3$	
The correct answer is $(2x+3)(x+1)$.	

Example	Student Practice
3. Factor. $4x^2 - 13x + 3$	**4.** Factor. $3x^2 - 4x + 1$

3. Factor. $4x^2 - 13x + 3$

The different factorizations of 4 are $(2)(2)$ and $(1)(4)$. The factorization of 3 is $(1)(3)$. Let us list the possible factoring combinations and compute the middle term by the FOIL method. Note that the signs of the constants in both factors will be negative.

Possible Factors   Middle Term

$(2x-3)(2x-1)$ $-8x$

$(4x-3)(x-1)$ $-7x$

$(4x-1)(x-3)$ $-13x$

The correct middle term is $-13x$. The correct answer is $(4x-1)(x-3)$.

---

5. Factor. $3x^2 - 2x - 8$ **6.** Factor. $4x^2 - 5x - 21$

The factorization of 3 is $(1)(3)$. The factorizations of 8 are $(8)(1)$ and $(4)(2)$. Let us list only one-half of the possibilities.

Possible Factors   Middle Term

$(x+8)(3x-1)$ $+23x$

$(x+1)(3x-8)$ $-5x$

$(x+4)(3x-2)$ $+10x$

$(x+2)(3x-4)$ $+2x$

The middle term $+2x$ is only incorrect because the sign is wrong. So we just reverse the signs of the constants.

The correct answer is $(x-2)(3x+4)$.

Example	Student Practice
7. Factor by grouping. $2x^2 + 5x + 3$	**8.** Factor by grouping. $2x^2 + 5x + 3$

First find the grouping number. The grouping number is $(2)(3) = 6$. The factors of 6 are $6 \cdot 1$ and $3 \cdot 2$. We choose the numbers 3 and 2 because their product is 6 and their sum is 5. Next, write $5x$ as the sum $2x + 3x$. Then, factor by grouping.

$$2x^2 + 5x + 3 = 2x^2 + 2x + 3x + 3$$
$$= 2x(x+1) + 3(x+1)$$
$$= (x+1)(2x+3)$$

Multiply to check.

$$(x+1)(2x+3) = 2x^2 + 3x + 2x + 3$$
$$= 2x^2 + 5x + 3$$

9. Factor by grouping. $3x^2 - 2x - 8$	**10.** Factor by grouping. $5x^2 - 13x - 6$

First find the grouping number. The grouping number is $(3)(-8) = -24$. Now we want two numbers whose product is -24 and whose sum is -2. They are -6 and 4. So we write $-2x$ as the sum $-6x + 4x$. Then, we factor by grouping.

$$3x^2 - 6x + 4x - 8 = 3x(x-2) + 4(x-2)$$
$$= (x-2)(3x+4)$$

Example	Student Practice
11. Factor. $9x^2 + 3x - 30$ We first factor out the common factor 3 from each term of the trinomial. $9x^2 + 3x - 30 = 3\left(3x^2 + 1x - 10\right)$ We then factor the trinomial by the grouping method or by the trial-and-error method. $3\left(3x^2 + 1x - 10\right) = 3(3x - 5)(x + 2)$	**12.** Factor. $8x^2 + 22x - 6$
13. Factor. $32x^2 - 40x + 12$ We first factor out the greatest common factor 4 from each term of the trinomial. $32x^2 - 40x + 12 = 4\left(8x^2 - 10x + 3\right)$ We then factor the trinomial by the grouping method or by the trial-and-error method. $4\left(8x^2 - 10x + 3\right) = 4(2x - 1)(4x - 3)$	**14.** Factor. $30x^2 - 65x + 30$

Extra Practice

1. Factor by the trial-and-error method. Check your answer by using FOIL.
$4x^2 - 12x + 5$

2. Factor by the grouping method. Check your answer by using FOIL. $7x^2 - 4x - 3$

3. Factor by any method. $5x^2 + 43x - 18$

4. Factor by first factoring out the greatest common factor. $15y^2 + 51y - 36$

Concept Check

Explain how you would factor $10x^3 + 18x^2y - 4xy^2$.

Chapter 6 Factoring
6.5 Special Cases of Factoring

Vocabulary
difference of two squares • negative • perfect-square trinomials • greatest common factor

1. The difference of two squares formula only works if the last term is _____.

2. A _____ is a trinomial where the first and last terms are perfect squares and the middle term is twice the product of the values whose squares are the first and last terms.

3. The _____ can be factored into the sum and difference of those values that were squared.

4. For some polynomials, we first need to factor out the _____.

Example	Student Practice
1. Factor. $9x^2 - 1$	**2.** Factor. $36x^2 - 1$

We see that the polynomial is in the form of the difference of two squares. $9x^2$ is a square and 1 is a square. $9x^2 = (3x)^2$ and $1 = (1)^2$. So using the formula we can write the following.

$$9x^2 - 1 = (3x+1)(3x-1)$$

Example	Student Practice
3. Factor. $25x^2 - 16$	**4.** Factor. $64x^2 - 9$

$25x^2 = (5x)^2$ and $16 = (4)^2$. Again we use the formula for the difference of two squares.

$$25x^2 - 16 = (5x+4)(5x-4)$$

Example	Student Practice
5. Factor. $81x^4 - 1$	**6.** Factor. $16x^4 - 1$

5. Factor. $81x^4 - 1$

Because $81x^4 = \left(9x^2\right)^2$ and $1 = \left(1\right)^2$, we see that

$$81x^4 - 1 = \left(9x^2 + 1\right)\left(9x^2 - 1\right).$$

The factoring is not complete. We can factor $9x^2 - 1$, because
$9x^2 - 1 = \left(3x + 1\right)\left(3x - 1\right).$

$$81x^4 - 1 = \left(9x^2 + 1\right)\left(3x + 1\right)\left(3x - 1\right)$$

6. Factor. $16x^4 - 1$

---

7. Factor. $x^2 + 6x + 9$

This is a perfect-square trinomial. The first and last terms are perfect squares because $x^2 = \left(x\right)^2$ and $9 = \left(3\right)^2$. The middle term, $6x,$ is twice the product of 3 and $x.$

Since $x^2 + 6x + 9$ is a perfect-square trinomial we can use the formula

$$a^2 + 2ab + b^2 = \left(a + b\right)^2$$

with $a = x$ and $b = 3.$ So we have

$$x^2 + 6x + 9 = \left(x + 3\right)^2.$$

8. Factor. $x^2 + 10x + 25$

Example	Student Practice
9. Factor.	**10.** Factor.
(a) $49x^2 + 42xy + 9y^2$	**(a)** $64x^2 + 80xy + 25y^2$
This is a perfect square trinomial, because $49x^2 = (7x)^2$, $9y^2 = (3y)^2$, and $42xy = 2(7x \cdot 3y)$.	
$49x^2 + 42xy + 9y^2 = (7x + 3y)^2$	
(b) $36x^4 - 12x^2 + 1$	**(b)** $49x^4 - 14x^2 + 1$
This is a perfect square trinomial, because $36x^4 = (6x^2)^2$, $1 = (1)^2$, and $12x^2 = 2(6x^2 \cdot 1)$.	
$36x^4 - 12x^2 + 1 = (6x^2 - 1)^2$	
11. Factor. $49x^2 + 35x + 4$	**12.** Factor. $36x^2 + 25x + 4$
This is not a perfect-square trinomial! Although the first and last terms are perfect squares since $(7x)^2 = 49x^2$ and $(2)^2 = 4$, the middle term, $35x$, is not double the product of 2 and $7x$. $35x \neq 28x$! So we must factor by trial and error or by grouping to obtain	
$49x^2 + 35x + 4 = (7x + 4)(7x + 1)$.	

189

Example	Student Practice
13. Factor. $12x^2 - 48$ We see that the greatest common factor is 12. First we factor out 12. Then we use the difference-of-two-squares formula, $a^2 - b^2 = (a+b)(a-b)$. $12x^2 - 48 = 12(x^2 - 4)$ $\qquad\qquad = 12(x+2)(x-2)$	**14.** Factor. $3x^2 - 75$
15. Factor. $24x^2 - 72x + 54$ First we factor out the greatest common factor, 6. Then we use the perfect-square-trinomial formula, $a^2 - 2ab + b^2 = (a-b)^2$. $24x^2 - 72x + 54 = 6(4x^2 - 12x + 9)$ $\qquad\qquad\qquad = 6(2x-3)^2$	**16.** Factor. $48x^2 - 72x + 27$

Extra Practice

1. Factor using the difference-of-two-squares formula. $25a^2 - 36b^2$

2. Factor by using the perfect-square trinomial formula. $36a^2 + 60ab + 25b^2$

3. Factor by using the difference-of-two-squares. $x^4 - 100$

4. Factor by using the perfect-square trinomial formula. $9x^4 - 30x^2 + 25$

Concept Check

Explain how to factor the polynomial $24x^2 + 120x + 150$.

Chapter 6 Factoring
6.6 A Brief Review of Factoring

Vocabulary

perfect-square trinomial • difference of two squares • factor by grouping • prime
trinomial of the form $x^2 + bx + c$ • trinomial of the form $ax^2 + bx + c$ • common factor

1. Some polynomials have a _____ consisting of a number, a variable, or both.

2. When we _____, we rearrange the order if the first two terms do not have a common factor.

3. If we cannot factor a polynomial by elementary methods, we will identify it as a _____ polynomial.

4. In a _____ there are three terms and the first and last terms are perfect squares.

Example	Student Practice
1. Factor.	**2.** Factor.
(a) $25x^3 - 10x^2 + x$	**(a)** $9y^3 + 6y^2 + y$
Factor out the common factor x. The other factor is a perfect-square trinomial.	
$25x^3 - 10x^2 + x = x(25x^2 - 10x + 1)$ $= x(5x-1)^2$	
(b) $20x^2y^2 - 45y^2$	**(b)** $-4x^3 + 28x^2 + 32x$
Factor out the common factor $5y^2$. The other factor is a difference of squares.	
$20x^2y^2 - 45y^2 = 5y^2(4x^2 - 9)$ $= 5y^2(2x+3)(2x-3)$	

Example	Student Practice
3. Factor. $ax^2 - 9a + 2x^2 - 18$	**4.** Factor. $bx^2 - 16b - 3x^2 + 48$

3. Factor. $ax^2 - 9a + 2x^2 - 18$

We factor by grouping since there are four terms. Factor out the common factor a from the first two terms and 2 from the second two terms.

$ax^2 - 9a + 2x^2 - 18$

$= a\left(x^2 - 9\right) + 2\left(x^2 - 9\right)$

Factor out the common factor $\left(x^2 - 9\right)$.

$a\left(x^2 - 9\right) + 2\left(x^2 - 9\right) = (a + 2)\left(x^2 - 9\right)$

Factor $x^2 - 9$ using the difference-of-two squares formula.

$(a + 2)\left(x^2 - 9\right) = (a + 2)(x - 3)(x + 3)$

---

4. Factor. $bx^2 - 16b - 3x^2 + 48$

---

5. Factor, if possible.

(a) $x^2 + 6x + 12$

The factors of 12 are

$(1)(12)$ or $(2)(6)$ or $(3)(4)$.

None of these pairs add up to 6, the coefficient of the middle term. Thus, the problem cannot be factored by the methods in this chapter.

(b) $25x^2 + 4$

We have a formula to factor the difference of two squares. There is no way to factor the sum of two squares. That is, $a^2 + b^2$ cannot be factored.

6. Factor, if possible.

(a) $x^2 + 5x + 17$

(b) $x^2 - x + 21$

Extra Practice

1. Factor, if possible. Be sure to factor completely. Factor out the greatest common factor first, if one exists.

   $x^2 + 9$

2. Factor, if possible. Be sure to factor completely. Factor out the greatest common factor first, if one exists.

   $63x - 7x^3$

3. Factor, if possible. Be sure to factor completely. Factor out the greatest common factor first, if one exists.

   $-x^3 + 12x^2 + 45x$

4. Factor, if possible. Be sure to factor completely. Factor out the greatest common factor first, if one exists.

   $x^2 + 2x + 3x + 6$

Concept Check

Explain how to completely factor $2x^2 + 6xw - 5x - 15w$.

Chapter 6 Factoring
6.7 Solving Quadratic Equations by Factoring

Vocabulary
quadratic equation   •   standard form   •   real roots   •   zero factor property

1. Many quadratic equations have two real number solutions, also called _____.

2. The _____ states that if $a \cdot b = 0$, then $a = 0$ or $b = 0$.

3. A _____ is a polynomial equation in one variable that contains a variable term of degree 2 and no terms of higher degree.

Example	Student Practice
1. Solve the equation to find the two roots. $2x^2 + 13x - 7 = 0$	**2.** Solve the equation to find the two roots. $3x^2 + 8x - 3 = 0$

The equation is in standard form. Factor. Then set each factor equal to 0 and solve the equations to find the two roots.

$$2x^2 + 13x - 7 = 0$$
$$(2x - 1)(x + 7) = 0$$
$$2x - 1 = 0 \quad x + 7 = 0$$
$$x = \frac{1}{2} \qquad x = -7$$

Check. If $x = \frac{1}{2}$ and if $x = -7$ then we have the following.

$$2\left(\frac{1}{2}\right)^2 + 13\left(\frac{1}{2}\right) - 7 = \frac{1}{2} + \frac{13}{2} - \frac{14}{2} = 0$$
$$2(-7)^2 + 13(-7) - 7 = 98 - 91 - 7 = 0$$

Thus $\frac{1}{2}$ and -7 are both roots of the equation $2x^2 + 13x - 7 = 0$.

Example	Student Practice
3. Solve the equation to find the two roots. $7x^2 - 3x = 0$	**4.** Solve the equation to find the two roots. $5x^2 - 4x = 0$

3. (continued)

The equation is in standard form. Here $c = 0$. Factor out the common factor. Then set each factor equal to 0 by the zero factor property. Solve the equations to find the two roots.

$$7x^2 - 3x = 0$$
$$x(7x - 3) = 0$$
$$x = 0 \quad 7x - 3 = 0$$
$$7x = 3$$
$$x = \frac{3}{7}$$

The two roots are 0 and $\frac{3}{7}$.

Check. Verify that 0 and $\frac{3}{7}$ are the roots of $7x^2 - 3x = 0$.

---

5. Solve. $x^2 = 12 - x$

The equation is not in standard form. Add x and -12 to both sides of the equation so that the left side is equal to zero. Then factor and set each factor equal to 0. Solve the equations for x.

$$x^2 = 12 - x$$
$$x^2 + x - 12 = 0$$
$$(x - 3)(x + 4) = 0$$
$$x - 3 = 0 \quad x + 4 = 0$$
$$x = 3 \qquad x = -4$$

The check is left to the student.

6. Solve. $x^2 = 63 - 2x$

Example	Student Practice

7. Carlos lives in Mexico City. He has a rectangular brick walkway in front of his house. The length of the walkway is 3 meters longer than twice the width. The area of the walkway is 44 square meters. Find the length and width of the rectangular walkway.

Let $w =$ the width in meters.

Then $2w + 3 =$ the length in meters.

Next, write an equation.

$$\text{area} = (\text{width})(\text{length})$$
$$44 = w(2w + 3)$$

Now, solve and state the answer.

$$44 = w(2w + 3)$$
$$44 = 2w^2 + 3w$$
$$0 = 2w^2 + 3w - 44$$
$$0 = (2w + 11)(w - 4)$$
$$2w + 11 = 0 \qquad w - 4 = 0$$
$$w = -5\frac{1}{2} \qquad w = 4$$

Since it would not make sense to have a rectangle with a negative number as a width, $-5\frac{1}{2}$ is not a valid solution.

Since $w = 4$, the width of the walkway is 4 meters. The length is $2w + 3$, so we have $2(4) + 3 = 8 + 3 = 11$. Thus, the length of the walkway is 11 meters.

The check is left to the student.

8. The length of a rectangle is 31 inches shorter than triple the width. The rectangle has an area of 60 square inches. Find the length and width of the rectangle.

Example	Student Practice

9. A tennis ball is thrown upward with an initial velocity of 8 meters/second. Suppose the initial height above the ground is 4 meters. At what time t will the ball hit the ground?

In this case $S = 0$ since the ball will hit the ground. The initial upward velocity is $v = 8$ meters/second. The initial height is 4 meters, so $h = 4$.

$$S = -5t^2 + vt + h$$
$$0 = -5t^2 + 8t + 4$$
$$5t^2 - 8t - 4 = 0$$
$$(5t + 2)(t - 2) = 0$$
$$5t + 2 = 0 \qquad t - 2 = 0$$
$$t = -\frac{2}{5} \qquad t = 2$$

We want a positive time for t in seconds; thus we do not use $t = -\frac{2}{5}$.

Therefore, the ball will strike the ground 2 seconds after it is thrown.

The check is left to the student.

10. A baseball is thrown upward with an initial velocity of 9 meters/second. Suppose the initial height above the ground is 18 meters. At what time t will the ball hit the ground?

Extra Practice

1. Solve for the roots of the quadratic equation. Check your answer.
$2x^2 - 5x - 3 = 0$

2. Solve for the roots of the quadratic equation. Check your answer.
$3x^2 = 27$

3. Solve for the roots of the quadratic equation. Check your answer.
$x^2 - 35 = 2x$

4. Solve for the roots of the quadratic equation. Check your answer.
$x^2 - 7x = -12$

Concept Check

Explain how you would solve the following problem: A rectangle has an area of 65 square feet. The length of the rectangle is 3 feet longer than double the width. Find the length and the width of the rectangle.

MATH COACH

Mastering the skills you need to do well on the test.

Watch the **MATH COACH** videos in MyMathLab® or on YouTube™ while you work the problems below. These helpful hints will help you avoid making common errors on test problems.

Factoring the Difference of Two Squares—Problem 2

Factor completely. $16x^2 - 81$

> **Helpful Hint:** It is important to learn the difference-of-two-squares formula: $a^2 - b^2 = (a+b)(a-b)$. Remember that the numerical values in both terms will be perfect squares. The first ten perfect squares are 1, 4, 9, 16, 25, 36, 49, 64, 81, and 100.

Did you remember that $(4x)^2 = 16x^2$? Yes _____ No _____

Did you remember that $9^2 = 81$? Yes _____ No _____

If you answered No to these questions, stop and review the list of the first ten perfect squares. Consider that $4^2 = 16$ and $x \cdot x = x^2$.

If you answered No, stop and review the difference-of-two-squares formula again. Make sure that one set of parentheses contains a + sign and the other set of parentheses contains a − sign.

If you answered Problem 2 incorrectly, go back and rework the problem using these suggestions.

Do you see how $16x^2 - 81$ can be factored using the formula $a^2 - b^2$? Yes _____ No _____

Factoring a Perfect-Square Trinomial—Problem 4   Factor completely. $9a^2 - 30a + 25$

> **Helpful Hint:** Remember the perfect-square-trinomial formula: $a^2 - 2ab + b^2 = (a-b)^2$. You must verify two things to determine if you can use this formula:
> (1) The numerical values in the first term and the last term must be perfect squares.
> (2) The middle term must equal "twice the product of the values whose squares are the first and last terms."

Did you remember that $(3a)^2 = 9a^2$? Yes _____ No _____

Did you remember that $5^2 = 25$? Yes _____ No _____

If you answered No to these questions, stop and review the first ten perfect squares. Consider that $3^2 = 9$ and $a \cdot a = a^2$.

If you answered No, check to see if the middle term, $30a$, equals twice the product of $3a$ and 5.

Now go back and rework the problem using these suggestions.

Do you see how $9a^2 - 30a + 25$ can be factored using the formula $(a-b)^2$? Yes _____ No _____

Factoring a Polynomial with Four Terms by Grouping—Problem 6

Factor completely. $10xy + 15by - 8x - 12b$

> **Helpful Hint:** Look for common factors first. We can find the greatest common factors of the first two terms and factor. Then we can find the greatest common factor of the second two terms and factor. Make sure that you obtain the same binomial factor for each step. Be careful with $+/-$ signs.

Did you identify $5y$ as the greatest common factor of the first two terms: $10xy + 15by$?

Yes _____ No _____

Did you identify 4 as the greatest common factor of the second two terms: $-8x - 12b$?

Yes _____ No _____

If you answered Yes to these questions, then you obtained $(2x + 3b)$ in the first term and $(-2x - 3b)$ in the second term. These are not the same binomial factor. Stop and consider how to get the same binomial factor of $(2x + 3b)$.

In your final answer, is the binomial factor of $(2x + 3b)$ listed once?

Yes _____ No _____

If you answered No, remember that to factor completely, we must remove all common factors from both terms. Stop now and complete this step.

If you answered Problem 6 incorrectly, go back and rework the problem using these suggestions.

Factoring a Polynomial with a Common Factor—Problem 19 Factor completely. $3x^2 - 3x - 90$

> **Helpful Hint:** Look for the greatest common factor of all three terms as your first step. Don't forget to include this common factor as part of your answer. Always check your final product to make sure that it matches the original polynomial. Do this by multiplying.

Did you obtain $(3x - 18)(x + 5)$ or $(3x - 15)(x - 6)$ as your answer?

Yes _____ No _____

If you answered Yes, then you forgot to factor out the greatest common factor 3 as your first step.

Do you see how to factor $x^2 - x - 30$?

Yes _____ No _____

If you answered No, remember that we are looking for two numbers with a product of -30 and a sum of -1.

Be sure to double-check your final answer to be sure there are no common factors and include 3 in your final answer.

Now go back and rework the problem using these suggestions.

Name: _____   Date: _____
Instructor: _____   Section: _____

Chapter 7 Rational Expressions and Equations
7.1 Simplifying Rational Expressions

Vocabulary
rational expression • fractional algebraic expression • basic rule of fractions
simplify the fraction • factors

1. Only _____ of both the numerator and the denominator can be divided out.

2. A _____ is a polynomial divided by another polynomial.

3. Dividing out common factors is how to _____.

4. The _____ states that for any rational expression $\dfrac{a}{b}$ and any polynomials a, b,

   and c, $\dfrac{ac}{bc} = \dfrac{a}{b}$.

Example	Student Practice
1. Reduce. $\dfrac{21}{39}$	**2.** Reduce. $\dfrac{56}{77}$
Use the rule $\dfrac{ac}{bc} = \dfrac{a}{b}$. Let $c = 3$. $\dfrac{21}{39} = \dfrac{7 \cdot \cancel{3}}{13 \cdot \cancel{3}} = \dfrac{7}{13}$	
3. Simplify. $\dfrac{4x+12}{5x+15}$ $\dfrac{4x+12}{5x+15} = \dfrac{4(x+3)}{5(x+3)}$ $= \dfrac{4\cancel{(x+3)}}{5\cancel{(x+3)}}$ $= \dfrac{4}{5}$	**4.** Simplify. $\dfrac{6x+8}{15x+20}$

Example	Student Practice
5. Simplify. $\dfrac{x^2+9x+14}{x^2-4}$	**6.** Simplify. $\dfrac{x^2+3x-28}{x^2-16}$

$$\frac{x^2+9x+14}{x^2-4}=\frac{(x+7)(x+2)}{(x-2)(x+2)}$$

$$=\frac{(x+7)\,\cancel{(x+2)}}{(x-2)\,\cancel{(x+2)}}$$

$$=\frac{x+7}{x-2}$$

Example	Student Practice
7. Simplify. $\dfrac{x^3-9x}{x^3+x^2-6x}$	**8.** Simplify. $\dfrac{x^3+8x^2+12x}{x^3-36x}$

$$\frac{x^3-9}{x^3+x^2-6x}=\frac{x(x^2-9x)}{x(x^2+x-6)}$$

$$=\frac{\cancel{x}\,\cancel{(x+3)}(x-3)}{\cancel{x}\,\cancel{(x+3)}(x-2)}$$

$$=\frac{x-3}{x-2}$$

Example	Student Practice
9. Simplify. $\dfrac{5x-15}{6-2x}$	**10.** Simplify. $\dfrac{8x-24}{42-14x}$

The variable terms in the numerator and in the denominator are opposite in sign. Likewise the numerical terms are opposite in sign. Factor out a negative number from the denominator.

$$\frac{5x-15}{6-2x}=\frac{5(x-3)}{-2(-3+x)}$$

$$=\frac{5\,\cancel{(x-3)}}{-2\,\cancel{(-3+x)}}$$

$$=-\frac{5}{2}$$

Example	Student Practice
11. Simplify. $\dfrac{2x^2-11x+12}{16-x^2}$	**12.** Simplify. $\dfrac{6x^2-7x-20}{25-4x^2}$

Factor the numerator and the denominator. Observe that $(x-4)$ and $(4-x)$ are opposites.

$$\frac{2x^2-11x+12}{16-x^2}=\frac{(x-4)(2x-3)}{(4-x)(4+x)}$$

Factor -1 out of $(+4-x)$ to obtain $-1(-4+x)$.

$$\frac{(x-4)(2x-3)}{(4-x)(4+x)}=\frac{(x-4)(2x-3)}{-1(-4+x)(4+x)}$$

$$=\frac{\cancel{(x-4)}(2x-3)}{-1\cancel{(-4+x)}(4+x)}$$

$$=\frac{(2x-3)}{-1(4+x)}$$

$$=-\frac{2x-3}{4+x}$$

13. Simplify. $\dfrac{x^2-7xy+12y^2}{2x^2-7xy-4y^2}$	**14.** Simplify. $\dfrac{10x^2-9xy-9y^2}{18x^2-13xy-21y^2}$

$$\frac{x^2-7xy+12y^2}{2x^2-7xy-4y^2}=\frac{(x-4y)(x-3y)}{(2x+y)(x-4y)}$$

$$=\frac{\cancel{(x-4y)}(x-3y)}{(2x+y)\cancel{(x-4y)}}$$

$$=\frac{x-3y}{2x+y}$$

Example	Student Practice
15. Simplify. $\dfrac{6a^2 + ab - 7b^2}{36a^2 - 49b^2}$	**16.** Simplify. $\dfrac{25a^2 - 16b^2}{10a^2 - 3ab - 4b^2}$

$$\frac{6a^2 + ab - 7b^2}{36a^2 - 49b^2} = \frac{(6a + 7b)(a - b)}{(6a + 7b)(6a - 7b)}$$

$$= \frac{\cancel{(6a + 7b)}(a - b)}{\cancel{(6a + 7b)}(6a - 7b)}$$

$$= \frac{a - b}{6a - 7b}$$

Extra Practice

1. Simplify. $\dfrac{3x + 12}{x^2 + 4x}$

2. Simplify. $\dfrac{9x^2 - 24x + 16}{9x^2 - 16}$

3. Simplify. $\dfrac{81 - x^2}{3x^2 - 21x - 54}$

4. Simplify. $\dfrac{25x^2 - 20xy + 4y^2}{15x^2 - xy - 2y^2}$

Concept Check

Explain why it is important to completely factor both the numerator and the denominator when simplifying $\dfrac{x^2 y - y^3}{x^2 y + xy^2 - 2y^3}$.

Chapter 7 Rational Expressions and Equations
7.2 Multiplying and Dividing Rational Expressions

Vocabulary
reciprocals   •   multiply   •   divide   •   greatest common factor

1. You should always check for the _____ as your first step when multiplying rational expressions.

2. Two numbers are _____ of each other if their product is 1.

3. To _____ two rational expressions, multiply the numerators and multiply the denominators.

4. To _____ two rational expressions, invert the second fraction and multiply it by the first fraction.

Example	**Student Practice**
1. Multiply. $\dfrac{x^2-x-12}{x^2-16} \cdot \dfrac{2x^2+7x-4}{x^2-4x-21}$	**2.** Multiply. $\dfrac{3x^2+34x+63}{2x^2+7x-15} \cdot \dfrac{x^2+9x+20}{x^2+13x+36}$

Factoring is always the first step.

$$\frac{x^2-x-12}{x^2-16} \cdot \frac{2x^2+7x-4}{x^2-4x-21}$$
$$=\frac{(x-4)(x+3)}{(x-4)(x+4)} \cdot \frac{(x+4)(2x-1)}{(x+3)(x-7)}$$

Apply the basic rule of fractions. Three pairs of factors divide out.

$$\frac{(x-4)(x+3)}{(x-4)(x+4)} \cdot \frac{(x+4)(2x-1)}{(x+3)(x-7)}$$
$$=\frac{\cancel{(x-4)}\,\cancel{(x+3)}}{\cancel{(x-4)}\,\cancel{(x+4)}} \cdot \frac{\cancel{(x+4)}\,(2x-1)}{\cancel{(x+3)}\,(x-7)}$$
$$=\frac{(2x-1)}{(x-7)}$$

Example	Student Practice
3. Multiply. $\dfrac{x^4 - 16}{x^3 + 4x} \cdot \dfrac{2x^2 - 8x}{4x^2 + 2x - 12}$	**4.** Multiply. $\dfrac{x^3 + x}{2x^4 - 2} \cdot \dfrac{4x^3 - 4x}{6x^3 + 38x^2 + 40x}$

Factor each numerator and denominator. Factoring out the greatest common factor first is very important.

$$\dfrac{x^4 - 16}{x^3 + 4x} \cdot \dfrac{2x^2 - 8x}{4x^2 + 2x - 12}$$

$$= \dfrac{(x^2 + 4)(x^2 - 4)}{x(x^2 + 4)} \cdot \dfrac{2x(x - 4)}{2(2x^2 + x - 6)}$$

$$= \dfrac{(x^2 + 4)(x + 2)(x - 2)}{x(x^2 + 4)} \cdot \dfrac{2x(x - 4)}{2(x + 2)(2x - 3)}$$

$$= \dfrac{\cancel{(x^2 + 4)}\ \cancel{(x + 2)}(x - 2)}{\cancel{x}\cancel{(x^2 + 4)}} \cdot \dfrac{\cancel{2}\cancel{x}(x - 4)}{\cancel{2}\cancel{(x + 2)}(2x - 3)}$$

$$= \dfrac{(x - 2)(x - 4)}{2x - 3} \text{ or } \dfrac{x^2 - 6x + 8}{2x - 3}$$

---

5. Divide. $\dfrac{6x + 12y}{2x - 6y} \div \dfrac{9x^2 - 36y^2}{4x^2 - 36y^2}$

$$= \dfrac{6x + 12y}{2x - 6y} \cdot \dfrac{4x^2 - 36y^2}{9x^2 - 36y^2}$$

$$= \dfrac{6(x + 2y)}{2(x - 3y)} \cdot \dfrac{4(x^2 - 9y^2)}{9(x^2 - 4y^2)}$$

$$= \dfrac{(3)(2)(x + 2y)}{2(x - 3y)} \cdot \dfrac{(2)(2)(x + 3y)(x - 3y)}{(3)(3)(x + 2y)(x - 2y)}$$

$$= \dfrac{\cancel{(3)}\ \cancel{(2)}\ \cancel{(x + 2y)}}{\cancel{2}\ \cancel{(x - 3y)}} \cdot \dfrac{(2)(2)(x + 3y)\ \cancel{(x - 3y)}}{\cancel{(3)}(3)\ \cancel{(x + 2y)}(x - 2y)}$$

$$= \dfrac{(2)(2)(x + 3y)}{3(x - 2y)}$$

$$= \dfrac{4(x + 3y)}{3(x - 2y)}$$

6. Divide.

$$\dfrac{5x^2 + 20x - 105}{4x^2 - 44x + 96} \div \dfrac{10x^2 + 75x + 35}{16x^2 - 140x + 96}$$

Example	Student Practice
7. Divide. $\dfrac{15-3x}{x+6} \div \left(x^2 - 9x + 20\right)$	**8.** Divide. $\dfrac{x+4}{x-2} \div \left(-x^2 + x + 20\right)$

Note that $x^2 - 9x + 20$ can be written as $\dfrac{x^2 - 9x + 20}{1}$.

$$\dfrac{15-3x}{x+6} \div \left(x^2 - 9x + 20\right)$$

$$= \dfrac{15-3x}{x+6} \cdot \dfrac{1}{x^2 - 9x + 20}$$

$$= \dfrac{-3(-5+x)}{x+6} \cdot \dfrac{1}{(x-5)(x-4)}$$

$$= \dfrac{-3\cancel{(-5+x)}}{x+6} \cdot \dfrac{1}{\cancel{(x-5)}(x-4)}$$

$$= \dfrac{-3}{(x+6)(x-4)}$$

or $-\dfrac{3}{(x+6)(x-4)}$ or $\dfrac{3}{(x+6)(4-x)}$

Extra Practice

1. Multiply. $\dfrac{32x^3}{8x^2 - 8} \cdot \dfrac{4x-4}{16x^2}$

2. Multiply.

$$\dfrac{2x^2 - 3x}{4x^3 + 4x^2 - 9x - 9} \cdot \dfrac{2x^2 + 5x + 3}{5x+7}$$

3. Divide. $\dfrac{9x^2 - 49}{3x^2 + 8x - 35} \div \left(3x^2 + x - 14\right)$

4. Divide. $\dfrac{9x^2 - 16}{9x^2 + 24x + 16} \div \dfrac{3x^2 - x - 4}{5x^2 - 5}$

Concept Check

Explain how you would divide $\dfrac{21x - 7}{9x^2 - 1} \div \dfrac{1}{3x+1}$.

Chapter 7 Rational Expressions and Equations
7.3 Adding and Subtracting Rational Expressions

Vocabulary
least common denominator   •   denominator   •   different   •   factor

1. If rational expressions have the same _____, they can be combined in the same way as arithmetic fractions.

2. If two rational expressions have _____ denominators, we first change them to equivalent rational expressions with the least common denominator.

3. The _____ is a product containing each different factor for each denominator of rational expressions.

4. The first step to find the LCD of two or more rational expressions is to _____ each denominator completely.

Example	Student Practice
1. Add. $\dfrac{5a}{a+2b}+\dfrac{6a}{a+2b}$	**2.** Add. $\dfrac{4x+3}{3x+4}+\dfrac{5-4x}{3x+4}$
Note that the denominators are the same. Only add the numerators. Do not change the denominator. $$\dfrac{5a}{a+2b}+\dfrac{6a}{a+2b}=\dfrac{5a+6a}{a+2b}=\dfrac{11a}{a+2b}$$	
3. Subtract. $$\dfrac{3x}{(x+y)(x-2y)}-\dfrac{8x}{(x+y)(x-2y)}$$ Write as one fraction and simplify. $$=\dfrac{3x}{(x+y)(x-2y)}-\dfrac{8x}{(x+y)(x-2y)}$$ $$=\dfrac{3x-8x}{(x+y)(x-2y)}=\dfrac{-5x}{(x+y)(x-2y)}$$	**4.** Subtract. $$\dfrac{4x+6}{(3x-2)(x+9)}-\dfrac{2x-5}{(3x-2)(x+9)}$$

Example	Student Practice
5. Find the LCD. $\dfrac{5}{2x-4}, \dfrac{6}{3x-6}$	**6.** Find the LCD. $\dfrac{5}{16x+20}, \dfrac{11}{28x+35}$

5. Factor each denominator.

$$2x - 4 = 2(x - 2)$$
$$3x - 6 = 3(x - 2)$$

The factors are 2, 3, and $(x-2)$. The LCD is the product of these factors.

$$LCD = (2)(3)(x-2) = 6(x-2)$$

---

7. Find the LCD.	**8.** Find the LCD.
$\dfrac{5}{12ab^2c}, \dfrac{13}{18a^3bc^4}$	$\dfrac{7}{24ab^2c^3}, \dfrac{13}{36a^2b^4c}$

7. The LCD will contain each factor repeated the greatest number of times that it occurs in any one denominator.

$$12ab^2x = 2\cdot 2\cdot 3 \quad\cdot a \qquad \cdot b\cdot b\cdot c$$
$$18a^3bc^4 = \quad 2\cdot 3\cdot 3\cdot a\cdot a\cdot a\cdot b\cdot\quad c\cdot c\cdot c\cdot c$$
$$LCD = 2\cdot 2\cdot 3\cdot 3\cdot a\cdot a\cdot a\cdot b\cdot b\cdot c\cdot c\cdot c\cdot c$$
$$LCD = 2^2\cdot 3^2\cdot a^3\cdot b^2\cdot c^4 = 36a^3b^2c^4$$

---

9. Add. $\dfrac{5}{xy} + \dfrac{2}{y}$	**10.** Add. $\dfrac{4}{xyz} + \dfrac{2}{y}$

9. Find the LCD. The two factors are x and y. Observe that the LCD is xy.

$$\frac{5}{xy} + \frac{2}{y} = \frac{5}{xy} + \frac{2}{y}\cdot\frac{x}{x}$$
$$= \frac{5}{xy} + \frac{2x}{xy}$$
$$= \frac{5+2x}{xy}$$

Example	Student Practice

11. Add. $\dfrac{3x}{x^2 - y^2} + \dfrac{5}{x+y}$

Factor the first denominator so that $x^2 - y^2 = (x+y)(x-y)$. Thus, the factors of the denominators are $(x+y)$ and $(x-y)$. Observe that the $LCD = (x+y)(x-y)$.

$$\dfrac{3x}{x^2 - y^2} + \dfrac{5}{x+y}$$

$$= \dfrac{3x}{(x+y)(x-y)} + \dfrac{5}{(x+y)} \cdot \dfrac{x-y}{x-y}$$

$$= \dfrac{3x}{(x+y)(x-y)} + \dfrac{5x-5y}{(x+y)(x-y)}$$

$$= \dfrac{8x-5y}{(x+y)(x-y)}$$

12. Add. $\dfrac{4}{2x-5} + \dfrac{3x+2}{4x^2 - 25}$

13. Add. $\dfrac{5}{x^2 - y^2} + \dfrac{3x}{x^3 + x^2 y}$

$$\dfrac{5}{x^2 - y^2} + \dfrac{3x}{x^3 + x^2 y}$$

$$= \dfrac{5}{(x+y)(x-y)} + \dfrac{3x}{x^2(x+y)}$$

$$= \dfrac{5}{(x+y)(x-y)} \cdot \dfrac{x^2}{x^2} + \dfrac{3x}{x^2(x+y)} \cdot \dfrac{x-y}{x-y}$$

$$= \dfrac{5x^2}{x^2(x+y)(x-y)} + \dfrac{3x^2 - 3xy}{x^2(x+y)(x-y)}$$

$$= \dfrac{x(8x - 3y)}{x^2(x+y)(x-y)}$$

$$= \dfrac{8x - 3y}{x(x+y)(x-y)}$$

14. Add. $\dfrac{6y}{3xy - y^2} + \dfrac{4y}{9x^2 - y^2}$

Example	Student Practice
15. Subtract. $\dfrac{3x+4}{x-2} - \dfrac{x-3}{2x-4}$	**16.** Subtract. $\dfrac{x+5}{2x+7} - \dfrac{x-2}{6x+21}$

Factor the second denominator.

$$\frac{3x+4}{x-2} - \frac{x-3}{2x-4} = \frac{3x+4}{x-2} - \frac{x-3}{2(x-2)}$$

$$= \frac{2}{2} \cdot \frac{3x+4}{x-2} - \frac{x-3}{2(x-2)}$$

$$= \frac{2(3x+4)-(x-3)}{2(x-2)}$$

$$= \frac{6x+8-x+3}{2(x-2)}$$

$$= \frac{5x+11}{2(x-2)}$$

Extra Practice

1. Find the LCD. Do not combine fractions.
$$\frac{7}{2x^2-11x+12}, \ \frac{13}{2x^2+7x-15}$$

2. Perform the operation indicated. Be sure to simplify. $\dfrac{5x}{x+1} - \dfrac{2x-5}{x+1}$

3. Perform the operation indicated. Be sure to simplify. $\dfrac{7}{x^2-y^2} + \dfrac{2y}{xy^2+y^3}$

4. Perform the operation indicated. Be sure to simplify. $\dfrac{1}{x^2+3x+2} - \dfrac{2}{x^2-x-6}$

Concept Check

Explain how to find the LCD of the fractions $\dfrac{3}{10xy^2z}$ and $\dfrac{6}{25x^2yz^3}$.

Chapter 7 Rational Expressions and Equations
7.4 Simplifying Complex Rational Expressions

Vocabulary

complex rational expression • complex fraction • numerator • denominator • LCD

1. A complex rational expression is also called a(n) _____.

2. A(n) _____ has a fraction in the numerator or in the denominator, or both.

3. A complex rational expression may contain two or more fractions in the _____ and denominator.

4. To simplify complex rational expressions, multiply the numerator and denominator of the complex fraction by the _____ of all the denominators appearing in the complex fraction.

Example	**Student Practice**
1. Simplify. $\dfrac{\dfrac{1}{x}}{\dfrac{2}{y^2}+\dfrac{1}{y}}$	2. Simplify. $\dfrac{\dfrac{1}{n^2}}{\dfrac{1}{m}+\dfrac{3}{4}}$

Add the two fractions in the denominator.

$$\frac{\dfrac{1}{x}}{\dfrac{2}{y^2}+\dfrac{1}{y}\cdot\dfrac{y}{y}}=\frac{\dfrac{1}{x}}{\dfrac{2+y}{y^2}}$$

Divide the fraction in the numerator by the fraction in the denominator.

$$\frac{1}{x}\div\frac{2+y}{y^2}=\frac{1}{x}\cdot\frac{y^2}{2+y}$$

$$=\frac{y^2}{x(2+y)}$$

Example	Student Practice
3. Simplify. $\dfrac{\dfrac{1}{x}+\dfrac{1}{y}}{\dfrac{3}{a}-\dfrac{2}{b}}$	**4.** Simplify. $\dfrac{\dfrac{1}{m}+\dfrac{1}{n}}{\dfrac{x}{5}-\dfrac{y}{3}}$

Observe that the LCD of the fractions in the numerator is xy. The LCD of the fractions in the denominator is ab.

$$\frac{\dfrac{1}{x}+\dfrac{1}{y}}{\dfrac{3}{a}-\dfrac{2}{b}}=\frac{\dfrac{1}{x}\cdot\dfrac{y}{y}+\dfrac{1}{y}\cdot\dfrac{x}{x}}{\dfrac{3}{a}\cdot\dfrac{b}{b}-\dfrac{2}{b}\cdot\dfrac{a}{a}}=\frac{\dfrac{y+x}{xy}}{\dfrac{3b-2a}{ab}}$$

$$=\frac{y+x}{xy}\cdot\frac{ab}{3b-2a}=\frac{ab(y+x)}{xy(3b-2a)}$$

| **5.** Simplify. $\dfrac{\dfrac{1}{x^2-1}+\dfrac{2}{x+1}}{x}$ | **6.** Simplify. $\dfrac{\dfrac{3x}{x^2+8x+12}-\dfrac{2}{x+6}}{5}$ |

We need to factor x^2-1.

$$\frac{\dfrac{1}{x^2-1}+\dfrac{2}{x+1}}{x}$$

$$=\frac{\dfrac{1}{(x+1)(x-1)}+\dfrac{2}{x+1}\cdot\dfrac{x-1}{x-1}}{x}$$

$$=\frac{\dfrac{1+2x-2}{(x+1)(x-1)}}{x}$$

$$=\frac{2x-1}{(x+1)(x-1)}\cdot\frac{1}{x}$$

$$=\frac{2x-1}{x(x+1)(x-1)}$$

Example	Student Practice

7. Simplify. $\dfrac{\dfrac{3}{a+b}-\dfrac{3}{a-b}}{\dfrac{5}{a^2-b^2}}$

$=\dfrac{\dfrac{3}{a+b}\cdot\dfrac{a-b}{a-b}-\dfrac{3}{a-b}\cdot\dfrac{a+b}{a+b}}{\dfrac{5}{a^2-b^2}}$

$=\dfrac{\dfrac{3a-3b}{(a+b)(a-b)}-\dfrac{3a+3b}{(a+b)(a-b)}}{\dfrac{5}{a^2-b^2}}$

$=\dfrac{\dfrac{-6b}{(a+b)(a-b)}}{\dfrac{5}{(a+b)(a-b)}}$

$=\dfrac{-6b}{(a+b)(a-b)}\cdot\dfrac{(a+b)(a-b)}{5}$

$=\dfrac{-6b}{5}$ or $-\dfrac{6b}{5}$

8. Simplify. $\dfrac{\dfrac{5}{x+y}+\dfrac{2}{x-y}}{\dfrac{7}{x^2-y^2}}$

---

9. Simplify by multiplying by the LCD.
$\dfrac{\dfrac{5}{ab^2}-\dfrac{2}{ab}}{3-\dfrac{5}{2a^2b}}$

$=\dfrac{2a^2b^2\left(\dfrac{5}{ab^2}-\dfrac{2}{ab}\right)}{2a^2b^2\left(3-\dfrac{5}{2a^2b}\right)}$

$=\dfrac{2a^2b^2\left(\dfrac{5}{ab^2}\right)-2a^2b^2\left(\dfrac{2}{ab}\right)}{2a^2b^2(3)-2a^2b^2\left(\dfrac{5}{2a^2b}\right)}$

$=\dfrac{10a-4ab}{6a^2b^2-5b}$

10. Simplify by multiplying by the LCD.
$\dfrac{\dfrac{4}{y}-\dfrac{7}{4x}}{5-\dfrac{9}{xy^2}}$

Example	Student Practice
11. Simplify by multiplying by the LCD. $$\dfrac{\dfrac{3}{a+b}-\dfrac{3}{a-b}}{\dfrac{5}{a^2-b^2}}$$	**12.** Simplify by multiplying by the LCD. $$\dfrac{\dfrac{5}{x+y}+\dfrac{2}{x-y}}{\dfrac{7}{x^2-y^2}}$$

The LCD of all individual fractions in the complex fraction is $(a+b)(a-b)$.

$$=\frac{(a+b)(a-b)\left(\dfrac{3}{a+b}\right)-(a+b)(a-b)\left(\dfrac{3}{a-b}\right)}{(a+b)(a-b)\left(\dfrac{5}{(a+b)(a-b)}\right)}$$

$$=\frac{3(a-b)-3(a+b)}{5}$$

$$=\frac{3a-3b-3a-3b}{5}$$

$$=-\frac{6b}{5}$$

Extra Practice

1. Simplify. $\dfrac{\dfrac{4}{a}+\dfrac{5}{b}}{\dfrac{2}{ab}}$

2. Simplify. $\dfrac{\dfrac{2}{x}+\dfrac{2}{y}}{x+y}$

3. Simplify. $\dfrac{\dfrac{2x}{x^2-25}}{\dfrac{4}{x+5}-\dfrac{3}{x-5}}$

4. Simplify. $\dfrac{\dfrac{3}{4x}-\dfrac{8}{3y}}{\dfrac{7}{a}+\dfrac{5}{b}}$

Concept Check

To simplify the following complex fraction, explain how you would add the two fractions in the numerator.

$$\frac{\dfrac{7}{x-3}+\dfrac{15}{2x-6}}{\dfrac{2}{x+5}}$$

Chapter 7 Rational Expressions and Equations
7.5 Solving Equations Involving Rational Expressions

Vocabulary
extraneous solution • no solution • LCD • exclude

1. The first step to solve an equation containing rational expressions is to determine the _____ of all the denominators.

2. If all of the apparent solutions of an equation are extraneous solutions, we say that the equation has _____.

3. A value that makes a denominator in the equation equal to zero is called a(n) _____.

4. _____ from your solution any value that would make the LCD equal to zero.

Example	**Student Practice**
1. Solve for x and check your solution. $$\frac{5}{x} + \frac{2}{3} = -\frac{3}{x}$$ The LCD is $3x$. Multiply each term by $3x$. $$3x\left(\frac{5}{x}\right) + 3x\left(\frac{2}{3}\right) = 3x\left(-\frac{3}{x}\right)$$ $$15 + 2x = -9$$ $$2x = -9 - 15$$ $$2x = -24$$ $$x = -12$$ Check. Replace each x by -12. $$\frac{5}{-12} + \frac{2}{3} \overset{?}{=} -\frac{3}{-12}$$ $$-\frac{5}{12} + \frac{8}{12} \overset{?}{=} \frac{3}{12}$$ $$\frac{3}{12} = \frac{3}{12}$$	**2.** Solve for x and check your solution. $$\frac{5}{x} + \frac{1}{4} = -\frac{6}{x}$$

Example	Student Practice
3. Solve and check. $\dfrac{6}{x+3} = \dfrac{3}{x}$	**4.** Solve and check. $\dfrac{2}{x+4} = \dfrac{6}{x}$

Observe that the LCD $= x(x+3)$.

$$x(x+3)\left(\dfrac{6}{x+3}\right) = x(x+3)\left(\dfrac{3}{x}\right)$$

$$6x = 3(x+3)$$

$$6x = 3x+9$$

$$3x = 9$$

$$x = 3$$

Check. Replace each x by 3.

$$\dfrac{6}{3+3} \overset{?}{=} \dfrac{3}{3}$$

$$\dfrac{6}{6} = \dfrac{3}{3}$$

---

5. Solve and check. $\dfrac{3}{x+5} - 1 = \dfrac{4-x}{2x+10}$

6. Solve and check.

$$\dfrac{3}{x+2} - \dfrac{5}{x+5} = \dfrac{x-2}{x^2+7x+10}$$

$$\dfrac{3}{x+5} - 1 = \dfrac{4-x}{2(x+5)}$$

$$2(x+5)\left(\dfrac{3}{x+5}\right) - 2(x+5)(1)$$

$$= 2(x+5)\left[\dfrac{4-x}{2(x+5)}\right]$$

$$2(3) - 2(x+5) = 4-x$$

$$6 - 2x - 10 = 4-x$$

$$-2x - 4 = 4-x$$

$$-4 = 4+x$$

$$-8 = x$$

The check is left to the student.

Example	Student Practice
7. Solve and check. $\dfrac{y}{y-2} - 4 = \dfrac{2}{y-2}$	**8.** Solve and check. $\dfrac{3x}{x-3} + 5 = \dfrac{3x}{x-3}$

Observe that the LCD is $y-2$.

$$(y-2)\left(\dfrac{y}{y-2}\right) - (y-2)(4)$$

$$= (y-2)\left(\dfrac{2}{y-2}\right)$$

$$y - 4(y-2) = 2$$

$$y - 4y + 8 = 2$$

$$-3y + 8 = 2$$

$$-3y = -6$$

$$\dfrac{-3y}{-3} = \dfrac{-6}{-3}$$

$$y = 2$$

This equation has no solution. We can see immediately that $y = 2$ is not a solution of the original equation. When we substitute 2 for y in a denominator, the denominator is equal to zero and the expression is undefined. Suppose that we tried to check the apparent solution by substituting 2 for y.

$$\dfrac{y}{y-2} - 4 = \dfrac{2}{y-2}$$

$$\dfrac{2}{2-2} - 4 \overset{?}{=} \dfrac{2}{2-2}$$

$$\dfrac{2}{0} - 4 = \dfrac{2}{0}$$

These expressions are not defined. There is no such number as $2 \div 0$. We see that 2 does not check. This equation has no solution.

219

Extra Practice

1. Solve and check. If there is no solution, so indicate. $\dfrac{2}{3x} + \dfrac{1}{6} = \dfrac{6}{x}$

2. Solve and check. If there is no solution, so indicate. $\dfrac{4}{a^2-1} = \dfrac{2}{a+1} + \dfrac{2}{a-1}$

3. Solve and check. If there is no solution, so indicate.

$$\dfrac{6}{5a+10} - \dfrac{1}{a-5} = \dfrac{4}{a^2-3a-10}$$

4. Solve and check. If there is no solution, so indicate.

$$\dfrac{x-2}{x^2-4x-5} + \dfrac{x+5}{x^2-25} = \dfrac{2x+13}{x^2+6x+5}$$

Concept Check

Explain how to find the LCD for the following equation. Do not solve the equation.

$$\dfrac{x}{x^2-9} + \dfrac{2}{3x-9} = \dfrac{5}{2x+6} + \dfrac{3}{2x^2-18}$$

Name: _____   Date: _____

Instructor: _____   Section: _____

Chapter 7 Rational Expressions and Equations
7.6 Ratio, Proportion, and Other Applied Problems

Vocabulary
ratio • proportion • cross multiplying • similar

1. A _____ is an equation that states that two ratios are equal.

2. _____ triangles are triangles that have the same shape, but may be different sizes.

3. If $\dfrac{a}{b} = \dfrac{c}{d}$, then $ad = bc$ is sometimes called _____.

4. A _____ is a comparison of two quantities.

Example	**Student Practice**
1. Michael took 5 hours to drive 245 miles on the turnpike. At the same rate, how many hours will it take him to drive a distance of 392 miles? Let x = the number of hours it will take to drive 392 miles. If 5 hours are needed to drive 245 miles, then x hours are needed to drive 392 miles. $$\frac{5 \text{ hours}}{245 \text{ miles}} = \frac{x \text{ hours}}{392 \text{ miles}}$$ $$5(392) = 245x$$ $$\frac{1960}{245} = x$$ $$8 = x$$ It will take Michael 8 hours to drive 392 miles. To check, do the computations and see if $\dfrac{5}{245} = \dfrac{8}{392}$.	**2.** Rose took 4 hours to drive 264 miles. How far will she drive in 9 hours?

Example	Student Practice

3. A ramp is 32 meters long and rises up 15 meters. A ramp at the same angle is 9 meters long. How high does the second ramp rise?

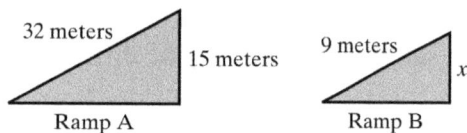

32 meters	
15 meters	9 meters
Ramp A	x
	Ramp B

$$\frac{32}{9} = \frac{15}{x}$$

$$32x = (9)(15)$$

$$32x = 135$$

$$x = \frac{135}{32} \text{ or } x = 4\frac{7}{32}$$

---

4. Triangle M is similar to Triangle N. Find the length of side x. Express your answer as a mixed number.

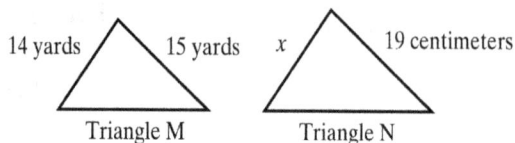

14 yards   15 yards x 19 centimeters

Triangle M   Triangle N

---

5. A French commuter airline flies from Paris to Avignon. Plane A flies at a speed that is 50 kilometers per hour faster than plane B. Plane A flies 500 kilometers in the amount of time that plane B flies 400 kilometers. Find the speed of each plane.

Let $s =$ the speed of plane B in kilometers per hour. Then $s + 50 =$ the speed of plane A in kilometers per hour. Each plane flies the same amount of time. That is, the time for plane A equals the time for plane B.

$$\frac{500}{s + 50} = \frac{400}{s}$$

$$500s = (s + 50)(400)$$

$$500s = 400s + 20,000$$

$$100s = 20,000$$

$$s = 200$$

Plane B travels 200 kilometers per hour. Since $s + 50 = 200 + 50 = 250$, plane A travels 250 kilometers per hour.

---

6. Two trains travel in opposite directions for the same amount of time. Train A traveled 200 kilometers, while train B traveled 175 kilometers. Train A traveled 15 kilometers per hour faster than train B. What is the speed of each train?

Example	Student Practice
7. Reynaldo can sort a huge stack of mail on an old sorting machine in 9 hours. His brother Carlos can sort the same amount of mail using a newer sorting machine in 8 hours. How long will it take them to do the job working together? Express your answer in hours and minutes. Round to the nearest minute.	**8.** Two people are painting a side of a house. The first person takes 4 hours to paint the side of the house, while the second person takes 6 hours. How long will it take both of them to paint the side of the house working together? Express your answer in hours and minutes. Round to the nearest minute.

If Reynaldo can do the job in 9 hours, then in 1 hour he could do $\frac{1}{9}$ of the job.

If Carlos can do the job in 8 hours, then in 1 hour he could do $\frac{1}{8}$ of the job. Let x = the number of hours it takes Reynaldo and Carlos to do the job together. In 1 hour together they could do $\frac{1}{x}$ of the job. The amount of work Reynaldo can do in 1 hour plus the amount of work Carlos can do in 1 hour must be equal to the amount of work they could do together in 1 hour.

$$\frac{1}{9}+\frac{1}{8}=\frac{1}{x}$$

$$72x\left(\frac{1}{9}\right)+72x\left(\frac{1}{8}\right)=72x\left(\frac{1}{x}\right)$$

$$8x+9x=72$$

$$x=4\frac{4}{17}$$

To change $\frac{4}{17}$ of an hour to minutes,

$$\frac{4}{17}\text{ hour}\times\frac{60\text{ minutes}}{1\text{ hour}}=\frac{240}{17}\text{ minutes},$$

which is approximately 14.118. Thus doing the job together will take 4 hours and 14 minutes.

Extra Practice

1. Solve. $\dfrac{3}{10} = \dfrac{11.4}{x}$

2. The scale on a map of Massachusetts is approximately $\dfrac{3}{4}$ inch to 25 miles. If the distance from a college to Boston measures 4 inches on the map, how far apart are the two locations?

3. Alicia is 4 feet tall and casts a shadow that is 9 feet long. At the same time of day, a tree casts a shadow that is 36 feet long. How tall is the tree?

4. Toshiro and Mathilda work in the library. Working alone, it takes Toshiro 7 hours to arrange and stack 1000 books, while it takes Mathilda 6.5 hours. To the nearest minute, how long will the job take if they work together to arrange and stack 1000 books?

Concept Check

Mike found that his car used 18 gallons of gas to travel 396 miles. He needs to take a trip of 450 miles and wants to know how many gallons of gas it will take. He set up the equation $\dfrac{18}{x} = \dfrac{450}{396}$. Explain what error he made and how he should correctly solve the problem.

MATH COACH

Mastering the skills you need to do well on the test.

Watch the **MATH COACH** videos in MyMathLab® or on You Tube™ while you work the problems below. These helpful hints will help you avoid making common errors on test problems.

Dividing Rational Expressions—Problem 5 $\dfrac{2a^2-3a-2}{a^2+5a+6} \div \dfrac{a^2-5a+6}{a^2-9}$

> **Helpful Hint:** The operation of division is performed by inverting the second fraction and multiplying it by the first fraction. This step should be done first, before you begin factoring.

Did you keep the first fraction the same as it is written and then multiply it by $\dfrac{a^2-9}{a^2-5a+6}$? Yes _____ No _____
If you answered No, stop and make this correction to your work.

Were you able to factor each expression so that you obtained $\dfrac{(2a+1)(a-2)}{(a+2)(a+3)} \cdot \dfrac{(a+3)(a-3)}{(a-2)(a-3)}$? Yes _____ No _____

If you answered No, go back and check each factoring step to see if you can get the same result. Complete the problem by dividing out any common factors.

If you answered Problem 5 incorrectly, go back and rework the problem using these suggestions.

Subtracting Rational Expressions with Different Denominators—Problem 8 $\dfrac{3x}{x^2-3x-18} - \dfrac{x-4}{x-6}$

> **Helpful Hint:** First factor the trinomial in the denominator so that you can determine the LCD of these two fractions. Then multiply by what is needed in the second fraction so that it becomes an equivalent fraction with the LCD as the denominator. When subtracting, it is a good idea to place brackets around the second numerator to avoid sign errors.

Did you factor the first denominator into $(x-6)(x+3)$?
Yes _____ No _____

Did you then determine that the LCD is $(x-6)(x+3)$?
Yes _____ No _____

If you answered No to these questions, stop and review how to factor the trinomial in the first denominator and how to find the LCD when working with polynomials as denominators.

Did you multiply the numerator and denominator of the second fraction by $(x+3)$ and place brackets around the product to obtain $\dfrac{3x}{(x-6)(x+3)} - \dfrac{[(x-4)(x+3)]}{(x-6)(x+3)}$?
Yes _____ No _____

Did you obtain $3x-x^2-x-12$ in the numerator? Yes _____ No _____

If you answered Yes, you forgot to distribute the negative sign. Rework the problem and make this correction.

As your final step, remember to combine like terms and then factor the new numerator before dividing out common factors.

Now go back and rework the problem using these suggestions.

Simplifying Complex Rational Expressions—Problem 10   Simplify. $\dfrac{\dfrac{6}{b}-4}{\dfrac{5}{bx}-\dfrac{10}{3x}}$

> **Helpful Hint:** There are two ways to simplify this expression:
> 1) combine the numerators and the denominators separately, or
> 2) multiply the numerator and denominator of the complex fraction by the LCD of all the denominators.
> Consider both methods and choose the one that seems easiest to you. We will show the steps of the first method for this particular problem.

Did you multiply 4 by $\dfrac{b}{b}$ and then subtract $\dfrac{6}{b}-\dfrac{4b}{b}$?

Yes _____ No _____

If you answered No, remember that 4 can be written as $\dfrac{4}{1}$, and to find the LCD, you must multiply numerator and denominator by the variable b.

In the denominator of the complex fraction, did you obtain $3bx$ as the LCD and multiply the first fraction by $\dfrac{3}{3}$ and the second fraction by $\dfrac{b}{b}$ before subtracting the two fractions?

Yes _____ No _____

If you answered No, stop and carefully change these two fractions into equivalent fractions with $3bx$ as the

denominator. Then subtract the numerators and keep the common denominator.

Were you able to rewrite the problem as follows: $\dfrac{6-4b}{b} \div \dfrac{15-10b}{3bx}$?

Yes _____ No _____

If you answered No, examine your steps carefully. Once you have one fraction in the numerator and one fraction in the denominator, you can rewrite the division as multiplication by inverting the fraction in the denominator.

If you answered Problem 10 incorrectly, go back and rework the problem using these suggestions.

Solving Equations Involving Rational Expressions—Problem 14

Solve for x. $\dfrac{x-3}{x-2}=\dfrac{2x^2-15}{x^2+x-6}-\dfrac{x+1}{x+3}$

> **Helpful Hint:** First factor any denominators that need to be factored so that you can determine the LCD of all denominators. Verify that you are solving an equation, then multiply each term of the equation by the LCD. Solve the resulting equation. Check your solution.

Did you factor x^2+x-6 to get $(x+3)(x-2)$ and identify the LCD as $(x+3)(x-2)$? Yes _____ No _____
If you answered No, go back and complete these steps again.

Did you notice that we are solving an equation and that we can multiply each term of the equation by the LCD and then multiply the binomials to obtain the equation

$x^2-9=(2x^2-15)-(x^2-x-2)$? Yes _____ No _____

If you answered No, look at the step $-\left[(x+1)(x-2)\right]$.

After you multiplied the binomials, did you distribute the negative sign and change the sign of all terms inside the grouping symbols? Yes _____ No _____

If you answered No, review how to multiply binomials and subtract polynomial expressions. Then combine like terms and solve the equation for x.

Now go back and rework the problem using these suggestions.

Chapter 8 Radicals
8.1 Square Roots

Vocabulary

square root • perfect square • principal square root • radical sign • radicand

1. The number beneath the radical sign is called the _____.

2. The _____ of a number is the nonnegative square root.

3. A square root of a _____ is an integer.

Example	Student Practice
1. Find.	**2.** Find.
(a) $\sqrt{144}$	**(a)** $\sqrt{25}$
$\sqrt{144} = 12$ since $(12)(12) = 144$ and $12 \geq 0$.	
	(b) $-\sqrt{64}$
(b) $-\sqrt{9}$	
Since $\sqrt{9} = 3$, $-\sqrt{9} = -3$.	
3. Find.	**4.** Find.
(a) $\sqrt{\dfrac{1}{4}}$	**(a)** $\sqrt{\dfrac{25}{144}}$
$\sqrt{\dfrac{1}{4}} = \dfrac{1}{2}$ since $\left(\dfrac{1}{2}\right)^2 = \left(\dfrac{1}{2}\right)\left(\dfrac{1}{2}\right) = \dfrac{1}{4}$.	
(b) $-\sqrt{\dfrac{4}{9}}$	**(b)** $-\sqrt{\dfrac{1}{16}}$
$-\sqrt{\dfrac{4}{9}} = -\dfrac{2}{3}$ since $\sqrt{\dfrac{4}{9}} = \dfrac{2}{3}$.	

Example	Student Practice
5. Find.	**6.** Find.
(a) $\sqrt{0.09}$	**(a)** $\sqrt{0.16}$
$\sqrt{0.09} = 0.3$ since $(0.3)(0.3) = 0.09$. Notice that $\sqrt{0.09}$ is not 0.03 because $(0.03)(0.03) = 0.0009$. Remember to count the decimal places when you multiply decimals.	**(b)** $-\sqrt{3600}$
(b) $\sqrt{1600}$	
$\sqrt{1600} = 40$ since $(40)(40) = 1600$.	**(c)** $\sqrt{361}$
(c) $\sqrt{225}$	
$\sqrt{225} = 15$	
7. Use a calculator or a table to approximate. Round to the nearest thousandth.	**8.** Use a calculator or a table to approximate. Round to the nearest thousandth.
(a) $\sqrt{28}$	**(a)** $\sqrt{12}$
Using a calculator, we have 28 $\boxed{\sqrt{x}}$ 5.291502622. Thus $\sqrt{28} \approx 5.292$.	
(b) $\sqrt{191}$	**(b)** $\sqrt{40}$
Using a calculator, we have 191 $\boxed{\sqrt{x}}$ 13.82027496. Thus $\sqrt{191} \approx 13.820$.	

Extra Practice

1. Find the two roots of 100.

2. Find the square root. Do not use a calculator or a table of square roots.

   $\sqrt{49}$

3. Find the square root. Do not use a calculator or a table of square roots.

   $-\sqrt{0.16}$

4. Find the square root. Do not use a calculator or a table of square roots.

   $\sqrt{\dfrac{36}{121}}$

Concept Check

Explain how you would go about determining $\sqrt{32,400}$ without using a calculator or a table of square roots.

Chapter 8 Radicals
8.2 Simplifying Radical Expressions

Vocabulary
complex number   •   multiplication   •   nonnegative   •   radicand

1. The _____ rule for square roots says $\sqrt{a} \cdot \sqrt{b} = \sqrt{ab}$.

2. In the _____ system, negative numbers have square roots.

3. All variable radicands must be assumed to be _____ in order to simplify radical expressions.

Example	**Student Practice**
1. Find.	**2.** Find.
(a) $\sqrt{9^4}$	**(a)** $\sqrt{5^2}$
Using the law of exponents, we rewrite 9^4 as a square.	
$\sqrt{9^4} = \sqrt{\left(9^2\right)^2}$	
$= 9^2$	
To check, multiply: $\left(9^2\right)\left(9^2\right) = 9^4$.	**(b)** $\sqrt{15^4}$
(b) $\sqrt{126^2}$	
$\sqrt{126^2} = 126$	
(c) $\sqrt{17^6}$	**(c)** $\sqrt{35^{18}}$
$\sqrt{17^6} = \sqrt{\left(17^3\right)^2}$	
$= 17^3$	

Example	Student Practice
3. Find.	**4.** Find.
(a) $\sqrt{x^6}$	**(a)** $\sqrt{n^{10}}$
By the law of exponents, $\left(x^3\right)^2 = x^6$.	
$$\sqrt{x^6} = \sqrt{\left(x^3\right)^2}$$ $$= x^3$$	
	(b) $\sqrt{b^{40}}$
(b) $\sqrt{y^{24}}$	
$$\sqrt{y^{24}} = y^{12}$$	
5. Find.	**6.** Find.
(a) $\sqrt{225x^2}$	**(a)** $\sqrt{289x^6}$
Use the multiplication rule for square roots.	
$$\sqrt{225x^2} = \sqrt{225}\sqrt{x^2} = 15x$$	**(b)** $\sqrt{x^8 y^{20}}$
(b) $\sqrt{x^6 y^{14}}$	
$$\sqrt{x^6 y^{14}} = x^3 y^7$$	
	(c) $\sqrt{196x^{14} y^{22}}$
(c) $\sqrt{169x^8 y^{10}}$	
$$\sqrt{169x^8 y^{10}} = 13x^4 y^5$$	
To check each answer, square it. The result should be the expression under the radical sign.	

Example	Student Practice
7. Simplify.	**8.** Simplify.
(a) $\sqrt{20}$	**(a)** $\sqrt{27}$
Look for perfect squares.	
$\sqrt{20} = \sqrt{4 \cdot 5}$ $= \sqrt{4}\sqrt{5}$ $= 2\sqrt{5}$	
(b) $\sqrt{48}$	**(b)** $\sqrt{32}$
$\sqrt{48} = \sqrt{16 \cdot 3}$ $= \sqrt{16}\sqrt{3}$ $= 4\sqrt{3}$	
9. Simplify.	**10.** Simplify.
(a) $\sqrt{x^3}$	**(a)** $\sqrt{x^7}$
To simplify a square root that has a variable with an exponent, use the form $\sqrt{x^n}\sqrt{x}$, where n is the largest possible even exponent.	
$\sqrt{x^3} = \sqrt{x^2}\sqrt{x}$ $= x\sqrt{x}$	
	(b) $\sqrt{x^{13}y^{15}}$
(b) $\sqrt{x^7 y^9}$	
$\sqrt{x^7 y^9} = \sqrt{x^6}\sqrt{x}\sqrt{y^8}\sqrt{y}$ $= x^3\sqrt{x}y^4\sqrt{y}$ $= x^3 y^4 \sqrt{xy}$	
Place the factors with exponents first and the radical factor second.	

Example	Student Practice
11. Simplify.	**12.** Simplify.
(a) $\sqrt{12y^5}$	**(a)** $\sqrt{75x^{17}}$
$\sqrt{12y^5} = \sqrt{4 \cdot 3 \cdot y^4 \cdot y}$ $= 2y^2\sqrt{3y}$	
(b) $\sqrt{18x^3 y^7 w^{10}}$	**(b)** $\sqrt{108x^8 y^9 z^{15}}$
$\sqrt{18x^3 y^7 w^{10}}$ $= \sqrt{9 \cdot 2 \cdot x^2 \cdot x \cdot y^6 \cdot y \cdot w^{10}}$ $= 3xy^3 w^5 \sqrt{2xy}$	

Extra Practice

1. Simplify. $\sqrt{6^4}$

2. Simplify. Assume that all variables represent positive numbers. $\sqrt{49r^6 s^4 t^8}$

3. Simplify. $\sqrt{90}$

4. Simplify. Assume that all variables represent positive numbers. $\sqrt{18x^3}$

Concept Check

Explain how you would simplify the following expression. $\sqrt{y^6 z^7}$

Chapter 8 Radicals
8.3 Adding and Subtracting Radical Expressions

Vocabulary
like radicals • simplify

1. Sometimes it is necessary to _____ one or more radicals before their terms can be combined.

2. To add or subtract square root radicals, the radicals must be _____.

Example	**Student Practice**
1. Combine. **(a)** $5\sqrt{2} - 8\sqrt{2}$ First check to see if the radicands are the same. If they are, you can combine the radicals. $5\sqrt{2} - 8\sqrt{2} = (5-8)\sqrt{2}$ $= -3\sqrt{2}$ **(b)** $7\sqrt{a} + 3\sqrt{a} - 5\sqrt{a}$ $7\sqrt{a} + 3\sqrt{a} - 5\sqrt{a} = (7+3-5)\sqrt{a}$ $= 5\sqrt{a}$	**2.** Combine. **(a)** $6\sqrt{5} - 2\sqrt{5}$ **(b)** $6\sqrt{x} - \sqrt{x} + 7\sqrt{x} - 8\sqrt{x}$
3. Combine. $5\sqrt{2a} + 3\sqrt{a} - 7\sqrt{2} + 3\sqrt{2a}$ The only terms that have the same radicand are $5\sqrt{2a}$ and $3\sqrt{2a}$. These terms may be combined. All other terms stay the same. $5\sqrt{2a} + 3\sqrt{a} - 7\sqrt{2} + 3\sqrt{2a}$ $= 8\sqrt{2a} + 3\sqrt{a} - 7\sqrt{2}$	**4.** Combine. $6\sqrt{x} - 4\sqrt{3x} + 8\sqrt{x} - 7\sqrt{3}$

Example	Student Practice
5. Combine.	**6.** Combine.
(a) $2\sqrt{3} + \sqrt{12}$	**(a)** $\sqrt{24} - \sqrt{54}$
Look for perfect square factors. Then simplify and combine like radicals.	
$2\sqrt{3} + \sqrt{12} = 2\sqrt{3} + \sqrt{4 \cdot 3}$ $= 2\sqrt{3} + 2\sqrt{3}$ $= 4\sqrt{3}$	**(b)** $\sqrt{32} + \sqrt{45} - \sqrt{18}$
(b) $\sqrt{12} - \sqrt{27} + \sqrt{50}$	
$\sqrt{12} - \sqrt{27} + \sqrt{50}$ $= \sqrt{4 \cdot 3} - \sqrt{9 \cdot 3} + \sqrt{25 \cdot 2}$ $= 2\sqrt{3} - 3\sqrt{3} + 5\sqrt{2}$ $= -\sqrt{3} + 5\sqrt{2} \text{ or } 5\sqrt{2} - \sqrt{3}$	
7. Combine. $\sqrt{2a} + \sqrt{8a} + \sqrt{27a}$	**8.** Combine. $\sqrt{32x} + \sqrt{36x} - \sqrt{16x} + \sqrt{72x}$
$\sqrt{2a} + \sqrt{8a} + \sqrt{27a}$ $= \sqrt{2a} + \sqrt{4 \cdot 2a} + \sqrt{9 \cdot 3a}$ $= \sqrt{2a} + 2\sqrt{2a} + 3\sqrt{3a}$ $= 3\sqrt{2a} + 3\sqrt{3a}$	
9. Combine. $3\sqrt{20} + 4\sqrt{45} - 2\sqrt{80}$	**10.** Combine. $2\sqrt{24} - 3\sqrt{54} + 4\sqrt{96}$
$3\sqrt{20} + 4\sqrt{45} - 2\sqrt{80}$ $= 3 \cdot \sqrt{4} \cdot \sqrt{5} + 4 \cdot \sqrt{9} \cdot \sqrt{5} - 2 \cdot \sqrt{16} \cdot \sqrt{5}$ $= 3 \cdot 2 \cdot \sqrt{5} + 4 \cdot 3 \cdot \sqrt{5} - 2 \cdot 4 \cdot \sqrt{5}$ $= 6\sqrt{5} + 12\sqrt{5} - 8\sqrt{5}$ $= 10\sqrt{5}$	

Example	Student Practice
11. Combine. $3a\sqrt{8a} + 2\sqrt{50a^3}$	**12.** Combine. $6\sqrt{20x^5} + 4x^2\sqrt{80x}$

$$3a\sqrt{8a} + 2\sqrt{50a^3}$$
$$= 3a\sqrt{4}\sqrt{2a} + 2\sqrt{25a^2}\sqrt{2a}$$
$$= 3a \cdot 2 \cdot \sqrt{2a} + 2 \cdot 5a \cdot \sqrt{2a}$$
$$= 6a\sqrt{2a} + 10a\sqrt{2a}$$
$$= 16a\sqrt{2a}$$

Extra Practice

1. Combine, if possible. Do not use a calculator or a table of square roots.
$2\sqrt{3} - 5\sqrt{3} + 4\sqrt{3}$

2. Combine, if possible. Do not use a calculator or a table of square roots.
$-\sqrt{90} + \sqrt{40} + \sqrt{10}$

3. Combine, if possible. Do not use a calculator or a table of square roots.
$3\sqrt{2x} + \sqrt{50x}$

4. Combine, if possible. Do not use a calculator or a table of square roots.
$3\sqrt{10y^3} - y\sqrt{40y}$

Concept Check

Explain how you would combine $3\sqrt{8x^3} + 5x\sqrt{98x}$.

Name: _____ Date: _____

Instructor: _____ Section: _____

Chapter 8 Radicals
8.4 Multiplying Radical Expressions

Vocabulary
FOIL method • multiplication

1. The basic rule for _____ of square root radicals is $\sqrt{a}\sqrt{b} = \sqrt{ab}$.

2. The _____ can be used for binomial radical expressions.

Example	Student Practice
1. Multiply. $\sqrt{7}\sqrt{14x}$	**2.** Multiply. $\sqrt{10x}\sqrt{15}$
$\sqrt{7}\sqrt{14x} = \sqrt{98x}$	
We do not stop here, because the radical $\sqrt{98x}$ can be simplified.	
$\sqrt{98x} = \sqrt{49 \cdot 2x} = \sqrt{49}\sqrt{2x} = 7\sqrt{2x}$	
3. Multiply.	**4.** Multiply.
(a) $\left(2\sqrt{3}\right)\left(5\sqrt{7}\right)$	**(a)** $\left(4\sqrt{2}\right)\left(5\sqrt{3}\right)$
Multiply the coefficients: $2 \cdot 5 = 10$. Multiply the radicals: $\sqrt{3}\sqrt{7} = \sqrt{21}$.	
$\left(2\sqrt{3}\right)\left(5\sqrt{7}\right) = 10\sqrt{21}$	
	(b) $\left(5\sqrt{15z}\right)\left(9\sqrt{35z}\right)$
(b) $\left(2a\sqrt{3a}\right)\left(3\sqrt{6a}\right)$	
$\left(2a\sqrt{3a}\right)\left(3\sqrt{6a}\right) = 6a\sqrt{18a^2}$	
$= 6a\sqrt{9}\sqrt{2}\sqrt{a^2}$	
$= 18a^2\sqrt{2}$	

Example	Student Practice
5. Find the area of a farm field that measures $\sqrt{9400}$ feet long and $\sqrt{4800}$ feet wide. Express your answer as a simplified radical. We multiply. $$\left(\sqrt{9400}\right)\left(\sqrt{4800}\right) = \left(10\sqrt{94}\right)\left(40\sqrt{3}\right)$$ $$= 400\sqrt{282} \text{ square feet } \left(\text{ft}^2\right)$$	**6.** Find the area of a swimming pool that measures $\sqrt{1200}$ feet long and $\sqrt{975}$ feet wide. Express your answer as a simplified radical.
7. Multiply and simplify. $\sqrt{5}\left(\sqrt{2} + 3\sqrt{7}\right)$ $$\sqrt{5}\left(\sqrt{2} + 3\sqrt{7}\right) = \sqrt{5}\sqrt{2} + 3\sqrt{5}\sqrt{7}$$ $$= \sqrt{10} + 3\sqrt{35}$$ In similar fashion, we can multiply a trinomial by another factor.	**8.** Multiply and simplify. $$\sqrt{2}\left(3\sqrt{6} + 7\sqrt{10}\right)$$
9. Multiply. $\sqrt{a}\left(3\sqrt{a} - 2\sqrt{5}\right)$ For any nonnegative real number a, $\sqrt{a} \cdot \sqrt{a} = a$. $$\sqrt{a}\left(3\sqrt{a} - 2\sqrt{5}\right) = 3\sqrt{a}\sqrt{a} - 2\sqrt{5}\sqrt{a}$$ $$= 3a - 2\sqrt{5a}$$	**10.** Multiply. $3\sqrt{x}\left(x\sqrt{5} + 7\sqrt{x}\right)$
11. Multiply. $\left(\sqrt{2} + 5\right)\left(\sqrt{2} - 3\right)$ $$\left(\sqrt{2} + 5\right)\left(\sqrt{2} - 3\right)$$ $$= \sqrt{4} - 3\sqrt{2} + 5\sqrt{2} - 15$$ $$= 2 + 2\sqrt{2} - 15$$ $$= -13 + 2\sqrt{2}$$	**12.** Multiply. $\left(\sqrt{5} + 3\right)\left(\sqrt{5} - 2\right)$

Example	Student Practice
13. Multiply. $\left(\sqrt{2}-3\sqrt{6}\right)\left(\sqrt{2}+\sqrt{6}\right)$	**14.** Multiply. $\left(\sqrt{5}+\sqrt{15}\right)\left(2\sqrt{5}-6\sqrt{15}\right)$

$$\left(\sqrt{2}-3\sqrt{6}\right)\left(\sqrt{2}+\sqrt{6}\right)$$
$$=\sqrt{4}+\sqrt{12}-3\sqrt{12}-3\sqrt{36}$$
$$=2-2\sqrt{12}-18$$
$$=-16-4\sqrt{3}$$

15. Multiply. $\left(2\sqrt{3}-\sqrt{6}\right)^2$	**16.** Multiply. $\left(4\sqrt{7}-\sqrt{14}\right)^2$

$$\left(2\sqrt{3}-\sqrt{6}\right)^2$$
$$=\left(2\sqrt{3}-\sqrt{6}\right)\left(2\sqrt{3}-\sqrt{6}\right)$$
$$=4\sqrt{9}-2\sqrt{18}-2\sqrt{18}+\sqrt{36}$$
$$=12-4\sqrt{18}+6$$
$$=18-12\sqrt{2}$$

Extra Practice

1. Multiply. Be sure to simplify any radicals in your answer. Do not use a calculator or a table of square roots. $\sqrt{6}\sqrt{10}$

2. Multiply. Be sure to simplify any radicals in your answer. Do not use a calculator or a table of square roots. $\sqrt{5}\left(\sqrt{10}+\sqrt{x}\right)$

3. Multiply. Be sure to simplify any radicals in your answer. Do not use a calculator or a table of square roots.
$\left(5-2\sqrt{3}\right)\left(2+3\sqrt{6}\right)$

4. Multiply. Be sure to simplify any radicals in your answer. Do not use a calculator or a table of square roots.
$\left(2\sqrt{3x}-7\right)\left(2\sqrt{3x}+7\right)$

Concept Check

Explain how to simplify $\left(3\sqrt{3}-2\right)^2$.

Name: _____   Date: _____

Instructor: _____   Section: _____

Chapter 8 Radicals
8.5 Dividing Radical Expressions

Vocabulary
rationalize the denominator   •   conjugates   •   quotient rule

1. The _____ for square roots can be used to divide square root radicals or to simplify square root radical expressions involving division.

2. Expressions like $5 - 3\sqrt{2}$ and $5 + 3\sqrt{2}$ are called _____.

3. If a fraction has a radical in the denominator, we _____ by multiplying to change the fraction to an equivalent one that has an integer in the denominator.

Example	**Student Practice**
1. Simplify.	**2.** Simplify.
(a) $\dfrac{\sqrt{75}}{\sqrt{3}}$	(a) $\dfrac{\sqrt{180}}{\sqrt{5}}$
3 is a factor of 75. Use the quotient rule to rewrite the expression.	
$\dfrac{\sqrt{75}}{\sqrt{3}} = \sqrt{\dfrac{75}{3}}$ $= \sqrt{25}$ $= 5$	
	(b) $\sqrt{\dfrac{4}{9}}$
(b) $\sqrt{\dfrac{25}{36}}$	
Both 25 and 36 are perfect squares, so rewrite the expression.	
$\sqrt{\dfrac{25}{36}} = \dfrac{\sqrt{25}}{\sqrt{36}}$ $= \dfrac{5}{6}$	

Example	Student Practice
3. Simplify. $\sqrt{\dfrac{20}{x^6}}$ $\sqrt{\dfrac{20}{x^6}} = \dfrac{\sqrt{20}}{\sqrt{x^6}} = \dfrac{\sqrt{4}\sqrt{5}}{x^3} = \dfrac{2\sqrt{5}}{x^3}$	**4.** Simplify. $\sqrt{\dfrac{54}{y^2}}$
5. Simplify. $\dfrac{3}{\sqrt{2}}$ Think, "What times 2 will make a perfect square?" $\dfrac{3}{\sqrt{2}} = \dfrac{3}{\sqrt{2}} \times 1 = \dfrac{3}{\sqrt{2}} \times \dfrac{\sqrt{2}}{\sqrt{2}} = \dfrac{3\sqrt{2}}{\sqrt{4}} = \dfrac{3\sqrt{2}}{2}$	**6.** Simplify. $\dfrac{7}{\sqrt{5}}$
7. Simplify. **(a)** $\dfrac{\sqrt{7}}{\sqrt{8}}$ Think, "What times 8 will make a perfect square?" $\dfrac{\sqrt{7}}{\sqrt{8}} = \dfrac{\sqrt{7}}{\sqrt{8}} \times \dfrac{\sqrt{2}}{\sqrt{2}} = \dfrac{\sqrt{14}}{\sqrt{16}} = \dfrac{\sqrt{14}}{4}$ **(b)** $\dfrac{3}{\sqrt{x^3}}$ Think, "What times x^3 will give an even exponent?" $\dfrac{3}{\sqrt{x^3}} \times \dfrac{\sqrt{x}}{\sqrt{x}} = \dfrac{3\sqrt{x}}{\sqrt{x^4}} = \dfrac{3\sqrt{x}}{x^2}$	**8.** Simplify. **(a)** $\dfrac{\sqrt{5}}{\sqrt{24}}$ **(b)** $\dfrac{5x}{\sqrt{x^9}}$

Example	Student Practice

9. Simplify. $\dfrac{\sqrt{2}}{\sqrt{27x}}$

Since it is not apparent what we should multiply 27 by to obtain a perfect square, we will begin by simplifying the denominator.

$$\dfrac{\sqrt{2}}{\sqrt{27x}} = \dfrac{\sqrt{2}}{3\sqrt{3x}}$$

$$= \dfrac{\sqrt{2}}{3\sqrt{3x}} \times \dfrac{\sqrt{3x}}{\sqrt{3x}}$$

$$= \dfrac{\sqrt{6x}}{3\sqrt{9x^2}}$$

$$= \dfrac{\sqrt{6x}}{9x}$$

10. Simplify. $\dfrac{\sqrt{3x}}{\sqrt{50x}}$

11. Simplify. $\dfrac{2}{\sqrt{3}-4}$

The conjugate of $\sqrt{3}-4$ is $\sqrt{3}+4$.

$$\dfrac{2}{\sqrt{3}-4} \cdot \dfrac{\sqrt{3}+4}{\sqrt{3}+4}$$

$$= \dfrac{2\sqrt{3}+8}{\left(\sqrt{3}\right)^2 + 4\sqrt{3} - 4\sqrt{3} - 4^2}$$

$$= \dfrac{2\sqrt{3}+8}{3-16}$$

$$= \dfrac{2\sqrt{3}+8}{-13}$$

$$= -\dfrac{2\sqrt{3}+8}{13}$$

12. Simplify. $\dfrac{6}{\sqrt{2}-\sqrt{7}}$

Example	Student Practice

13. Rationalize the denominator. $\dfrac{\sqrt{3}+\sqrt{2}}{\sqrt{3}-\sqrt{2}}$

The conjugate of $\sqrt{3}-\sqrt{2}$ is $\sqrt{3}+\sqrt{2}$.

$$\frac{\sqrt{3}+\sqrt{2}}{\sqrt{3}-\sqrt{2}}\cdot\frac{\sqrt{3}+\sqrt{2}}{\sqrt{3}+\sqrt{2}}$$

$$=\frac{\sqrt{9}+\sqrt{6}+\sqrt{6}+\sqrt{4}}{\left(\sqrt{3}\right)^{2}-\left(\sqrt{2}\right)^{2}}$$

$$=\frac{3+2\sqrt{6}+2}{3-2}$$

$$=\frac{5+2\sqrt{6}}{1}$$

$$=5+2\sqrt{6}$$

14. Rationalize the denominator. $\dfrac{\sqrt{6}+\sqrt{5}}{\sqrt{6}-\sqrt{5}}$

Extra Practice

1. Simplify. Be sure to rationalize all denominators. Do not use a calculator or a table of square roots. $\dfrac{\sqrt{48}}{\sqrt{12}}$

2. Simplify. Be sure to rationalize all denominators. Do not use a calculator or a table of square roots. $\sqrt{\dfrac{3}{15}}$

3. Rationalize the denominator. Simplify your answer. Do not use a calculator or a table of square roots. $\dfrac{x+4}{\sqrt{x}-2}$

4. Rationalize the denominator. Simplify your answer. Do not use a calculator or a table of square roots. $\dfrac{3\sqrt{5}+2}{\sqrt{8}-\sqrt{6}}$

Concept Check

Explain how you would simplify $\dfrac{x-4}{\sqrt{x}+2}$.

Chapter 8 Radicals
8.6 The Pythagorean Theorem and Radical Equations

Vocabulary
hypotenuse • legs • Pythagorean Theorem • radical equation • extraneous root

1. A(n) _____ is an equation with a variable in one or more of the radicands.

2. An apparent solution that does not satisfy the original equation is called a(n) _____.

3. The shorter sides of a right triangle are referred to as the _____.

4. The longest side of a right triangle is called the _____.

Example	Student Practice
1. The ramp to Tony Pitkin's barn rises 5 feet over a horizontal distance of 12 feet. How long is the ramp? That is, find the length of the hypotenuse of a right triangle whose legs are 5 feet and 12 feet.	2. Find the length of the hypotenuse of a right triangle with legs of 24 yards and 32 yards.

Write the Pythagorean Theorem. Substitute the known values into the equation.

$$c^2 = a^2 + b^2$$
$$c^2 = 5^2 + 12^2$$
$$c^2 = 25 + 144$$
$$c^2 = 169$$
$$c = \pm\sqrt{169}$$
$$= \pm 13$$

The hypotenuse is 13 feet. We do not use -13 because length is not negative.

The check is left to the student.

Example	Student Practice
3. A 25-foot ladder is placed against a building. The foot of the ladder is 8 feet from the wall. At approximately what height does the top of the ladder touch the building? Round to the nearest tenth.	**4.** A kite is on a string that is 50 feet long and is fastened to the ground at the other end. Assuming the string is a straight line, how high is the kite if it is flying 8 feet away from where it is fastened to the ground? Round to the nearest tenth.

$$c^2 = a^2 + b^2$$
$$25^2 = 8^2 + b^2$$
$$625 = 64 + b^2$$
$$561 = b^2$$
$$\pm\sqrt{561} = b$$

We want only the positive value for the distance, so $b = \sqrt{561}$. Using a calculator, we have $561\;\boxed{\sqrt{}}\;23.685438$. Rounding, we obtain $b \approx 23.7$. The ladder touches the building at a height of approximately 23.7 feet.

Example	Student Practice
5. Solve and check. $1 + \sqrt{5x - 4} = 5$	**6.** Solve and check. $-3 + \sqrt{2x - 5} = 6$

$$1 + \sqrt{5x - 4} = 5$$
$$\sqrt{5x - 4} = 4$$
$$\left(\sqrt{5x - 4}\right)^2 = (4)^2$$
$$5x - 4 = 16$$
$$5x = 20$$
$$x = 4$$

Check.

$$1 + \sqrt{5(4) - 4} \overset{?}{=} 5$$
$$1 + \sqrt{20 - 4} \overset{?}{=} 5$$
$$1 + \sqrt{16} \overset{?}{=} 5$$
$$1 + 4 = 5$$

Example	Student Practice
7. Solve and check. $\sqrt{x+3} = -7$	**8.** $\sqrt{6x+8} = -3$

7. Solve and check. $\sqrt{x+3} = -7$

$$\sqrt{x+3} = -7$$
$$x+3 = 49$$
$$x = 46$$

Check.

$$\sqrt{46+3} \overset{?}{=} -7$$
$$\sqrt{49} \overset{?}{=} -7$$
$$7 \neq -7$$

There is no solution.

8. $\sqrt{6x+8} = -3$

---

9. Solve and check. $2 + \sqrt{2x-1} = x$

$$2 + \sqrt{2x-1} = x$$
$$\left(\sqrt{2x-1}\right)^2 = (x-2)^2$$
$$2x-1 = x^2 - 4x + 4$$
$$0 = x^2 - 6x + 5$$
$$0 = (x-5)(x-1)$$
$$x = 5 \text{ or } x = 1$$

Check.
If $x = 5$:

$$2 + \sqrt{2(5)-1} \overset{?}{=} 5$$
$$\sqrt{9} \overset{?}{=} 3$$
$$3 = 3$$

If $x = 1$:

$$2 + \sqrt{2(1)-1} \overset{?}{=} 1$$
$$\sqrt{1} \overset{?}{=} -1$$
$$1 \neq -1$$

Thus only 5 is a solution to this equation.

10. Solve and check. $\sqrt{4x+1} + 3 = x - 2$

Example	Student Practice
11. Solve and check. $\sqrt{3x+3} = \sqrt{5x-1}$	**12.** Solve and check. $\sqrt{3x+4} = \sqrt{2x+8}$

Here there are two radicals. Each radical is already isolated.

$$\sqrt{3x+3} = \sqrt{5x-1}$$
$$\left(\sqrt{3x+3}\right)^2 = \left(\sqrt{5x-1}\right)^2$$
$$3x+3 = 5x-1$$
$$3 = 2x-1$$
$$4 = 2x$$
$$2 = x$$

Check.

$$\sqrt{3(2)+3} \overset{?}{=} \sqrt{5(2)-1}$$
$$\sqrt{6+3} \overset{?}{=} \sqrt{10-1}$$
$$\sqrt{9} = \sqrt{9}$$

Extra Practice

1. A right triangle has legs a and b and hypotenuse c. Find the exact length of the missing side c if $a=9$ and $b=9$.

2. A right triangle has legs a and b and hypotenuse c. Find the exact length of the missing side b if $a=\sqrt{72}$ and $c=10$.

3. Solve for the variable. Check your solutions. $\sqrt{2x+5} = 7$

4. Solve for the variable. Check your solutions. $\sqrt{4x-13} = 2x-6$

Concept Check

A student attempted to solve the equation $2+\sqrt{1-8x} = x$ and found values of $x=-3$ and $x=-1$. Explain how you would check these values to see if they are solutions to the original equation.

Chapter 8 Radicals
8.7 Word Problems Involving Radials: Direct and Inverse Variation

Vocabulary
varies directly   •   constant of variation   •   vary inversely

1. The constant is often called the _____.

2. If y _____ as x, then $y = kx$, where k is a constant.

3. If one variable is a constant multiple of the reciprocal of another, the two variables are said to _____.

Example	Student Practice
1. Cliff's salary varies directly as the number of hours worked. Last week he earned \$33.60 for working 7 hours. This week he earned \$52.80. How many hours did he work?	**2.** Mary's salary varies directly as the number of hours worked. Last week she earned \$41.84 for working 8 hours. This week she earned \$78.45. How many hours did she work?

Let S = salary, h = number of hours worked, and k = constant of variation. Since his salary varies directly as the number of hours he worked, $S = k \cdot h$. Find the constant k by substituting the known values of $S = 33.60$ and $h = 7$.

$$33.60 = 7k$$
$$\frac{33.60}{7} = \frac{7k}{7}$$
$$4.80 = k$$

How many hours did he work to earn \$52.80?

$$S = 4.80h$$
$$52.80 = 4.80h$$
$$11 = h$$

Cliff worked 11 hours this week.

Example	Student Practice

3. In a certain class of racing cars, the maximum speed varies directly as the square root of the horsepower of the engine. If a car with 225 horsepower can achieve a maximum speed of 120 mph, what speed could it achieve with 256 horsepower?

Let $V =$ the maximum speed, $h =$ the horsepower of the engine, and $k =$ the constant of variation.

Since the maximum speed (V) varies directly as the square root of the horsepower of the engine, we can write the following.

$$V = k\sqrt{h}$$
$$120 = k\sqrt{225}$$
$$120 = 15k$$
$$8 = k$$

Now we can write the direct variation equation with the known value for k.

$$V = 8\sqrt{h}$$
$$= 8\sqrt{256}$$
$$= (8)(16)$$
$$= 128$$

Thus, a car with 256 horsepower could achieve a maximum speed of 128 mph.

4. The volume of a certain type of balloon varies directly as the cube of the amount of air put in the balloon. If the volume was 32 centimeters cubed when 2 milliliters of air was in the balloon, what is the volume when 7 milliliters of air is in the balloon?

Example	Student Practice
5. A car manufacturer is thinking of reducing the size of the wheel used in a subcompact car. The number of times a car wheel must turn to cover a given distance varies inversely as the radius of the wheel. (Notice that this says that the smaller the wheel, the more times it must turn to cover a given distance.) A wheel with a radius of 0.35 meter must turn 400 times to cover a specified distance on a test track. How many times would it have to turn if the radius were reduced to 0.30 meter?	**6.** A car manufacturer is thinking of reducing the size of the wheel used in a subcompact car. The number of times a car wheel must turn to cover a given distance varies inversely as the radius of the wheel. (Notice that this says that the smaller the wheel, the more times it must turn to cover a given distance.) A wheel with a radius of 0.40 meter must turn 600 times to cover a specified distance on a test track. How many times would it have to turn if the radius were reduced to 0.32 meter?

Let $n =$ the number of times the car wheel turns, $r =$ the radius of the wheel, and $k =$ the constant of variation. Since the number of turns varies inversely as the radius, we can write the following.

$$n = \frac{k}{r}$$

$$400 = \frac{k}{0.35}$$

$$140 = k$$

How many times must the wheel turn if the radius is 0.30 meter?

$$n = \frac{140}{r}$$

$$n = \frac{140}{0.30}$$

$$n = 466\frac{2}{3}$$

The wheel would have to turn $466\frac{2}{3}$ times to cover the same distance if the radius were only 0.30 meter.

Example	Student Practice
7. The illumination of a light source varies inversely as the square of the distance from the source. The illumination measures 25 candlepower when a certain light is 4 meters away. Find the illumination when the light is 8 meters away. Since the illumination varies inversely as the square of the distance, $I = \dfrac{k}{d^2}$. $I = \dfrac{k}{d^2}$ $25 = \dfrac{k}{4^2}$ $400 = k$ $I = \dfrac{400}{d^2} = \dfrac{400}{8^2} = \dfrac{25}{4} = 6.25$ The illumination is 6.25 candlepower when the light source is 8 meters away.	**8.** The resistance to the flow of electric current in a wire of fixed length varies inversely as the square of the diameter of the wire. When the resistance measured is 0.45 ohm, the diameter of the wire is 0.02 centimeter. If we use wire that is 0.04 centimeter in diameter, what will the resistance be?

Extra Practice

1. If y varies directly as the cube of x, and $y = 3$ when $x = \dfrac{1}{2}$, find y when $x = 4$.

2. If y varies inversely as the square of x, and $y = 3$ when $x = \dfrac{1}{3}$, find y when $x = 3$.

3. The cost of shipping a package varies directly as the weight of the package. If the cost of shipping a 15 pound package is \$11.25, what is the cost of shipping a package that weighs 180 pounds?

4. The amount of time in minutes it takes for an ice cube to melt varies inversely with the temperature of the water that the ice cube is placed in. When an ice cube is placed in 45°F water, it takes 4.3 minutes to melt. How long would it take an ice cube of this size to melt if it were placed in 75°F water?

Concept Check

The distance a body falls from rest varies directly as the square of the time it falls. If an object falls 64 feet in two seconds, explain how you would write an equation to describe this relationship. How would you find the constant k in this equation?

MATH COACH

Mastering the skills you need to do well on the test.

Watch the **MATH COACH** videos in MyMathLab®or on You Tube
while you work the problems below. These helpful hints will
help you avoid making common errors on test problems.

Adding Radical Expressions That Require Simplifying—

Problem 6   Combine and simplify. $\sqrt{4a}+\sqrt{8a}+\sqrt{36a}+\sqrt{18a}$

> **Helpful Hint:** You must simplify each radical first. Terms can only be
> combined when the expression inside the radical is exactly the same.

Did you identify that 4 and 36 are perfect squares and then
simplify the first radical to $2\sqrt{a}$ and the third radical to
$6\sqrt{a}$? Yes _____ No _____

If you answered No, notice that the two radicals can be
rewritten as $\sqrt{4}\cdot\sqrt{a}$ and $\sqrt{36}\cdot\sqrt{a}$. You can then take the
square root of 4 and the square root of 36. Go back and
complete this step again.

Did you look for perfect squares and then rewrite the second
radical as $\sqrt{4\cdot2\cdot a}$ and the fourth radical as $\sqrt{9\cdot2\cdot a}$?

Yes _____ No _____

If you answered No, remember that we want
to factor the radicand into products of perfect
squares whenever possible. Now the square
roots of 4 and 9 can be taken to simplify these
radicals further.

The last step is to combine any radicals where
the expression inside the radical is exactly the
same.

If you answered Problem 6 incorrectly, go
back and rework the problem using these
suggestions.

Multiplying Two Binomial Radical Expressions—Problem 10   Multiply $\left(4\sqrt{2}-\sqrt{5}\right)\left(3\sqrt{2}+\sqrt{5}\right)$.

> **Helpful Hint:** Remember that you can use the FOIL method to multiply any two binomials. This also applies
> to binomial radical expressions. Be careful to separate numbers outside the radical and numbers inside the
> radical when completing the multiplication.

When you multiplied the first two terms and simplified that
product, did you get 24?
Yes _____ No _____

If you answered No, remember we must separately multiply
the numbers outside the radicals and then multiply the
numbers inside the radical signs. Be sure to simplify radical
expressions such as $\sqrt{4}$.

When you multiplied the outer terms, did you get $4\sqrt{10}$?
Yes _____ No _____

If you answered No, notice that $4\sqrt{2}\cdot\sqrt{5}=4\cdot1\sqrt{2\cdot5}$
$=4\sqrt{10}$.

The next steps are to multiply the inner terms,
multiply the last two terms, and then combine
any like terms.

Now go back and rework the problem using
these suggestions.

Simplifying the Quotient of Two Binomials That Contain Radicals—Problem 13

Simplify. $\dfrac{\sqrt{3}+4}{5+\sqrt{3}}$

Helpful Hint: When the denominator of a fraction contains a radical expression, we multiply both the numerator and the denominator of the fraction by the conjugate of the denominator.

Did you identify the conjugate of the denominator as $5-\sqrt{3}$? Yes ____ No ____

Did you multiply the numerator and denominator by $5-\sqrt{3}$

to obtain the expression $\dfrac{\left(\sqrt{3}+4\right)\left(5-\sqrt{3}\right)}{\left(5+\sqrt{3}\right)\left(5-\sqrt{3}\right)}$?

Yes ____ No ____

If you answered No to these questions, go back and review the definition of conjugate and try these two steps again.

After completing the multiplication, did you get the

unsimplified expression $\dfrac{5\sqrt{3}-\sqrt{9}+20-4\sqrt{3}}{25-5\sqrt{3}+5\sqrt{3}-\sqrt{9}}$?

Yes ____ No ____

If you answered No, slowly go through the step of multiplying the two binomials in the numerator.

Then carefully multiply the two binomials in the denominator.

After completing these steps, you can look for the square roots to evaluate. Then combine any like terms separately in both the numerator and denominator. Your final result will still be a fraction.

If you answered Problem 13 incorrectly, go back and rework the problem using these suggestions.

Solving and Verifying the Solution of a Radical Equation—Problem 19

Solve. Verify your solutions. $x=5+\sqrt{x+7}$

Helpful Hint: Always isolate the radical on one side of the equation before squaring each side of the equation. Be sure to check your results for extraneous roots.

Did you first isolate the radical to get $x-5=\sqrt{x+7}$?
Yes ____ No ____

When you squared each side of this equation, did you multiply and then simplify to obtain the equation $x^2-10x+25=x+7$? Yes ____ No ____

If you answered No to these questions, notice that $(x-5)^2=(x-5)(x+5)$ on the left side. Also remember that $\left(\sqrt{x+7}\right)^2=\sqrt{x+7}\cdot\sqrt{x+7}=x+7$.
Stop now and complete these calculations.

Did you transform the equation to $x^2-11x+18=0$?
Yes ____ No ____

If you answered No, try to get all the terms on the left and only 0 on the right. Then factor the resulting quadratic equation.

When you check both possible answers in the original equation, did both answers work?
Yes ____ No ____

If you answered Yes, carefully replace x by 2 in the equation. Remember that when you obtain $2=5+3$ you can see immediately that this is an invalid equation. The value $x=2$ does not check. It is an extraneous root.

Now go back and rework the problem using these suggestions.

256

Name: _____   Date: _____

Instructor: _____   Section: _____

Chapter 9 Quadratic Equations
9.1 Introduction to Quadratic Equations

Vocabulary
quadratic equation   •   standard form   •   greatest common factor   •   zero factor property

1. The _____ of a quadratic equation is $ax^2 + bx + c = 0$, where a, b, and c are real numbers and $a \neq 0$.

2. If it is possible to factor the quadratic equation, then we use the _____ to find the solution.

3. To solve equations of the form $ax^2 + bx = 0$, begin by factoring out the _____ .

4. A _____ is a polynomial of degree two.

Example	Student Practice
1. Place each quadratic equation in standard form and identify the real numbers a, b, and c.	**2.** Place each quadratic equation in standard form and identify the real numbers a, b, and c.
(a) $5x^2 - 6x + 3 = 0$ This equation is in standard form, $ax^2 + bx + c = 0$. Match each term to the standard form. $5x^2 - 6x + 3 = 0$ $a = 5,\ b = -6,\ c = 3$	**(a)** $3x^2 + 4x - 6 = 0$
(b) $-2x^2 + 15x + 4 = 0$ It is easier to work with quadratic equations if the first term is not negative. Multiply each term on both sides of the equation by -1. $2x^2 - 15x - 4 = 0$ $a = 2,\ b = -15,\ c = -4$	**(b)** $5x + 7x^2 - 6 = 0$

Example	Student Practice
3. Solve. $7x^2 + 9x - 2 = -8x - 2$	**4.** Solve. $3x^2 - 9x + 7 = 7 - 6x$

3. Solve. $7x^2 + 9x - 2 = -8x - 2$

The equation is not in standard form.
Add $8x + 2$ to each side and simplify.
$$7x^2 + 9x - 2 = -8x - 2$$
$$7x^2 + 9x - 2 + 8x + 2 = 0$$
$$7x^2 + 17x = 0$$

Factor. Then set each factor equal to zero and solve for x.

$$7x^2 + 17x = 0$$
$$x(7x + 17) = 0$$
$$x = 0 \qquad 7x + 17 = 0$$
$$7x = -17$$
$$x = -\frac{17}{7}$$

The solutions are 0 and $-\dfrac{17}{7}$.

4. Solve. $3x^2 - 9x + 7 = 7 - 6x$

---

5. Solve and check. $8x - 6 + \dfrac{1}{x} = 0$

Multiply each term by the LCD, x.
$$x(8x) - x(6) + x\left(\frac{1}{x}\right) = x(0)$$
$$8x^2 - 6x + 1 = 0$$

Factor the equation and solve for x.
$$8x^2 - 6x + 1 = 0$$
$$(4x - 1)(2x - 1) = 0$$
$$4x - 1 = 0 \qquad 2x - 1 = 0$$
$$x = \frac{1}{4} \qquad\qquad x = \frac{1}{2}$$
The check is left to the student.

6. Solve and check. $12x + 7 + \dfrac{1}{2x} = 0$

Example	Student Practice

7. A truck delivery company can handle a maximum of 36 truck routes in one day, and every two cities have a distinct truck route between them. How many separate cities can the truck company service?

Use the equation $t = \dfrac{n^2 - n}{2}$.

Substitute 36 for t, the number of truck routes.

$$t = \frac{n^2 - n}{2}$$

$$36 = \frac{n^2 - n}{2}$$

Multiply both sides by 2. Then write the equation in standard form.

$$36 = \frac{n^2 - n}{2}$$

$$2(36) = 2\left(\frac{n^2 - n}{2}\right)$$

$$72 = n^2 - n$$

$$0 = n^2 - n - 72$$

Factor.

$$0 = n^2 - n - 72$$

$$0 = (n-9)(n+8)$$

Set each factor equal to zero and solve.

$$n - 9 = 0 \qquad n + 8 = 0$$

$$n = 9 \qquad\quad n = -8$$

We reject -8 as a meaningless solution, because we cannot have a negative number of cities. Thus, the trucking company can service 9 cities.

8. A truck delivery company can handle a maximum of 21 truck routes in one day, and every two cities have a distinct truck route between them. How many separate cities can the truck company service?

Use the equation $t = \dfrac{n^2 - n}{2}$.

Extra Practice

1. Place the quadratic equation in standard form and identify the real numbers a, b, and c.

$$2x^2 - 27 = -3x^2 + 12$$

2. Solve. $6x^2 + 3x = 0$

3. Solve and check. $x^2 = 10x - 21$

4. Solve and check. $\dfrac{8}{x} = \dfrac{x}{3} + \dfrac{5}{3}$

Concept Check

In the following problem, explain how you would place the quadratic equation in standard form. $2 = \dfrac{5}{x+1} + \dfrac{3}{x-1}$

Chapter 9 Quadratic Equations
9.2 Using the Square Root Property and Completing the Square to Find Solutions

Vocabulary
square root property   •   completing the square

1. Rewriting an equation so that it has the form $(ax+b)^2 = c$ is called _____.

2. The _____ states that if $x^2 = a$, then $x = \sqrt{a}$ or $x = -\sqrt{a}$, for all nonnegative real numbers a.

Example	Student Practice
1. Solve.	**2.** Solve.
(a) $x^2 = 49$	**(a)** $x^2 = 100$
$x^2 = 49$ $x = \pm\sqrt{49}$ $x = \pm 7$	
(b) $x^2 = 20$	**(b)** $x^2 = 72$
$x^2 = 20$ $x = \pm\sqrt{20}$ $x = \pm 2\sqrt{5}$	
(c) $5x^2 = 125$	**(c)** $6x^2 = 120$
Divide both sides by 5 before taking the square root. $\dfrac{5x^2}{5} = \dfrac{125}{5}$ $x^2 = 25$ $x = \pm\sqrt{25}$ $x = \pm 5$	

Example	Student Practice
3. Solve. $3x^2 + 5x = 18 + 5x + x^2$	**4.** Solve. $5x^2 + 2x - 13 = 2x + 35 + 2x^2$

3. Solve. $3x^2 + 5x = 18 + 5x + x^2$

Simplify the equation by placing all the variable terms on the left and the constants on the right.

$$3x^2 + 5x = 18 + 5x + x^2$$
$$2x^2 = 18$$
$$x^2 = 9$$
$$x = \pm 3$$

---

5. Solve. $(3x+1)^2 = 8$

6. Solve. $(2x-5)^2 = 20$

Take the square root of both sides and simplify as much as possible.

$$(3x+1)^2 = 8$$
$$3x + 1 = \pm\sqrt{8}$$
$$3x + 1 = \pm 2\sqrt{2}$$

Now we must solve the two equations expressed by the plus or minus statement.

$$3x + 1 = +2\sqrt{2} \qquad 3x + 1 = -2\sqrt{2}$$
$$3x = -1 + 2\sqrt{2} \qquad 3x = -1 - 2\sqrt{2}$$
$$x = \frac{-1 + 2\sqrt{2}}{3} \qquad x = \frac{-1 - 2\sqrt{2}}{3}$$

The roots of this quadratic equation are irrational numbers. They are
$$\frac{-1 + 2\sqrt{2}}{3} \text{ and } \frac{-1 - 2\sqrt{2}}{3} \text{ or } \frac{-1 \pm 2\sqrt{2}}{3}.$$
We cannot simplify these roots further, so we leave them in this form.

Example	Student Practice
7. Solve. $4x^2 + 4x - 3 = 0$	**8.** Solve. $3x^2 - 6x - 24 = 0$

7. Solve. $4x^2 + 4x - 3 = 0$

Place the constant on the right. This puts the equation in the form $ax^2 + bx = c$. Then divide all terms by 4.

$$4x^2 + 4x - 3 = 0$$
$$4x^2 + 4x = 3$$
$$x^2 + x = \frac{3}{4}$$

Take one-half the coefficient of x and square it, $\left(\frac{1}{2}\right)^2 = \frac{1}{4}$. Add $\frac{1}{4}$ to each side.

$$x^2 + x + \frac{1}{4} = \frac{3}{4} + \frac{1}{4}$$

Factor the left side. Then use the square root property to solve for x.

$$x^2 + x + \frac{1}{4} = \frac{3}{4} + \frac{1}{4}$$
$$\left(x + \frac{1}{2}\right)^2 = 1$$
$$x + \frac{1}{2} = \pm\sqrt{1}$$
$$x + \frac{1}{2} = \pm 1$$
$$x + \frac{1}{2} = 1 \qquad x + \frac{1}{2} = -1$$
$$x = \frac{1}{2} \qquad x = -\frac{3}{2}$$

Thus the two roots are $\frac{1}{2}$ and $-\frac{3}{2}$.

The check is left to the student.

Extra Practice

1. Solve using the square root property.
   $$3x^2 = 192$$

2. Solve using the square root property.
   $$(2x - 4)^2 = 60$$

3. Solve by completing the square.
   $$x^2 - 8x = 11$$

4. Solve by completing the square.
   $$2x^2 - 5x - 4 = 0$$

.

Concept Check

Explain the first three steps of how you would solve the following problem by completing the square. $5x^2 - 10x + 2 = 0$

Chapter 9 Quadratic Equations
9.3 Using the Quadratic Formula to Find Solutions

Vocabulary
quadratic formula   •   discriminant

1. If the _____ is a negative number, the roots are not real numbers.

2. Use the _____ to find the roots of any quadratic equation of the form $ax^2 + bx + c = 0$, where a, b, and c are real numbers and $a \neq 0$.

Example	Student Practice
1. Solve using the quadratic formula. $3x^2 + 10x + 7 = 0$	**2.** Solve using the quadratic formula. $2x^2 + 3x - 5 = 0$

1. Solve using the quadratic formula.
$3x^2 + 10x + 7 = 0$

In our given equation, we have $a = 3$, $b = 10$, and $c = 7$. Write the quadratic formula and substitute the values for a, b, and c. Then, simplify.

$$x = \frac{-b \pm \sqrt{b^2 - 4ac}}{2a}$$

$$= \frac{-10 \pm \sqrt{(10)^2 - 4(3)(7)}}{2(3)}$$

$$= \frac{-10 \pm \sqrt{100 - 84}}{6}$$

$$= \frac{-10 \pm \sqrt{16}}{6} = \frac{-10 \pm 4}{6}$$

Solve using the positive sign.
$$x = \frac{-10 + 4}{6} = \frac{-6}{6} = -1$$

Solve using the negative sign.
$$x = \frac{-10 - 4}{6} = \frac{-14}{6} = -\frac{7}{3}$$

2. Solve using the quadratic formula.
$2x^2 + 3x - 5 = 0$

Example	Student Practice
3. Solve. $x^2 = 5 - \dfrac{3}{4}x$	**4.** Solve. $x^2 = \dfrac{2}{3}x + 4$

First, we obtain an equivalent equation that does not have fractions. Multiply each term by the LCD, 4, and simplify.

$$x^2 = 5 - \frac{3}{4}x$$

$$4(x^2) = 4(5) - 4\left(\frac{3}{4}x\right)$$

$$4x^2 = 20 - 3x$$

Add $3x - 20$ to each side to write the equation in standard form.

$$4x^2 = 20 - 3x$$

$$4x^2 + 3x - 20 = 0$$

$$a = 4, \ b = 3, \ \text{and} \ c = -20$$

Substitute the values for a, b, and c, into the quadratic formula.

$$x = \frac{-b \pm \sqrt{b^2 - 4ac}}{2a}$$

$$x = \frac{-3 \pm \sqrt{(3)^2 - 4(4)(-20)}}{2(4)}$$

$$x = \frac{-3 \pm \sqrt{9 + 320}}{8}$$

$$x = \frac{-3 \pm \sqrt{329}}{8}$$

Example	Student Practice
5. Find the roots of $3x^2 - 5x = 7$. Approximate to the nearest thousandth.	**6.** Find the roots of $4x^2 - 7x = 12$. Approximate to the nearest thousandth.

Place the equation in standard form.

$$3x^2 - 5x = 7$$
$$3x^2 - 5x - 7 = 0$$

$a = 3$, $b = -5$, and $c = -7$

Substitute the values for a, b, and c into the quadratic formula, then simplify.

$$x = \frac{-b \pm \sqrt{b^2 - 4ac}}{2a}$$

$$x = \frac{-(-5) \pm \sqrt{(-5)^2 - 4(3)(-7)}}{2(3)}$$

$$x = \frac{5 \pm \sqrt{25 + 84}}{6}$$

$$x = \frac{5 \pm \sqrt{109}}{6}$$

Look up $\sqrt{109}$ in the square root table. $\sqrt{109} \approx 10.440$. Use this result to solve for x.

$$x \approx \frac{5 + 10.440}{6} = \frac{15.440}{6} \approx 2.573$$

$$x \approx \frac{5 - 10.440}{6} = \frac{-5.440}{6} \approx -0.907$$

The two roots are 2.573 and -0.907.

Example	Student Practice

7. Determine whether $3x^2 = 5x - 4$ has real number solution(s).

First, we place the equation in standard form.
$$3x^2 = 5x - 4$$

$$3x^2 - 5x + 4 = 0$$
$$a = 3,\ b = -5,\ c = 4$$

Substitute the values for a, b, and c into the discriminant, $b^2 - 4ac$.

$$b^2 - 4ac = (-5)^2 - 4(3)(4)$$
$$= 25 - 48$$
$$= -23$$

The discriminant is negative. Thus $3x^2 = 5x - 4$ has no real number solution(s).

8. Determine whether $4x^2 = 6x - 5$ has real number solution(s).

Extra Practice

1. Solve. $x^2 - 2x = 4$

2. Solve. $3x^2 - 2x + 4 = 0$

3. Solve. $\dfrac{1}{5} + \dfrac{4}{x} = \dfrac{x}{5}$

4. Solve. $3x^2 + 8x + 12 = 0$

Concept Check

Explain how you would determine if the quadratic equation $5x^2 - 8x + 9 = 0$ has real solutions or has no real solutions?

Chapter 9 Quadratic Equations
9.4 Graphing Quadratic Equations

Vocabulary

parabola   •   vertex

1. The _____ is the highest point of a parabola that opens downward or the lowest point of a parabola that opens upward.

2. The graph of a quadratic equation is shaped like a _____.

Example	**Student Practice**
1. Graph.	**2.** Graph.

1. Graph.

(a) $y = x^2$

Make a table of values.

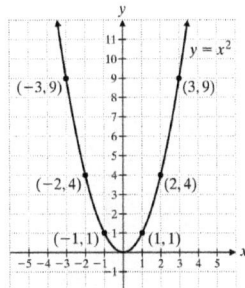

x	y
−3	9
−2	4
−1	1
0	0
1	1
2	4
3	9

(b) $y = -x^2$

x	y
−3	−9
−2	−4
−1	−1
0	0
1	−1
2	−4
3	−9

Notice that when the coefficient of x^2 is negative, the graph flips upside down.

2. Graph.

(a) $y = \dfrac{1}{4}x^2$

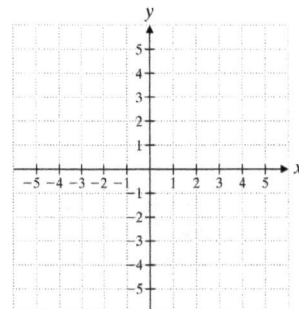

x	y
4	
2	
0	
−2	
−4	

(b) $y = -\dfrac{1}{4}x^2$

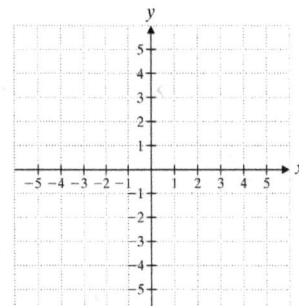

x	y
4	
2	
0	
−2	
−4	

Example	Student Practice

3. Graph $y = x^2 - 2x$. Identify the coordinates of the vertex.

First, make a table of values.

x	y
-2	8
-1	3
0	0
1	-1
2	0
3	3
4	8

Next, plot the points.

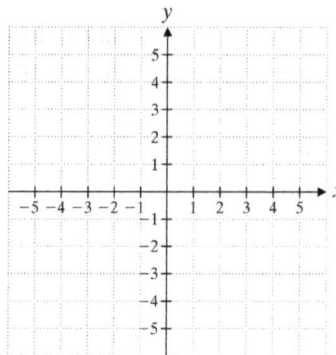

Notice that the x-intercepts are at $(0,0)$ and $(2,0)$. The x-coordinates are the solutions to the equation $x^2 - 2x = 0$.

$$x^2 - 2x = 0$$
$$x(x - 2) = 0$$
$$x = 0, \quad x = 2$$

4. Graph $y = x^2 + 4x$. Identify the coordinates of the vertex.

x	y
1	
0	
-1	
-2	
-3	
-4	
-5	

270

Example	Student Practice

5. $y = -x^2 - 2x + 3$. Determine the vertex and the x-intercepts. Then sketch the graph.

We will first determine the coordinates of the vertex. We begin by finding the x-coordinate.

$$x = \frac{-b}{2a} = \frac{-(-2)}{2(-1)} = \frac{2}{-2} = -1$$

$$y = -x^2 - 2x + 3$$
$$y = -(-1)^2 - 2(-1) + 3 = 4$$

The point $(-1, 4)$ is the vertex.

Find the x-intercepts, the solutions to the equation $-x^2 - 2x + 3 = 0$. Since it is easier to factor if the first term is positive, add $x^2 + 2x - 3$ to each side. Then factor and solve for x.

$$-x^2 - 2x + 3 = 0$$
$$0 = x^2 + 2x - 3$$
$$0 = (x-1)(x+3)$$
$$x = 1 \qquad x = -3$$

The x-intercepts are $(1, 0)$ and $(-3, 0)$.

Graph. Since $a = -1$ the parabola opens down and the vertex is the highest point.

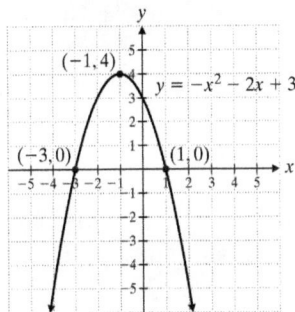

6. $y = x^2 - 6x + 5$. Determine the vertex and the x-intercepts, then sketch the graph.

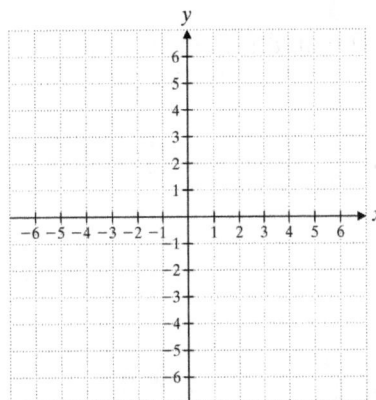

Extra Practice

1. Graph $y = x^2 - 3$.

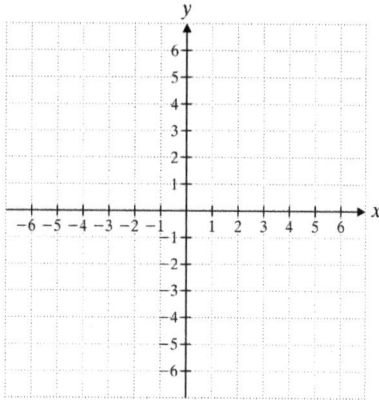

2. Graph $y = 3(x+1)^2$.

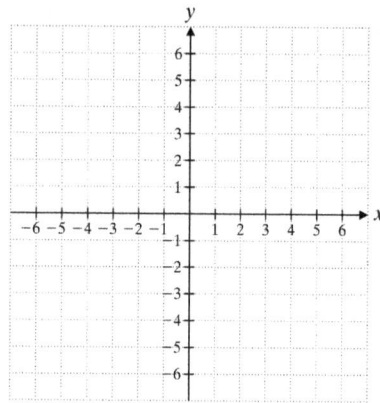

3. Graph $y = -x^2 + 2x + 4$.

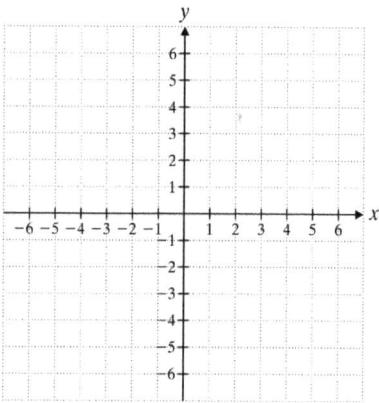

4. Graph $y = x^2 + 3x + 4$. Identify the coordinates of the vertex.

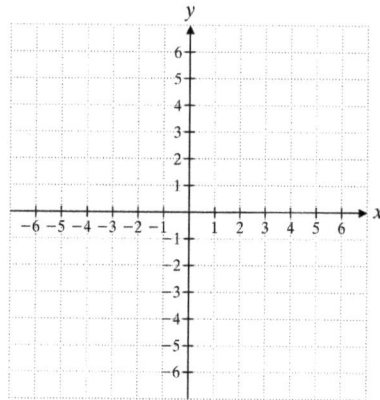

.

Concept Check

Explain how you would find the vertex of the equation $y = 4x^2 + 16x - 2$.

Name: _____   Date: _____

Instructor: _____   Section: _____

Chapter 9 Quadratic Equations
9.5 Formulas and Applied Problems

Vocabulary
quadratic equation   •   substitution

1. If there are two equations in two variables use _____ to solve for one variable, then use that result in the other equation to solve for the other variable.

2. The Pythagorean Theorem is a _____

Example	Student Practice
1. The hypotenuse of a right triangle is 25 meters in length. One leg is 17 meters longer than the other. Find the length of each leg.	**2.** The hypotenuse of a right triangle is 30 meters in length. One leg is 6 meters shorter than the other. Find the length of each leg.

Example 1 (continued):

Understand the problem. Draw a picture.

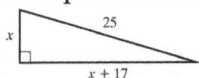

Since the problem involves a right triangle, use the Pythagorean Theorem.
$$a^2 + b^2 = c^2$$
$$x^2 + (x+17)^2 = 25^2$$

Solve and state the answer.
$$x^2 + (x+17)^2 = 25^2$$
$$x^2 + x^2 + 34x + 289 = 625$$
$$x^2 + 17x - 168 = 0$$
$$(x+24)(x-7) = 0$$
$$x + 24 = 0 \qquad\qquad x - 7 = 0$$
$$x = -24 \qquad\qquad x = 7$$

Note $x = -24$ is not a valid solution. Thus, one leg is 7 meters and the other leg is $7 + 17 = 24$ meters in length. The check is left to the student.

Example	Student Practice
3. The ski club rented a bus to travel to Mount Snow. The members agreed to share the cost of $180 equally. On the day of the ski trip, three members were sick with the flu and could not go. This increased the share of each person going on the trip by $10. How many people originally planned to attend?	**4.** A group of friends wants to charter a fishing boat for a day. The fee for the boat is $750 per day. On the day of the charter, 5 of the friends cancelled, and the cost for each person going on the boat went up $25. How many friends were originally supposed to go?

Let $s = $ the number of students and $c = $ the cost for each student in the original group. Write an equation.

$$s \cdot c = 180 \quad (1)$$

If three people were sick, the number of students dropped by three, but the cost for each increased by $10. The total is still $180

$$(s-3)(c+10) = 180 \quad (2)$$

Solve equation (1) for c to get $c = \dfrac{180}{s}$.

Substitute the result into equation (2).

$$(s-3)(c+10) = 180$$

$$(s-3)\left(\frac{180}{s} + 10\right) = 180$$

$$(s-3)\left(\frac{180 + 10s}{s}\right) = 180$$

$$(s-3)(180 + 10s) = 180s$$

$$10s^2 + 150s - 540 = 180s$$

$$10s^2 - 30s - 540 = 0$$

$$s^2 - 3s - 54 = 0$$

$$(s-9)(s+6) = 0$$

$$s = 9, \qquad s = -6$$

The number of students originally going on the ski trip was 9.

Example	Student Practice

5. Minette is fencing in a garden that borders the back of a large barn. She has 120 feet of fencing. She would like a rectangular garden that measures 1350 square feet in area. She wants to use the back of the barn, so she needs to use fencing on three sides only. What dimensions should she use for her garden?

Draw a picture.

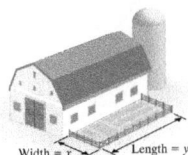

Width = x   Length = y

Write the equations. Look at the drawing. The 120 feet of fencing will be needed for the width twice and the length once.

$120 = 2x + y$ (1)

The area is $A = (\text{width})(\text{length})$.

$1350 = xy$ (2)

Solve for y in equation (1). Substitute $y = 120 - 2x$ in (2) and solve for x.

$$1350 = xy$$
$$1350 = x(120 - 2x)$$
$$1350 = 120x - 2x^2$$
$$2x^2 - 120x + 1350 = 0$$
$$x^2 - 60x + 675 = 0$$
$$(x - 15)(x - 45) = 0$$
$$x = 15, \quad x = 45$$

To find the dimensions of the garden we can substitute $x = 15$ and $x = 45$ into equation (1). If the width is 15 feet, the length is 90 feet. If the width is 45 feet, the length is 30 feet.

6. Larry needs a new pen for his goats. The pen needs to be 7500 square feet. He has 275 feet of fence and his land is along a river that he will use as one side of his pen. What dimensions should he use for the fence?

Extra Practice

1. A lot is in the shape of a right triangle. The shorter leg measures 150 meters. The hypotenuse is 50 meters longer than the length of the longer leg. How long is the longer leg?

2. The area of a rectangle is 180 square meters. If the length is 3 meters more than the width, what are the dimensions of the rectangle?

3. A number of computer club members pitched in to purchase a computer for the class. The computer cost $850. The members planned to share the cost equally. Nine new members joined the club, and they chipped in to share in the purchase of the computer, which lowered the cost per person by $9. How many members were in the club initially?

4. Leslie wants to enclose a rectangular area of 1800 square feet in her backyard. She will use fencing for three sides of the area, and the fourth side of the area will be formed by the back of her house. She wants to use exactly 125 feet of fencing. What are the possible dimensions of the area?

.

Concept Check

The hiking club needs to raise $720 through dues paid by each member. However, 4 people dropped out of the club so the dues went up by $6 for each member. Explain how you could set up an equation to find out how many members were in the club prior to the dropout.

MATH COACH

Mastering the skills you need to do well on the test.

Watch the **MATH COACH** videos in MyMathLab® or on YouTube™ while you work the problems below. These helpful hints will help you avoid making common errors on test problems.

Solving an Equation Using the Quadratic Formula—

Problem 1 Solve by any desired method. If there is no real number solution, say so. $5x^2 + 7x = 4$

> **Helpful Hint:** We often use the quadratic formula when other methods of factoring, square roots, or completing the square are not practical. Make sure that you rewrite the equation in $ax^2 + bx + c = 0$ form before completing any other steps.

Did you rewrite the equation as $5x^2 + 7x - 4 = 0$?
Yes _____ No _____
Can you see that this equation cannot be factored?
Yes _____ No _____
If you answered No to these questions, review how to write a quadratic equation in standard form. Try to break the equation into possible factors to discover that the equation cannot be factored.
Did you identify $a = 5$, $b = 7$, and $c = -4$?
Yes _____ No _____
Did you substitute these values into the quadratic formula to

obtain $x = \dfrac{-7 \pm \sqrt{(7)^2 - 4(5)(-4)}}{2(5)}$? Yes _____ No _____

If you answered No to these questions, go back and substitute the correct values into the quadratic formula. Be careful with signs. Simplify the expression further to find the possible solutions.

If you answered Problem 1 incorrectly, go back and rework the problem using these suggestions.

Determining Whether a Quadratic Equation Has No Real Solutions—Problem 3

Solve by any desired method. If there is no real solution, say so. $2x^2 - 2x + 5 = 0$?

> **Helpful Hint:** If you use the quadratic formula and obtain the square root of a negative number, then you know that there is no real number solution to the quadratic equation.

Did you rewrite the equation as $2x^2 - 2x + 5 = 0$?
Yes _____ No _____
Did you identify $a = 2$, $b = -2$, and $c = 5$?
Yes _____ No _____
If you answered No to these questions, stop and review how to write a quadratic equation in standard form and how to identify the values of a, b, and c. Be careful with signs.

When you substituted the values for a, b, and c in the quadratic formula, did you get

$x = \dfrac{-(-2) \pm \sqrt{(-2)^2 - 4(2)(5)}}{2(2)}$? Yes _____ No _____

If you answered No, stop, carefully make the required substitutions, and then simplify. You should get a negative value for the radicand.

Now go back and rework the problem using these suggestions.

Solving a Quadratic Equation That Requires Simplification First—Problem 9

Solve by any desired method. If there is no real number solution, say so. $2x(x-6)=6-x$

> **Helpful Hint:** Be sure to remove the parentheses and combine like terms first. Then write the equation in $ax^2+bx+c=0$ form before choosing your method of solution.

When you removed the parentheses, did you get
$2x^2-12x=6-x$? Yes _____ No _____

Next, did you combine like terms and write the equation as
$2x^2-11x-6=0$? Yes _____ No _____

If you answered No to these questions, go back and complete these steps again. Be careful to avoid sign errors.

Did you determine that you could factor $2x^2-11x-6=0$?
Yes _____ No _____

Did you choose the first term in one set of parentheses as $2x$ and the first term in the other set of parentheses as x?
Yes _____ No _____

If you answered No to these questions, try to factor the expression on the left side of the equation again.

Remember to set each set of parentheses equal to 0 and then solve for x.

If you answered Problem 9 incorrectly, go back and rework this problem using these suggestions.

Graphing a Quadratic Equation and Finding the Vertex—Problem 12

Graph the quadratic equation. Locate the vertex. $y=-x^2+8x-12$

> **Helpful Hint:** Make sure that the quadratic equation is written in $y=ax^2+bx+c$ form. Use the vertex formula to find the vertex. Next, find the intercepts by setting $y=0$ and $x=0$. Notice that when $a<0$, the graph shows a parabola opening downward.

Did you identify that the equation is written in
$y=ax^2+bx+c$ form and that $a=-1$, $b=8$, and $c=-12$?
Yes _____ No _____

Did you determine that the x-coordinate of the vertex is 4?
Yes _____ No _____

If you answered No to these questions, remember that the vertex formula for the x-coordinate of the vertex point is
$x=\dfrac{-b}{2a}$. You must then substitute this value for x in the original equation to find the y-coordinate for the vertex point. Graph this point on the coordinate plane.

Did you let $y=0$ to find the x-intercepts and let $x=0$ to find the y-intercepts? Yes _____ No _____

If you answered No, remember to follow these steps and solve for the other variable in each case to find the intercept

points. Graph these points on the same coordinate plane with the vertex.

Once you know that the parabola opens downward and you have located the vertex and intercept points, you can graph the quadratic equation. For this problem, your vertex point should be the highest point on the graph.

Now go back and rework this problem using these suggestions.

Worksheet Answers Chapter 0

Section 0.1

Vocabulary
1. natural numbers
2. denominator
3. lowest terms
4. prime numbers

Student Practice
2. (a) $\dfrac{11}{13}$

   (b) $\dfrac{3}{5}$
4. $\dfrac{1}{13}$
6. 9
8. $\dfrac{3}{5}$
10. 9
12. $\dfrac{77}{12}$
14. (a) 11
    (b) 20

Extra Practice
1. 4
2. $55\dfrac{3}{5}$
3. $\dfrac{29}{17}$
4. 64

Concept Check
Answers may vary. Possible solution: First, multiply the whole number by the denominator. Then, add this to the numerator. The result is the new numerator. The denominator does not change.

Section 0.2

Vocabulary
1. least common denominator
2. denominator
3. perimeter
4. mixed number

Student Practice
2. (a) 1

   (b) $\dfrac{3}{2}$ or $1\dfrac{1}{2}$
4. $\dfrac{1}{4}$
6. 70
8. 126
10. $\dfrac{95}{108}$
12. $47\dfrac{1}{3}$ yd

Extra Practice
1. 100
2. $\dfrac{5}{9}$
3. $3\dfrac{7}{36}$
4. $3\dfrac{1}{2}$

Concept Check
Answers may vary. Possible solution: Write each denominator as the product of prime factors. The LCD is a product containing each different factor. If a factor occurs more than once in any one denominator, the LCD will contain that factor repeated the greatest number of times that it occurs in any one denominator.

Section 0.3

Vocabulary
1. complex fraction
2. numerators
3. invert and multiply method
4. mixed numbers

Student Practice
2. (a) $\dfrac{2}{17}$

   (b) $\dfrac{11}{27}$

279

4. 20

6. $44\dfrac{5}{8}$ m^2

8. $\dfrac{2}{15}$

10. 35

12. $\dfrac{3}{8}$

14. $3\dfrac{4}{7}$ or $\dfrac{25}{7}$

16. $\dfrac{4}{5}$ image per hour

Extra Practice

1. $\dfrac{2}{49}$

2. 18

3. $\dfrac{3}{2}$ or $1\dfrac{1}{2}$

4. 14 pieces

Concept Check

Answers may vary. Possible solution: Change each mixed number to an improper fraction. Invert the second fraction and multiply the result by the first fraction. Simplify.

Section 0.4

Vocabulary

1. decimal
2. divisor
3. decimal places
4. decimal points

Student Practice

2. (a) $\dfrac{9}{100,000}$; 5 decimal places; nine hundred-thousandths

   (b) $4\dfrac{25}{1000}$; 3 decimal places, four and twenty-five thousandths

4. $0.8666\ldots$ or $0.8\overline{6}$

6. $\dfrac{18}{25}$

8. 37.11

10. 0.000425

12. 84,000

14. 93

16. 713.83

Extra Practice

1. $32\dfrac{41}{500}$; thirty-two and eighty-two thousandths

2. 6.1368

3. 19.32

4. 1678.5

Concept Check

Answers may vary. Possible solution: Move the decimal point over 4 places to the right on the divisor $(0.0035 \Rightarrow 35)$.

Then move the decimal point the same number of places on the dividend $(0.252 \Rightarrow 2520)$.

Section 0.5

Vocabulary

1. percent
2. nonzero digit
3. estimation
4. decimal point

Student Practice

2. (a) 0.019%
   (b) 60%

4. (a) 595%
   (b) 790%

6. (a) 0.008
   (b) 1.012

8. 31.95

10. (a) $79.95
    (b) $43.05

12. 87.5%

14. 6%

16. 1000 square meters

Extra Practice

1. 0.576%

2. 1

3. 260%

4. 0.02

Concept Check

Answers may vary. Possible solution:
Move the decimal point two places to the
right and add the % symbol.

Section 0.6

Vocabulary

1. mathematics blueprint
2. check

Student Practice

2. $650
4. (a) 18%
   (b) 28%
   (c) The Northwest

Extra Practice

1. 492 people
2. $701.25
3. 1.7 kilometers
4. $3380

Concept Check

Answers may vary. Possible solution:
Multiply the number of kilometers by
0.62 to obtain miles.

Section 1.1

Vocabulary

1. absolute value
2. opposite numbers
3. real numbers
4. negative numbers

Student Practice

2. (a) Irrational, real
   (b) Rational, real
   (c) Rational, real
4. (a) −6.23
   (b) +109.4
   (c) −345
   (d) +15
6. (a) $\dfrac{7}{8}$
   (b) 30 feet above sea level
8. (a) 5.34
   (b) $\sqrt{5}$
   (c) 0
10. −14
12. −6
14. 0.1
16. −8

Extra Practice

1. Irrational, real
2. $\dfrac{7}{8}$
3. −25
4. 15

Concept Check

Answers may vary. Possible solution: Adding a negative number to a negative number will always result in a number further in the negative direction. However, adding numbers of opposite sign could result in a negative number if the absolute value of the negative number is larger than that of the positive number. Further, adding numbers of opposite sign could result in a positive number, if the absolute value of the positive number is greater than that of the negative number. Finally, adding numbers of opposite sign could result in 0 if their absolute values are equal.

Section 1.2

Vocabulary

1. subtract
2. additive inverse property

Student Practice

2. 12
4. −6
6. (a) $-\dfrac{5}{9}$
   (b) $-\dfrac{3}{20}$
8. 0
10. (a) −20
    (b) −5
    (c) 25
    (d) $-\dfrac{26}{5}$
12. 629 feet

Extra Practice

1. −3.29
2. $-\dfrac{7}{30}$
3. −11
4. $179-(-22)$; 201 feet

Concept Check

Answers may vary. Possible solution: The result could be positive if the absolute value of the number subtracted is greater than the other number $\left[-2-(-3)=-2+3=1\right]$.

The result could be zero if the two numbers are the same $\left[-2-(-2)=-2+2=0\right]$.

The result could be negative if the absolute value of the number being subtracted is less than the other number

$$\left[-2-(-1) = -2+1 = -1\right].$$

Section 1.3
Vocabulary
1. undefined
2. negative
3. division
4. multiplication

Student Practice
2. (a) 20

   (b) $-\dfrac{25}{12}$ or $-2\dfrac{1}{12}$
4. (a) 32.4
   (b) −60
6. (a) 8
   (b) −5
8. (a) 1100
   (b) −0.6
10. $\dfrac{5}{12}$
12. −40

Extra Practice
1. −462.94
2. 16
3. −175
4. $-\dfrac{5}{6}$

Concept Check
Answers may vary. Possible solution:
Multiplying an even number of
negative numbers results in a positive
number, whereas multiplying an odd
number of negative numbers results in
a negative number.

Section 1.4
Vocabulary
1. exponent
2. cubed
3. base
4. variable

Student Practice
2. (a) $(-3)^4$

(b) $(-6)^8$

(c) $(n)^6$

4. (a) 64
   (b) 37
6. (a) −64
   (b) 256
   (c) −256
   (d) −256
8. (a) $\dfrac{1}{125}$
   (b) 0.027
   (c) $\dfrac{64}{343}$
   (d) 400
   (e) −24

Extra Practice
1. $(-ab)^2$
2. $-\dfrac{1}{8}$
3. −40
4. 57

Concept Check
Answers may vary. Possible solution:
If you have parentheses surrounding the
−2, then the base is −2 and the
exponent is 6. The result is 64. If you do
not have parentheses, then the base is 2.
You evaluate to obtain 64 and then take
the opposite of 64, which is −64. Thus,
$(-2)^6 = 64$ and $-2^6 = -64$.

Section 1.5
Vocabulary
1. add and subtract
2. parentheses
3. order of operations
4. multiply and divide

Student Practice
2. 60
4. 24
6. 23
8. $-\dfrac{5}{48}$

283

Extra Practice

1. 1
2. 29
3. 2
4. $-\dfrac{31}{5}$

Concept Check

Answers may vary. Possible solution: Evaluate the operations within the parentheses. Then raise the result to a power of 3. Then divide. Finally, add and subtract from left to right.

Section 1.6

Vocabulary

1. distributive property
2. algebraic expression
3. factor
4. term

Student Practice

2. (a) $6x + 18y$

   (b) $-8m - 20n$

4. $-m + 6n$

6. (a) $\dfrac{3}{4}m^2 - 6m + \dfrac{15}{4}$

   (b) $1.3x^2 + 1.69x + 0.91$

8. $-5x^2 - 15xy + 30x$

10. $-8y + 6y^2$

12. $3500x + 4000y$

Extra Practice

1. $-12x - 6$
2. $\dfrac{1}{6}x + \dfrac{1}{5}y$
3. $-18x + 12y - 6z$
4. $30(15 + 7n);\ 450 + 210n$

Concept Check

Answers may vary. Possible solution:

Distribute $\left(-\dfrac{3}{7}\right)$ to each term within

the parentheses.

$$\left(-\dfrac{3}{7}\right)\!\left(21x^2 - 14x + 3\right)$$

$$= \left(-\dfrac{3}{7}\right)\!\left(21x^2\right) + \left(-\dfrac{3}{7}\right)\!\left(-14x\right) + \left(-\dfrac{3}{7}\right)\!\left(3\right)$$

$$= -9x^2 + 6x - \dfrac{9}{7}$$

Section 1.7

Vocabulary

1. term
2. combining
3. combine like terms
4. like terms

Student Practice

2. (a) $4x$ and $2x$

   (b) $4x^2$ and $-6x^2$; $5x$ and $9x$

4. (a) $7z^5$

   (b) $2z$

6. $3x^3$

8. (a) $7.6m + 7.2n$

   (b) $8a^2b - 4ab^3$

   (c) $6xy + 2x^2y - 9xy^2 + 5x^2y^2$

10. $14x - 5c - 2t$

12. $\dfrac{17}{5}x - \dfrac{5}{14}y$

14. $-17c + 47cd$

Extra Practice

1. $-2a + 12 + 4a^2$
2. $\dfrac{9}{14}x^2 - \dfrac{5}{8}y$
3. $-22a + 22b$
4. $2ab + 13ac - 3bc + 30cd$

Concept Check

Answers may vary. Possible solution: Use the distributive property to remove the parentheses. Then simplify by combining like terms.

$$1.2(3.5x - 2.2y) - 4.5(2.0x + 1.5y)$$

$$= 4.2x - 2.64y - 9x - 6.75y$$

$$= -4.8x - 9.39y$$

Section 1.8

Vocabulary

1. square
2. circumference
3. area
4. evaluate

Student Practice

2. -25
4. (a) 75
   (b) 225
6. 33
8. 118
10. 13.5 yd^2
12. 50.24 yd^2
14. $-40°\text{C}$
16. You are driving at approximately 112.7 km/h. You are exceeding the speed limit.

Extra Practice

1. -3
2. 19
3. -1
4. 113 m^2

Concept Check

Answers may vary. Possible solution: The formula for the area of a circle is $A = \pi r^2$. Therefore, first find the radius from the diameter $\left(r = \dfrac{d}{2} \right)$.

Use $\pi \approx 3.14$.

$A = 3.14 (6 \text{ m})^2$

$\quad = 3.14 (36 \text{ m}^2) = 113.04 \text{ m}^2$

Section 1.9

Vocabulary

1. fraction bars
2. distributive property
3. grouping symbols

Student Practice

2. $4x - 60$
4. $-32x + 8y + 8$
6. $32x - 42$
8. $-8x - 8$

Extra Practice

1. $-5a - 9b$
2. $10y - 2x$
3. $a - 6b$
4. $-10x^2 - 16x + 20$

Concept Check

Answers may vary. Possible solution: First combine like terms within the square brackets. Then use the distributive property to remove grouping symbols and combine like terms after each step.

$3\left\{ 2 - 3 \left[4x - 2(x+3) + 5x \right] \right\}$

$= 3 \left\{ 2 - 3 \left[9x - 2(x+3) \right] \right\}$

$= 3 \left\{ 2 - 3 \left[9x - 2x - 6 \right] \right\}$

$= 3 \left\{ 2 - 3 \left[7x - 6 \right] \right\}$

$= 3 \left\{ 2 - 21x + 18 \right\}$

$= 3 \left\{ 20 - 21x \right\}$

$= 60 - 63x$

Section 2.1

Vocabulary
1. additive inverse
2. equation
3. identity
4. addition principle

Student Practice
2. $x = 34$
4. $x = 2.7$
6. No. $x = 36$
8. $x = -\dfrac{5}{21}$

Extra Practice
1. $x = 6$
2. $x = 3.53$
3. No. $x = -6$
4. Yes.

Concept Check
Answers may vary. Possible solution: Substitute the value 3.8 for x in the equation. Simplify. If the resultant equation is true, $x = 3.8$ is the solution. If the resultant equation is not true, $x = 3.8$ is not the solution.

Section 2.2

Vocabulary
1. multiplication principle
2. division principle
3. multiplicative inverse
4. terminating decimal

Student Practice
2. $x = 21$
4. $x = 7$
6. $x = \dfrac{15}{8}$
8. $x = -7$
10 $x = -15$
12 $x = -6$
14 $x = 7$

Extra Practice
1. $x = 13$
2. $x = -25$

3. Yes
4. No; $x = -0.12$

Concept Check
Answers may vary. Possible solution:

Change $36\dfrac{2}{3}$ to an improper fraction.

Substitute that value for x in the equation. Simplify. If the resultant equation is true, $x = 36\dfrac{2}{3}$ is the solution.

Section 2.3

Vocabulary
1. multiplication principle
2. addition principle
3. distributive property
4. like terms

Student Practice
2. $x = 2$
4. $x = -\dfrac{2}{3}$
6. $x = \dfrac{17}{4}$
8. $x = -\dfrac{1}{7}$
10 $z = 20$

Extra Practice
1. $x = 5$
2. $x = 12$
3. $x = \dfrac{1}{4}$
4. $z = 2$

Concept Check
Answers may vary. Possible solution: Use the distributive property to remove parentheses. Combine like terms on the left side of the equation. Move the variable terms to the left side of the equation and the constants to the right side of the equation. Simplify.

Section 2.4

Vocabulary
1. decimal
2. no solution
3. LCD

Student Practice

2. $x = \dfrac{6}{7}$

4. $x = -15$

6. $x = 112$

8. $x = \dfrac{1}{3}$

Extra Practice

1. $p = \dfrac{3}{2}$

2. $x = -2$

3. $x = -5$

4. $x = -5$

Concept Check

Answers may vary. Possible solution:
Multiply both sides of the equation by
the LCD, 12. Add or subtract terms on
both sides of the equation to get all terms
containing x on one side of the
equation. Add or subtract a constant
value to both sides of the equation to get
all terms not containing x on the other
side of the equation. Divide both sides
by the coefficient of x and simplify the
solution if necessary. Finally, check the
solution.

Section 2.5

Vocabulary
1. sum of
2. difference between
3. product of
4. quotient of

Student Practice

2. (a) $x - 5$
   (b) $3x$
   (c) $\dfrac{x}{5}$ or $\dfrac{1}{5}x$
   (d) $x + 10$

4. a. $3x + 8$

b. $3(x + 8)$

c. $\dfrac{1}{4}(x + 6)$

6. $c = $ Chuck's car's mileage;
   $c + 12{,}500 = $ Tom's car's mileage

8. $w = $ the width; $3w - 7 = $ the length

$$3w - 7$$
$$w$$

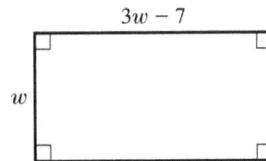

10. $s = $ the second angle; $5s = $ the first
    angle; $s + 40 = $ the third angle

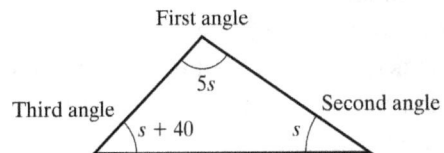

First angle

$5s$

Third angle   Second angle

$s + 40$ s

Extra Practice

1. $x + 12$

2. $\dfrac{1}{4}x - 1$

3. $x = $ value of Allison's car;
   $x + \$2300 = $ value of Alicia's car

4. $s = $ number of Scott's comic
   books; $s + 17 = $ number of
   Patricia's comic books;
   $3s = $ number of Walter's comic
   books; $4s - 5 = $ number of
   Adrienne's comic books

Concept Check

Answers may vary. Possible solution:
"one-third of the sum" means you

multiply $\dfrac{1}{3}$ times the sum. The sum

will be the quantity in the parentheses

because the $\dfrac{1}{3}$ is multiplied by the

whole sum; $\dfrac{1}{3}(x + 7)$.

Section 2.6

Vocabulary
1. check

287

2. solve and state the answer
3. write an equation

Student Practice

2. 85
4. 65
6. 32°
8. 12 hours for John; 13 hours for Yuri

Extra Practice

1. 546
2. 15
3. 18 energy bars
4. 300 minutes

Concept Check

Answers may vary. Possible solution: Since we want to know how many pairs of socks, let $x =$ the number of pairs of socks. Set up an equation to represent the total amount spent.

$2(\$23) + \$0.75x = \$60.25$

Section 2.7

Vocabulary

1. interest
2. simple interest
3. percent

Student Practice

2. 16 hours
4. $800
6. $2000 invested at 6%; $1500 invested at 5%
8. 21 nickels; 17 dimes

Extra Practice

1. $16,800
2. $975
3. $54,000
4. 129 bills

Concept Check

Answers may vary. Possible solution: Since each amount is given in terms of quarters, let $x =$ the number of quarters. Then $2x =$ the number of dimes, and $x + 1 =$ the number of nickels. Now set up an equation and solve, given the value of each coin and the total.

$0.25x + 0.10(2x) + 0.05(x + 1) = 2.55$

Section 2.8

Vocabulary

1. solution set
2. inequalities
3. graph
4. is greater than

Student Practice

2. (a) $>$
   (b) $>$
   (c) $<$
   (d) $>$
   (e) $<$
4. (a) x is less than 3.

   (b) five halves is greater than or equal to x.

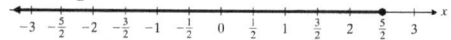

6. (a) $c < \$125$
   (b) $p \le 35$
8. $x > \dfrac{3}{2}$

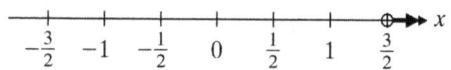

10. $x > -\dfrac{1}{3}$

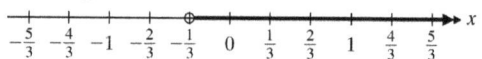

12. $x > -\dfrac{23}{72}$

Extra Practice

1. $<$
2.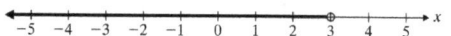
3. $w \le 1,000,000$
4. $x < 3$

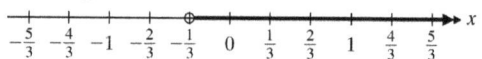

Concept Check

Answers may vary. Possible solution: $12 < x$ is the same as $x > 12$, but written in different ways. The graphs will be exactly the same.

Section 3.1

Vocabulary

1. coordinates
2. graphs
3. y-axis
4. x-coordinate

Student Practice

2.

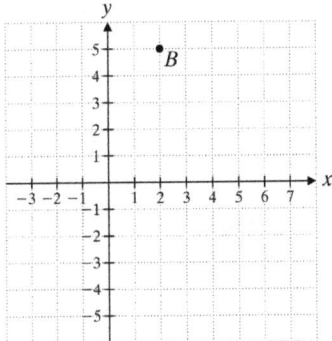

4.

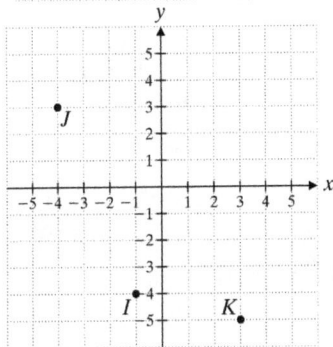

6. $(-3,6)$

8. No

10. $(-2,3)$

Extra Practice

1.

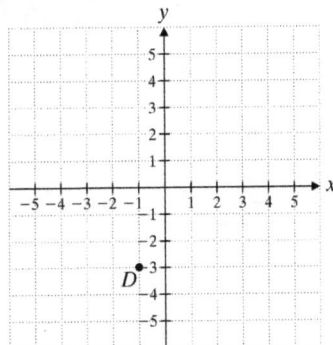

2. $(3,5)$

3. (a) $(-1,5)$

   (b) $(3,-3)$

4. (a) $(-1,-8)$

   (b) $(4,2)$

Concept Check

Answers may vary. Possible solution: Isolate x on the left side of the equation. Substitute the given value for y and solve for x.

Section 3.2

Vocabulary

1. y-intercept
2. linear equation
3. x-intercept

Student Practice

2.

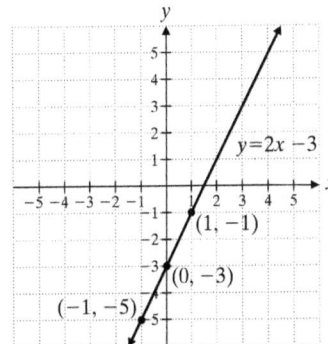

4.

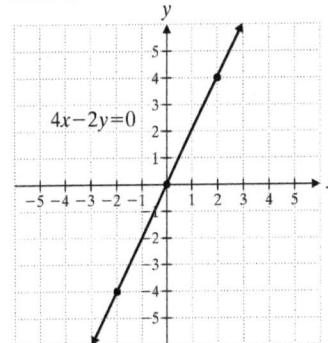

6. (a) x-intercept:$(-1,0)$

   y-intercept:$(0,-2)$

(b)

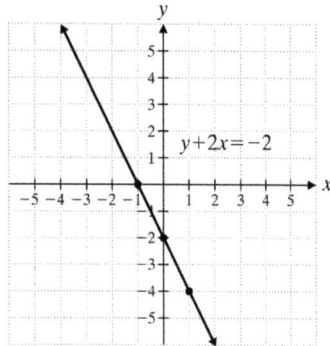

$y + 2x = -2$

8.

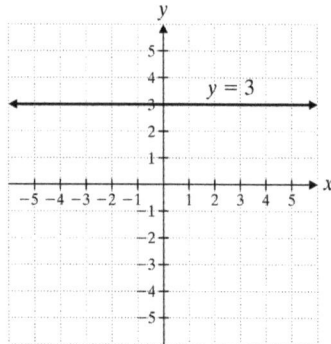

$y = 3$

Extra Practice

1.

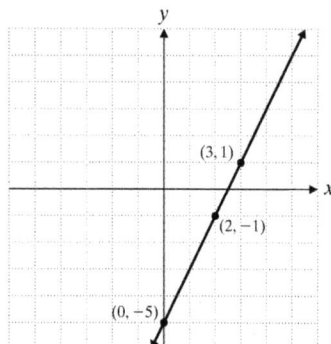

$(3, 1)$
$(2, -1)$
$(0, -5)$

2.

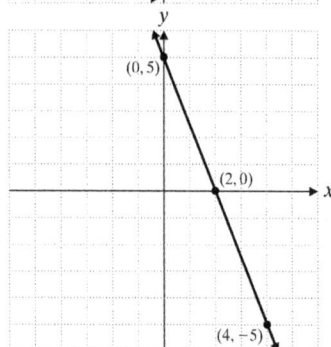

$(0, 5)$
$(2, 0)$
$(4, -5)$

3.

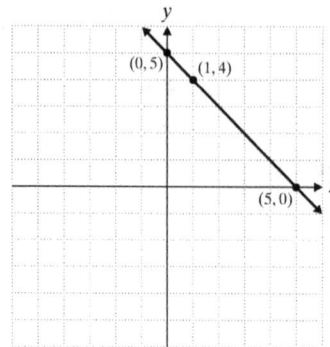

$(0, 5)$ $(1, 4)$
$(5, 0)$

4.

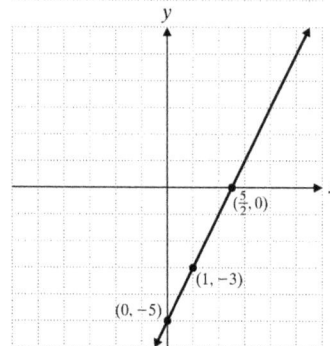

$\left(\frac{5}{2}, 0\right)$
$(1, -3)$
$(0, -5)$

Concept Check

Answers may vary. Possible solution:
The most important ordered pair is
$(0, 0)$, since it is both the x- and

y- intercept of the equation.

Section 3.3

Vocabulary
1.   slope
2.   undefined slope
3.   slope-intercept form

Student Practice

2. $m = \dfrac{2}{3}$

4.   (a) No slope
     (b) $m = 0$

6. $m = -3$, y-intercept: $(0, 4)$

8.
     (a) $y = \dfrac{3}{4}x - 7$
     (b) $3x - 4y = 28$

10.

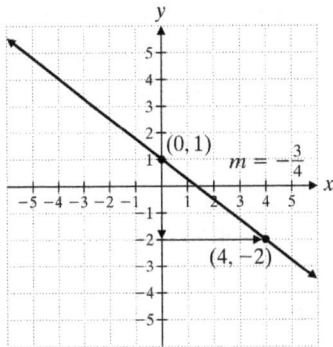

Extra Practice

1. $-\dfrac{1}{3}$

2. $m = 5$, y-intercept: $(0,0)$

3. $y = -4x + \dfrac{4}{5}$

4. $y = -2$

Concept Check

Answers may vary. Possible solution: Slope measures the vertical change per one unit of horizontal change.

Section 3.4

Vocabulary

1. y-intercept
2. vertical units
3. perpendicular lines
4. parallel lines

Student Practice

2. $y = -\dfrac{1}{2}x - 4$

4. $y = -\dfrac{3}{5}x + 7$

6. $y = -\dfrac{5}{2}x + 5$

8.
   (a) $\dfrac{5}{7}$

   (b) $-\dfrac{7}{5}$

Extra Practice

1. $y = -\dfrac{1}{2}x + 2.5$

2. $y = 5x - 11$

3. $y = \dfrac{7}{6}x - 1$

4. $y = -5x + 8$

Concept Check

Answers may vary. Possible solution: Zero slope indicates a horizontal line. Since the line is horizontal, and it passes through $(-2, -3)$, the equation must be $y = -3$.

Section 3.5

Vocabulary

1. solution
2. test point
3. solid line

Student Practice

2.

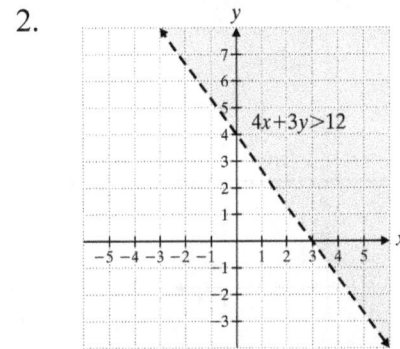

4.

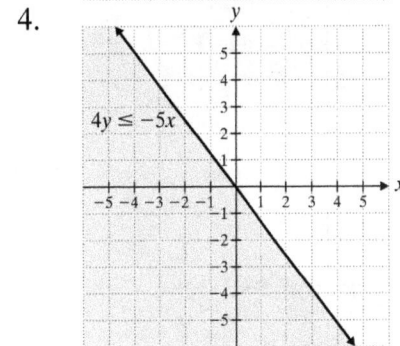

6.

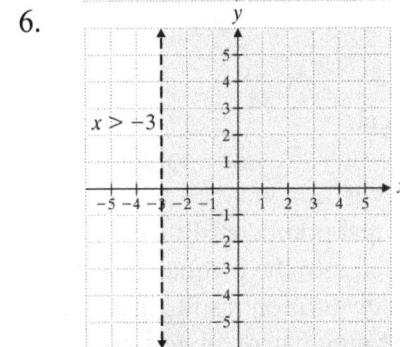

Extra Practice

1.
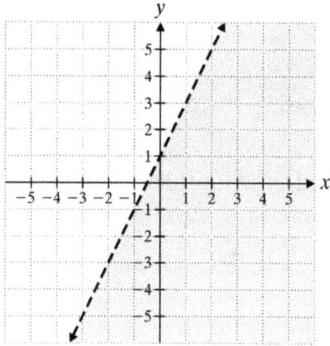

2.
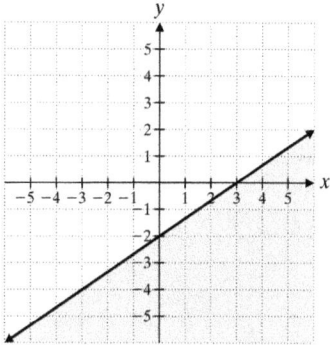

3.

4.

Concept Check
Answers may vary. Possible solution:
The inequality is first graphed without
shading. The test point coordinates are
then substituted into the inequality. If the
result of the substitution results in a true
statement, the area where the test point

lies is shaded. If false, the opposite area
is shaded.

Section 3.6
Vocabulary
1. relation
2. function
3. vertical line test
4. domain

Student Practice
2. The domain is $\{-3, -2, 4, 7\}$.

    The range is $\{1, 6\}$.

4. (a) Not a Function
    (b) Function

6.
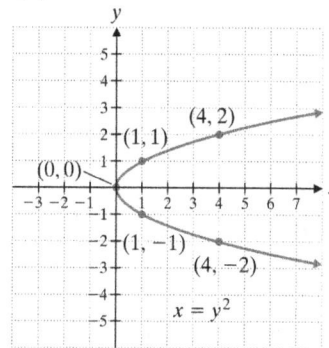

8. (a) Not a function
    (b) Function
10. (a) 49
    (b) 19
    (c) 7

Extra Practice
1. The domain is $\{2.5, 3.5, 5.5, 8.5\}$.

    The range is $\{-6, -2, 0, 3\}$.

    Function
2. (a) 1
    (b) 28
    (c) 10
3. Not a function

292

4.

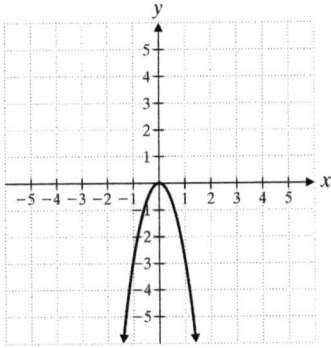

Concept Check

Answers may vary. Possible solution: Duplicate elements are recorded only once in the domain and in the range.

Section 4.1

Vocabulary

1. consistent
2. system of equations
3. dependent
4. inconsistent

Student Practice

2. $(1,1)$

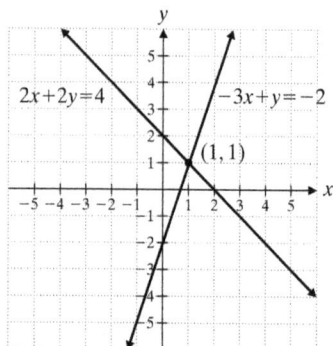

4. No solution; inconsistent system

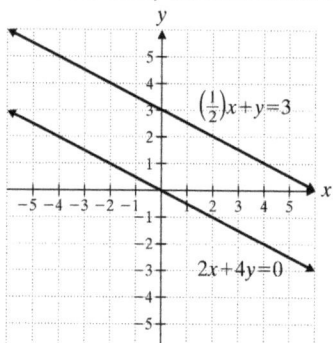

6. Infinite number of solutions; dependent system

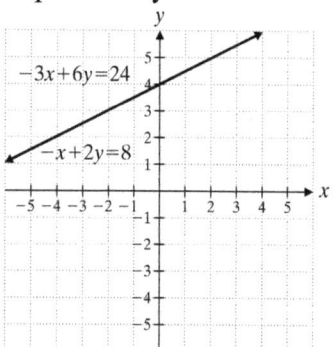

8. (a)
$$y = 50 + 40x$$
$$y = 90 + 30x$$

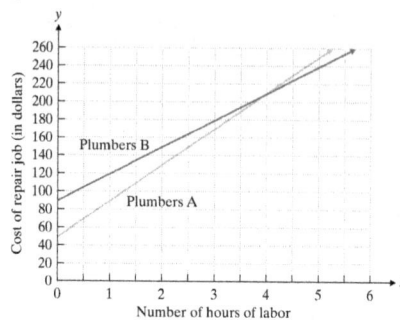

(b)
(c) 4 hours
(d) Plumbers B

Extra Practice

1. $(-2,-6)$

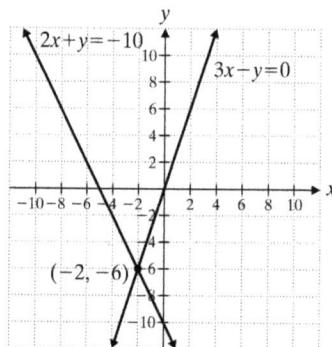

2. No solution; inconsistent system

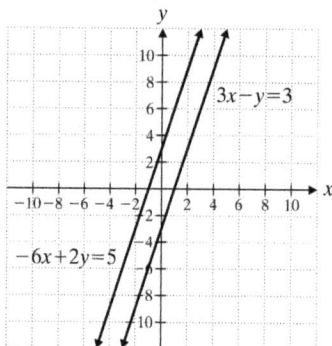

3. $(6,2)$

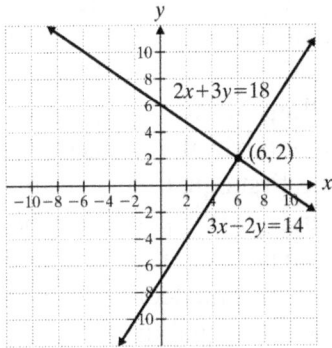

4. Infinite number of solutions; dependent system

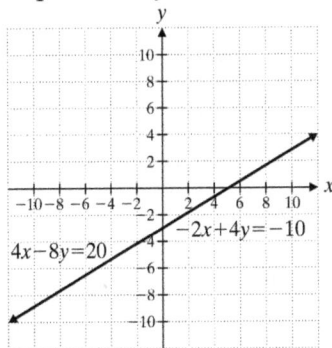

Concept Check

Answers may vary. Possible solution: The two equations represent lines that are parallel lines. There is, therefore, no solution.

Section 4.2

Vocabulary

1. linear equation
2. substitution method

Student Practice

2. $(-2,3)$

4. $\left(\dfrac{3}{8},\dfrac{5}{8}\right)$

Extra Practice

1. $(0,-3)$

2. $\left(\dfrac{54}{5},\dfrac{6}{5}\right)$

3. $(-6,6)$

4. $(-45,33)$

Concept Check

Answers may vary. Possible solution: Label the equations (1) and (2).

Solve (1) for y and label this equation (3). Substitute the value of y in (3) into (2) and solve (2) for x. Substitute the found value of x back into (3) and solve for y. Check that the values of x and y satisfy both (1) and (2).

Section 4.3

Vocabulary

1. elimination method
2. addition method

Student Practice

2. $(4,-1)$

4. $(2,3)$

6. $(-3,1)$

Extra Practice

1. $(29,23)$

2. $(-7,-4)$

3. $\left(1,\dfrac{4}{3}\right)$

4. $\left(\dfrac{1}{2},-2\right)$

Concept Check

Answers may vary. Possible solution: Multiplying both equations by 100 will eliminate decimals. Or, multiply the first equation by 10 and the second equation by 100.

Section 4.4

Vocabulary

1. identity
2. substitution method
3. inconsistent
4. dependent

Student Practice

2. $(15,8)$

4. No solution; inconsistent system

6. Infinite number of solutions; dependent system

Extra Practice

1. $(-4,-5)$

2. No solution; inconsistent system

3. $(3,5)$

4. $(-3,-5)$

Concept Check

Answers may vary. Possible solution: Label the equations (1) and (2).

Multiply (1) by 20 and (2) by -48.

Then add equations to eliminate fractions and x.

Section 4.5

Vocabulary

1. system of equations

2. substitution method

Student Practice

2. The first employee plants 15 bulbs per hour and the second employee plants 20 bulbs per hour.

4. The student should use 50 milliliters of the 25% solution and 100 milliliters of the 10% solution.

6. The plane's speed in still air was 80 nautical miles per hour and the wind speed was 20 nautical miles per hour.

Extra Practice

1. They have 17 quarters and 9 nickels.

2. They should include 5 pounds of large snack bars and 5 pounds of small snack bars.

3. Ronaldo hammers 85 nails per hour and Lee hammers 75 nails per hour.

4. It will be 25 years until the populations are the same. Each population will be 55,600 people.

Concept Check

Answers may vary. Possible solution: The system will consist of two equations in terms of x and y. x will equal the number of gallons of 50% pure juice. y will equal the number of gallons of 30% pure juice. The first equation states that the sum of x and y will equal 500. The second equation states that each amount of the mixture times its decimal fraction of juice will sum to equal the total mixture quantity times its decimal fraction of juice.

Worksheet Answers Chapter 5

Section 5.1

Vocabulary
1. base
2. exponent
3. quotient rule
4. numerical coefficient

Student Practice
2. (a) z^5
   (b) 2^8
4. $-8x^6 y^3$
6. x^4
8. $\dfrac{1}{n^4}$
10. $\dfrac{x^2}{2y^5}$
12. (a) y^{32}
    (b) -1
14. $\dfrac{y^3}{z^6}$
16. $-\dfrac{x^{15}}{64y^6}$

Extra Practice
1. $24x^6 y^5$
2. 9
3. $\dfrac{1}{b^4}$
4. $\dfrac{9^2}{14^2 x^6}$, or $\dfrac{81}{196x^6}$

Concept Check
Answers may vary. Possible solution:
In the numerator and the denominator, raise each factor inside the parentheses to the power.

$$\dfrac{4^2 \left(x^3\right)^2}{2^3 \left(x^4\right)^3}$$

Evaluate the constants, and multiply exponents on the variable expressions.

$$\dfrac{16x^6}{8x^{12}}$$

Divide the numbers and subtract exponents on the variable expressions. Because the larger exponent is in the denominator, the variable expression will be in the denominator, $\dfrac{2}{x^6}$.

Section 5.2

Vocabulary
1. significant digits
2. base
3. scientific notation
4. negative exponent

Student Practice
2. $\dfrac{1}{2401}$
4. $\dfrac{y^8}{16x^4 z^{12}}$
6. 2.564×10^3
8. (a) 1.3×10^{-3}
   (b) 1.0×10^{-6}
10. (a) 0.000528
    (b) $3,221,400$
12. 3.86×10^{11} hr

Extra Practice
1. $\dfrac{1}{343}$
2. $\dfrac{b^8 c^{12}}{81a^{12}}$
3. 6340
4. 2.0×10^{-1}

Concept Check
Answers may vary. Possible solution:
Raise each factor inside the parentheses to the power.

$$4^{-3} \left(x^{-3}\right)^{-3} \left(y^4\right)^{-3}$$

Multiply exponents on the variable expressions.

$$4^{-3} x^9 y^{-12}$$

Rewrite as a fraction.

$$\frac{x^9}{4^3 y^{12}}$$

Evaluate 4^3.

$$\frac{x^9}{64 y^{12}}$$

Section 5.3
Vocabulary
1. binomial
2. degree of a term
3. decreasing order
4. degree of a polynomial

Student Practice
2. (a) Degree 7; monomial
   (b) Degree 2; trinomial
   (c) Degree 7; binomial
4. $4x^2 - 8x + 5$
6. $\frac{7}{4}x^2 - \frac{8}{5}x + \frac{5}{8}$
8. $-3x^2 - 10x + 5$
10. 26.4 miles per gallon

Extra Practice
1. Degree 6; trinomial
2. $10.8x - 14$
3. $5r^4 - r^2 + 8$
4. 9.01 miles per gallon

Concept Check
Answers may vary. Possible solution: To determine the degree of the polynomial, first determine the degree of each term by finding the sum of the exponents on the variables in each term. The degree of the first term, $2xy^2$, is $1 + 2 = 3$, and the degree of the second term, $-5x^3y^4$, is $3 + 4 = 7$. The degree of the polynomial is the greater of these, which is 7. To determine whether the polynomial is a monomial, a binomial, or a trinomial, we must count the number of terms in the polynomial. The polynomial has two terms, $2xy^2$ and $-5x^3y^4$, so it is a binomial.

Section 5.4
Vocabulary
1. monomial
2. FOIL
3. distributive property

Student Practice
2. $-5y^3 + 20y^2$
4. $2x^4y - 14x^3y - 20x^2y$
6. $15x^2 + 4x - 3$
8. $7x^2 + 14xz - 3xy - 6yz$
10. $16x^2 + 24xy + 9y^2$
12. $20x^4 - 20x^2y^4 - 15y^8$
14. $8x^2 - 2$

Extra Practice
1. $-10x^4 + 3x^2$
2. $x^2 - 14x + 33$
3. $15x^2 - 11xy - 56y^2$
4. $8x^4 - 2x^2y^3 - 15y^6$

Concept Check
Answers may vary. Possible solution: First write the square of the binomial as the product of the binomial and itself.
$$(7x - 3)(7x - 3)$$
Then use FOIL and collect like terms.
$$49x^2 - 21x - 21x + 9 = 49x^2 - 42x + 9$$

Section 5.5
Vocabulary
1. vertical multiplication
2. square of a difference
3. polynomial

Student Practice
2. $64y^2 - 16$
4. $4x^2 - 81y^2$
6. $49x^2 + 14xy + y^2$
8. $2x^4 + 2x^3 - 29x^2 + 9x + 30$
10. $x^4 - 7x^3 + 6x^2 - 22x + 40$
12. $2x^3 + 6x^2 - 8x - 24$

Extra Practice

1. $x^2 - 100$

2. $25a^2 - 9b^2$

3. $24x^3 + 2x^2 + 7x + 3$

4. $x^3 - 4x^2 - 7x + 10$

Concept Check

Answers may vary. Possible solution:
To use the formula
$(a+b)^2 = a^2 + 2ab + b^2$ to multiply

$(6x - 9y)^2$, first identify a and b: $a = 6x$

and $b = -9y$. Then substitute these
values for a and b in the formula and
simplify.
Math displayed below.

$(6x - 9y)^2$

$= (6x)^2 + 2(6x)(-9y) + (-9y)^2$

$= 36x^2 - 108xy + 81y^2$

Concept Check

Answers may vary. Possible solution:
Multiply the quotient and the divisor.
Then add the remainder. You should
get the original dividend.

$(x-2)(x^2 + 2x + 8) + 13$

$= x^3 + 2x^2 + 8x - 2x^2 - 4x - 16 + 13$

$= x^3 + 4x - 3$

Yes, the answer checks.

Section 5.6

Vocabulary

1. monomial
2. long division
3. descending order
4. subtraction

Student Practice

2. $3x^3 + x - 4$

4. $2x^2 + 2x + 1 + \dfrac{5}{x+1}$

6. $4x^2 - 2x - 3 + \dfrac{8}{2x+1}$

8. $4x^2 + 4x + 6 + \dfrac{24}{2x-3}$

Extra Practice

1. $6a^5 - 2a^3 + 4a - 1$

2. $4y^2 - 6y + 9$

3. $3x^2 - 2x + 5$

4. $y^2 + y - 1 - \dfrac{5}{y-1}$

Worksheet Answers Chapter 6

Section 6.1

Vocabulary
1. factor
2. common factor
3. to factor
4. greatest common factor

Student Practice

2. (a) $5(y+3z)$

    (b) $y(12-5z)$

4. $11y(4y^2+5y-x)$

6. (a) $7(3m^2-4n^2)$

    (b) $m^2(mn^2+9n^2+3m^2)$

8. $9xy^2(3y-4x-x^2y)$

10. $11m^3n^2(m+1)$

12. $(3x+4y)(5y^2-1)$

14. $\pi(m^2+4n^2+9z^2)$

Extra Practice

1. $3x(x^2+4x-7)$

2. $6x(10xy+3y-4)$

3. $(x+3y)(8a-b)$

4. $(5a-1)(4x-3)$

Concept Check

Answers may vary. Possible solution:
Determine that the largest integer that will divide into the coefficient of all terms is 36. Determine that the variables common to all terms are a^2 and b^2.
Write the above common factors as the first part of the answer (the first factor).
Remove common factors, and what remains is the second part of the answer (the second factor).

Section 6.2

Vocabulary
1. common factor
2. factoring by grouping
3. commutative property
4. FOIL

Student Practice

2. $(2z+5)(z-4)$

4. $(3x+1)(4x+5)$

6. $(y+3z)(7+x)$

8. $(z+y)(n+6)$

10. $(2x+3)(x-3)$

12. $(5z-1)(n-m)$

14. $(5x-3w)(7y-3z)$

Extra Practice

1. $(x-2)(x+1)$

2. $(x+5)(x-2)$

3. $(3x-2)(2x+5)$

4. $(3a+2b)(4x+5y)$

Concept Check

Answers may vary. Possible solution:
Start by grouping terms $10ax$ with $5ab$ and $2bx$ with b^2.
$(2x+b)$ can be factored out of both groups leaving the second factor to be $(5a+b)$.

Section 6.3

Vocabulary
1. first terms
2. second terms
3. outer and inner terms
4. last terms

Student Practice

2. $(x+3)(x+6)$

4. $(x+11)(x+3)$

6. $(x-4)(x-7)$

8. $(x-6)(x+4)$

10. $(z-2)(z+14)$

12. $(x^2+2)(x^2-5)$

14. $4(x-7)(x+3)$

16. $(x+10)(x-7)$

Extra Practice

1. $(x+2)(x+4)$

2. $(a-3)(a-4)$

3. $(x-4)(x+3)$

4. $2(x+2)(x+7)$

Concept Check

Answers may vary. Possible solution:
The first step is to factor out the greatest common factor of 4, leaving $4(x^2-x-30)$. Next write the expression in factored form using variables m and n, $4(x+m)(x+n)$.

Next determine that the product of m and n is -30, and the sum is -1. m and n may equal 5 and -6. Substitute the values of m and n, then check.

Section 6.4

Vocabulary

1. trial-and-error method
2. multiplying
3. greatest common factor
4. grouping method

Student Practice

2. $(2x+5)(x+4)$

4. $(3x-1)(x-1)$

6. $(x-3)(4x+7)$

8. $(x+1)(2x+3)$

10. $(x-3)(5x+2)$

12. $2(4x-1)(x+3)$

14. $5(2x-3)(3x-2)$

Extra Practice

1. $(2x-1)(2x-5)$

2. $(7x+3)(x-1)$

3. $(5x-2)(x+9)$

4. $3(5y-3)(y+4)$

Concept Check

Answers may vary. Possible solution:
First step is to factor out common coefficients and variables.

$2x(5x^2+9xy-2y^2)$

Next step is to factor the inside expression using grouping.
The grouping number is -10.

$2x\ 5x^2+10xy-xy-2y^2$

$=2x\ 5x(x+2y)-y(x+2y)$

$=2x(5x-y)(x+2y)$

Lastly, check the solution by multiplying the factors.

Section 6.5

Vocabulary

1. negative
2. perfect-square trinomials
3. difference of two squares
4. greatest common factor

Student Practice

2. $(6x+1)(6x-1)$

4. $(8x+3)(8x-3)$

6. $(4x^2+1)(2x+1)(2x-1)$

8. $(x+5)^2$

10. (a) $(8x+5y)^2$

   (b) $(7x^2-1)^2$

12. $(4x+1)(9x+4)$

14. $3(x+5)(x-5)$

16. $3(4x-3)^2$

Extra Practice

1. $(5a-6b)(5a+6b)$

2. $(6a+5b)^2$

3. $(x^2-10)(x^2+10)$

4. $(3x^2-5)^2$

Concept Check

Answers may vary. Possible solution:
First factor out the common factor of 2.

$2(12x^2+60x+75)$

Next use grouping to factor the inside
expression. The grouping number is 900.

$2(12x^2+30x+30x+75)$

$=2[6x(2x+5)+15(2x+5)]$

$=2(6x+15)(2x+5)$

Lastly, check by multiplying.

Section 6.6

Vocabulary

1. common factor
2. factor by grouping
3. prime
4. perfect-square trinomial

Student Practice

2.
   (a) $y(3y+1)^2$

   (b) $-4x(x-8)(x+1)$

4. $(b-3)(x-4)(x+4)$

6. (a) prime
   (b) prime

Extra Practice

1. prime

2. $7x(3-x)(3+x)$

3. $-x(x+3)(x-15)$

4. $(x+3)(x+2)$

Concept Check

Answers may vary. Possible solution:
The first step is to group terms and
factor out common factors.

$2x(x+3w)-5(x+3w)$

Next, factor out $(x+3w)$.

$(x+3w)(2x-5)$

Finally, check by multiplying.

Section 6.7

Vocabulary

1. real roots
2. zero factor property
3. quadratic equation

Student Practice

2. $\dfrac{1}{3}$ and -3

4. 0 and $\dfrac{4}{5}$

6. $x=7$, $x=-9$

8. width $=12$ in., length $=5$ in.

10. 3 seconds

Extra Practice

1. $3, -\dfrac{1}{2}$

2. $-3, 3$

3. $7, -5$

4. $3, 4$

Concept Check

Answers may vary. Possible solution:
Let $x=$ width of rectangle, then
length $=(2x+3)$.

$A=(\text{width})(\text{length})$

$65=x(2x+3)$

$65=2x^2+3x$

$0=2x^2+3x-65$

$0=(2x+13)(x-5)$

Set each factor equal to 0 and solve for
x.

$2x+13=0 \qquad x-5=0$

$\quad\ 2x=-13 \qquad\quad x=5$

$\quad\quad\ x=-\dfrac{13}{2}$

x cannot be negative in this case,
because it describes a length, So $x=5$.

length $=2(5)+3=13$ feet

width $=5$ feet

Section 7.1

Vocabulary
1. factors
2. rational expression
3. simplify the fraction
4. basic rule of fractions

Student Practice

2. $\dfrac{8}{11}$

4. $\dfrac{2}{5}$

6. $\dfrac{x+7}{x+4}$

8. $\dfrac{x+2}{x-6}$

10. $-\dfrac{4}{7}$

12. $-\dfrac{3x+4}{5+2x}$

14. $\dfrac{5x+3y}{9x+7y}$

16. $\dfrac{5a+4b}{2a+b}$

Extra Practice

1. $\dfrac{3}{x}$

2. $\dfrac{3x-4}{3x+4}$

3. $-\dfrac{x+9}{3(x+2)}$

4. $\dfrac{5x-2y}{3x+y}$

Concept Check
Answers may vary. Possible solution: Completely factoring both numerator and denominator is the only way to see what factors are shared, and may consequently be eliminated. In this case, it can be seen that $(x-y)$ is a common factor.

Section 7.2

Vocabulary
1. greatest common factor
2. reciprocals
3. multiply
4. divide

Student Practice

2. $\dfrac{3x+7}{2x-3}$

4. $\dfrac{x}{(x+5)(3x+4)}$ or $\dfrac{x}{3x^2+19x+20}$

6. $\dfrac{4x-3}{2x+1}$

8. $-\dfrac{1}{(x-5)(x-2)}$

Extra Practice

1. $\dfrac{x}{x+1}$

2. $\dfrac{x}{5x+7}$

3. $\dfrac{1}{(x-2)(x+5)}$

4. $\dfrac{5(x-1)}{3x+4}$

Concept Check
Answers may vary. Possible solution: The first step is to change the operation from division to multiplication, by changing the operator and inverting the second fraction. Secondly, all terms must be factored completely. Next, common factors in the numerators and denominators may be canceled. Lastly the multiplication operation is performed.

Section 7.3

Vocabulary
1. denominator
2. different
3. least common denominator

303

4. factor

Student Practice

2. $\dfrac{8}{3x+4}$

4. $\dfrac{2x+11}{(3x-2)(x+9)}$

6. $28(4x+5)$

8. $72a^2b^4c^3$

10. $\dfrac{2xz+4}{xyz}$

12. $\dfrac{11x+22}{4x^2-25}$ or $\dfrac{11(x+2)}{(2x+5)(2x-5)}$

14. $\dfrac{18x+10y}{9x^2-y^2}$ or $\dfrac{2(9x+5y)}{(3x+y)(3x-y)}$

16. $\dfrac{2x+17}{6x+21}$ or $\dfrac{2x+17}{3(2x+7)}$

Extra Practice

1. $(2x-3)(x-4)(x+5)$

2. $\dfrac{3x+5}{x+1}$

3. $\dfrac{2x+5y}{x^2y-y^3}$ or $\dfrac{2x+5y}{y(x+y)(x-y)}$

4. $\dfrac{-x-5}{x^3-7x-6}$ or $-\dfrac{x+5}{(x-3)(x+1)(x+2)}$

Concept Check
Answers may vary. Possible solution:
First factor each denominator completely. The LCD will be the product containing each different factor. If a factor occurs more than once in any one denominator, the LCD will contain that factor repeated the greatest number of times that it occurs in any one denominator.

Section 7.4

Vocabulary
1. complex fraction
2. complex rational expression
3. numerator
4. LCD

Student Practice

2. $\dfrac{4m}{n^2(3m+4)}$

4. $\dfrac{15(m+n)}{mn(3x-5y)}$

6. $\dfrac{x-4}{5(x+2)(x+6)}$

8. $\dfrac{7x-3y}{7}$

10. $\dfrac{16xy-7y^2}{20xy^2-36}$

12. $\dfrac{7x-3y}{7}$

Extra Practice

1. $\dfrac{5a+4b}{2}$

2. $\dfrac{2}{xy}$

3. $\dfrac{2x}{x-35}$

4. $\dfrac{-ab(32x-9y)}{12xy(7b+5a)}$

Concept Check
Answers may vary. Possible solution: The first step is to find the LCD for the fractions in the numerator.

$x-3=(x-3)$

$2x-6=2\cdot(x-3)$

$LCD=2(x-3)$

Next, multiply the first fraction in the numerator by the 2 to obtain common denominators in the numerator. Lastly, add the two fractions in the numerator.

Section 7.5

Vocabulary
1. LCD
2. no solution
3. extraneous solution
4. exclude

Student Practice

2. $x = -44$

4. $x = -6$

6. $x = \dfrac{7}{3}$

8. no solution

Extra Practice

1. $x = 32$

2. no solution

3. $a = 60$

4. $x = -10$

Concept Check

Answers may vary. Possible solution:
To find the LCD first factor each
denominator, then multiply one instance
of each factor.
The math is displayed below.

$$x^2 - 9 = (x - 3)(x + 3)$$

$$3x - 9 = 3(x - 3)$$

$$2x + 6 = 2(x + 3)$$

$$2x^2 - 18 = 2(x - 3)(x + 3)$$

$$\text{LCD} = 2 \cdot 3 \cdot (x - 3)(x + 3)$$

Concept Check

Answers may vary. Possible solution:
One of the fractions needs to be
inverted in order for the equation to be
an accurate statement.

Section 7.6

Vocabulary

1. proportion

2. similar

3. cross multiplying

4. ratio

Student Practice

2. 594 miles

4. $17\dfrac{11}{15}$ centimeters

6. Train A traveled 120 kilometers
per hour. Train B traveled 105
kilometers per hour.

8. 2 hours and 24 minutes

Extra Practice

1. $x = 38$

2. $133\dfrac{1}{3}$ miles

3. 16 feet

4. 3 hours and 22 minutes

Worksheet Answers Chapter 8

Section 8.1

Vocabulary
1. radicand
2. principal square root
3. perfect square

Student Practice
2. (a) 5
   (b) −8
4. (a) $\dfrac{5}{12}$
   (b) $-\dfrac{1}{4}$
6. (a) 0.4
   (b) −60
   (c) 19
8. (a) 3.464
   (b) 6.325

Extra Practice
1. 10 and −10
2. 7
3. −0.4
4. $\dfrac{6}{11}$

Concept Check
Answers may vary. Possible solution:
Start by factoring the radicand by 10^2, or 100 leaving a more manageable $10\sqrt{324}$. For the remaining radicand, guess logically what its square root might be. For example, $15^2 = 225$ is too small, but $20^2 = 400$ is too big. But $18^2 = 324$.

Therefore $10\sqrt{324} = 10\sqrt{18^2} = 180$.

Section 8.2

Vocabulary
1. multiplication
2. complex number
3. place-value system

Student Practice
2. (a) 5
   (b) 15^2

4. (a) n^5
   (b) b^{20}

   (c) 35^9

6. (a) $17x^3$
   (b) $x^4 y^{10}$
   (c) $14x^7 y^{11}$

8. (a) $3\sqrt{3}$
   (b) $4\sqrt{2}$

10. (a) $x^3\sqrt{x}$
    (b) $x^6 y^7 \sqrt{xy}$

12. (a) $5x^8\sqrt{3x}$
    (b) $6x^4 y^4 z^7 \sqrt{3yz}$

Extra Practice
1. 6^2
2. $7r^3 s^2 t^4$
3. $3\sqrt{10}$
4. $3x\sqrt{2x}$

Concept Check
Answers may vary. Possible solution:
Factor the radicand into variables with the largest possible even exponents. All variables with even exponents may then be removed from the radicand by halving the exponent, leaving only the variables with exponents of 1 in the radicand.

Section 8.3

Vocabulary
1. simplify
2. like radicals

Student Practice
2. (a) $4\sqrt{5}$
   (b) $4\sqrt{x}$
4. $14\sqrt{x} - 4\sqrt{3x} - 7\sqrt{3}$
6. (a) $-\sqrt{6}$
   (b) $\sqrt{2} + 3\sqrt{5}$
8. $2\sqrt{x} + 10\sqrt{2x}$

306

10. $11\sqrt{6}$

12. $28x^2\sqrt{5x}$

Extra Practice

1. $\sqrt{3}$

2. 0

3. $8\sqrt{2x}$

4. $y\sqrt{10y}$

Concept Check

Answers may vary. Possible solution:
Factor $4x^2$ out of the radicand of the
first term and 49 out of the second.
Combining the two terms results in
$41x\sqrt{2x}$.

Section 8.4

Vocabulary

1. multiplication
2. FOIL method

Student Practice

2. $5\sqrt{6x}$

4. (a) $20\sqrt{6}$

(b) $225z\sqrt{21}$

6. $300\sqrt{13}$ square feet

8. $6\sqrt{3}+14\sqrt{5}$

10. $21x+3x\sqrt{5x}$

12. $-1+\sqrt{5}$

14. $-80-20\sqrt{3}$

16. $126-56\sqrt{2}$

Extra Practice

1. $2\sqrt{15}$

2. $5\sqrt{2}+\sqrt{5x}$

3. $10-4\sqrt{3}-18\sqrt{2}+15\sqrt{6}$

4. $12x-49$

Concept Check

Answers may vary. Possible solution:
Expand and use the FOIL method to
multiply terms. Combine terms with
equal radicands.

Section 8.5

Vocabulary

1. quotient rule
2. conjugates
3. rationalize the denominator

Student Practice

2. (a) 6

(b) $\dfrac{2}{3}$

4. $\dfrac{3\sqrt{6}}{y}$

6. $\dfrac{7\sqrt{5}}{5}$

8. (a) $\dfrac{\sqrt{30}}{12}$

(b) $\dfrac{5\sqrt{x}}{x^4}$

10. $\dfrac{\sqrt{6}}{10}$

12. $-\dfrac{6\sqrt{7}+6\sqrt{2}}{5}$

14. $11+2\sqrt{30}$

Extra Practice

1. 2

2. $\dfrac{\sqrt{5}}{5}$

3. $\dfrac{2x+8+4\sqrt{x}+x\sqrt{x}}{x-4}$

4. $\dfrac{3\sqrt{30}+6\sqrt{10}+2\sqrt{6}+4\sqrt{2}}{2}$

Concept Check

Answers may vary. Possible solution:
Rationalize the denominator by
multiplying both the numerator and the
denominator by the conjugate of the
denominator.

Section 8.6

Vocabulary

1. radical equation
2. extraneous root

3. legs
4. hypotenuse

Student Practice

2. 40 yards
4. 49.4 feet
6. $x = 43$
8. No solution
10. $x = 12$
12. $x = 4$

Extra Practice

1. $c = 9\sqrt{2}$
2. $b = 2\sqrt{7}$
3. $x = 22$
4. $x = \dfrac{7}{2}$

Concept Check

Answers may vary. Possible solution:
Substitute found values into the original
equation to test the logic of each value.

$$k = \frac{d}{t^2} = \frac{64}{2^2} = 16$$

Section 8.7

Vocabulary

1. constant of variation
2. varies directly
3. vary inversely

Student Practice

2. 15 hours
4. 1372 centimeters cubed
6. 750 times
8. 0.1125 ohm

Extra Practice

1. $y = 1536$
2. $y = \dfrac{1}{27}$
3. $135
4. 2.58 minutes

Concept Check

Answers may vary. Possible solution:
$t = $ time
$d = $ distance
$k = $ constant
When $d = 64$, $t = 2$.
Substitute the values for d and t into the
equation, and solve for k.

Worksheet Answers Chapter 9

Section 9.1

Vocabulary
1. standard form
2. zero factor property
3. greatest common factor
4. quadratic equation

Student Practice
2. (a) $3x^2 + 4x - 6 = 0$
   $a = 3,\ b = 4,\ c = -6$
   (b) $7x^2 + 5x - 6 = 0$
   $a = 7,\ b = 5,\ c = -6$
4. $x = 0,\ x = 1$
6. $x = -\dfrac{1}{12},\ x = -\dfrac{1}{2}$
8. $n = 7$

Extra Practice
1. $5x^2 - 39 = 0$
   $a = 5,\ b = 0,\ c = -39$
2. $x = 0,\ x = -\dfrac{1}{2}$
3. $x = 7,\ x = 3$
4. $x = -8,\ x = 3$

Concept Check
Answers may vary. Possible solution:
Multiply each term by the LCD.
Simplify and move all terms to the left side of the equation.

Section 9.2

Vocabulary
1. completing the square
2. square root property

Student Practice
2. (a) $x = \pm 10$
   (b) $x = \pm 6\sqrt{2}$
   (c) $x = \pm 2\sqrt{5}$
4. $x = \pm 4$
6. $x = \dfrac{5 \pm 2\sqrt{5}}{2}$
8. $x = -2,\ x = 4$

Extra Practice
1. $x = \pm 8$
2. $x = 2 \pm \sqrt{15}$
3. $x = 4 \pm 3\sqrt{3}$
4. $x = \dfrac{5 \pm \sqrt{57}}{4}$

Concept Check
Answers may vary. Possible solution:
First, subtract 2 from both sides of the equation. Then divide all terms of the equation by 5. Then add the square of the quantity, −2 divided by 2, to both sides of the equation. Then add the square of quantity ten divided by two to both sides of the equation. $\left(\dfrac{10}{2}\right)^2 = 25$.

Lastly, combine like terms on both sides of the equation.

Section 9.3

Vocabulary
1. discriminant
2. quadratic formula

Student Practice
2. $x = 1,\ x = -\dfrac{5}{2}$
4. $x = \dfrac{1 \pm \sqrt{37}}{3}$
6. $x = -1.066,\ x = 2.816$
8. No real number solution

Extra Practice
1. $x = 1 \pm \sqrt{5}$
2. No real solutions
3. $x = 5$ or $x = -4$
4. No real solutions

Concept Check
Answers may vary. Possible solution:
Attempt to solve using the quadratic formula. If the resulting radicand is negative there are no real solutions.

Section 9.4

Vocabulary
1. vertex
2. parabola

Student Practice

2. (a)

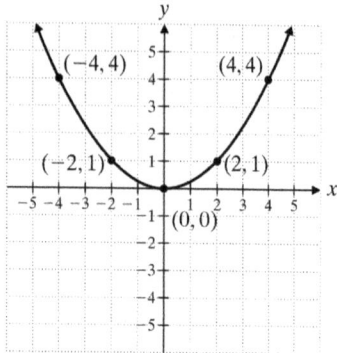

(b)

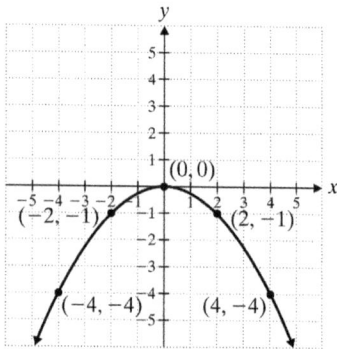

4. Vertex: $(-2,-4)$

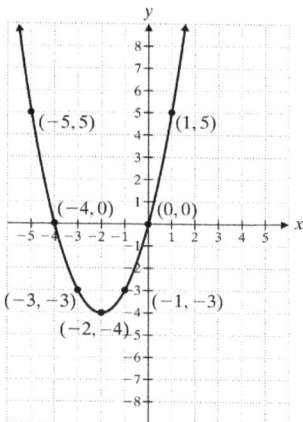

6. Vertex: $(3,-4)$

   x-intercepts: $(1,0), \ (5,0)$

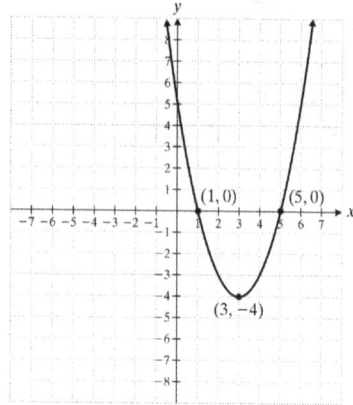

Extra Practice

1.

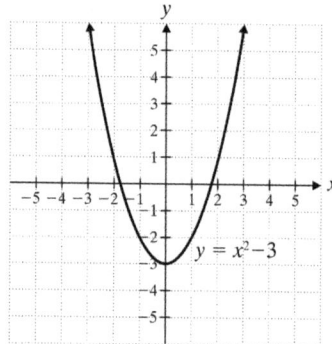

$y = x^2 - 3$

2.

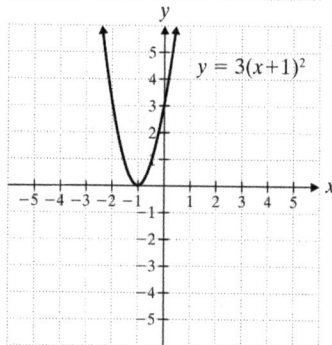

$y = 3(x+1)^2$

3.

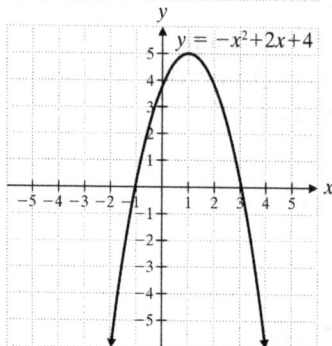

$y = -x^2 + 2x + 4$

4. Vertex: $\left(-\dfrac{3}{2}, \dfrac{7}{4}\right)$

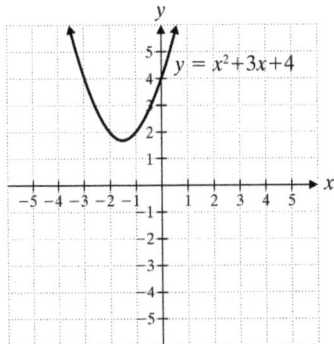

$y = x^2 + 3x + 4$

Two equations with two unknowns can be solved by substitution.

Concept Check
Answers may vary. Possible solution:
The x-coordinate of the vertex of the
equation $y = 4x^2 + 16x - 2$ is found by
using $x = \dfrac{-b}{2a}$ when the equation is in the
form $y = ax^2 + ax + c$. After the
x-coordinate has been found, it may be
substituted back into the original
equation to find the y-coordinate.

Section 9.5
Vocabulary
1. substitution
2. quadratic equation
Student Practice
2. 24 meters and 18 meters
4. 15 friends
6. Length $= 100$ feet
   Width $= 75$ feet
Extra Practice
1. 200 meters
2. 15 meters by 12 meters
3. 25 members
4. 40 feet by 45 feet or

   22.5 feet by 80 feet
Concept Check
Answers may vary. Possible solution:
Set $x =$ original number of people,
$d =$ original due per person.
$$x \cdot d = 750 \quad (1)$$
$$(x-4)(d+6) = 750 \quad (2)$$

GUIDED LEARNING
VIDEO WORKSHEETS

Pearson

Name: _____   Date: _____

Instructor: _____   Section: _____

Guided Learning Video Worksheet: Reducing Fractions to Lowest Terms

Text: *Beginning Algebra: Early Graphing*
Student Learning Objective 0.1.2: Simplify fractions to lowest terms using prime numbers.

Follow along with *Guided Learning Video 0.1.2, Reducing Fractions to Lowest Terms*.

Understanding the Big Picture: As you listen to the first part of the video, when the pencil icon appears, pause and fill in the blanks, choosing from the words listed.

factor • divide • equivalent • reduced

1. Fractions are said to be _____ when the values of the fractions are the same.

2. A fraction is _____ to lowest terms when its numerator and denominator have no common _____ other than one.

3. To reduce a fraction, _____ out any common factor in the numerator and denominator.

Follow along with the two Guided Learning Video examples, and fill in the blanks on the left as you learn with the instructor. When the pencil icon appears, pause and try the Student Practice on your own.

Guided Learning Video	Pause: Student Practice
1. a. Reduce the fraction $\dfrac{6}{14}$.	**2. a.** Reduce the fraction $\dfrac{12}{20}$.
b. Reduce the fraction $\dfrac{32}{40}$.	**b.** Reduce the fraction $\dfrac{30}{54}$.
Find the largest number that is a common factor of both the numerator and denominator. Then, divide the numerator and denominator by that number.	
a. $\dfrac{6 \div \boxed{}}{14 \div \boxed{}} = \dfrac{\boxed{}}{\boxed{}}$	
b. $\dfrac{32 \div \boxed{}}{40 \div \boxed{}} = \dfrac{\boxed{}}{\boxed{}}$	

Guided Learning Video ⏵ GUIDED LEARNING VIDEO	Pause: Student Practice
3. Simplify $\frac{30}{75}$ using the method of prime factors.	4. Simplify $\frac{42}{60}$ using the method of prime factors.

Rewrite the numerator and denominator as the product of prime factors. Then, cancel out common prime factors.

$$\frac{30}{75} = \frac{\square \times \square \times \square}{\square \times \square \times \square} = \frac{\square}{\square}$$

Helpful Hint: Pause the video and write the helpful hint in your own words.

For each Active Video Lesson, when the pencil icon appears, pause and work the problem. Then press play to check your work.

Active Video Lesson 1	Active Video Lesson 2
5. Reduce $\frac{28}{70}$ to lowest terms.	6. Reduce $\frac{24}{39}$ to lowest terms.

Guided Learning Video Worksheet: Changing Mixed Numbers to Improper Fractions

Text: *Beginning Algebra: Early Graphing*
Student Learning Objective 0.1.3: Convert between improper fractions and mixed numbers.

Follow along with *Guided Learning Video 0.1.3, Changing Mixed Numbers to Improper Fractions.*

Understanding the Big Picture: As you listen to the first part of the video, when the pencil icon appears, pause and fill in the blanks, choosing from the words listed.

improper • mixed • proper • one • zero

1. The value of a(n) _____ fraction is less than one.

2. When the value of a fraction's numerator is greater than or equal to its denominator, the fraction is called _____.

3. The sum of a whole number greater than _____ and a proper fraction is a _____ number.

Follow along with the two Guided Learning Video examples, and fill in the blanks on the left as you learn with the instructor. When the pencil icon appears, pause and try the Student Practice on your own.

Guided Learning Video ⏵ GUIDED LEARNING VIDEO	Pause: Student Practice ✏️
1. Change $9\frac{5}{6}$ to an improper fraction.	2. Change $4\frac{3}{5}$ to an improper fraction.
Use the procedure for changing a mixed number to an improper fraction to write the improper fraction. $9\frac{5}{6} = \dfrac{\boxed{} \times \boxed{} + \boxed{}}{\boxed{}} = \dfrac{\boxed{}}{\boxed{}}$	

3. Change $14\dfrac{7}{8}$ to an improper fraction.

   Use the procedure for changing a mixed number to an improper fraction to write the improper fraction.

   $$14\dfrac{7}{8} = \dfrac{\boxed{} \times \boxed{} + \boxed{}}{\boxed{}} = \dfrac{\boxed{}}{\boxed{}}$$

4. Change $17\dfrac{2}{9}$ to an improper fraction.

Helpful Hint: Pause the video and write the helpful hint in your own words.

For each Active Video Lesson, when the pencil icon appears, pause and work the problem. Then press play to check your work.

Active Video Lesson 1 ✏ | **Active Video Lesson 2** ✏

5. Change $19\dfrac{4}{7}$ to an improper fraction.

6. Change $8\dfrac{2}{7}$ to an improper fraction.

Name: _____   Date: _____

Instructor: _____   Section: _____

Guided Learning Video Worksheet: Change Improper Fractions to Mixed Numbers

Text: *Beginning Algebra: Early Graphing*
Student Learning Objective 0.1.3: Convert between improper fractions and mixed numbers.

Follow along with *Guided Learning Video 0.1.3, Change Improper Fractions to Mixed Numbers.*

Understanding the Big Picture: As you listen to the first part of the video, when the pencil icon appears, pause and fill in the blanks, choosing from the words listed.

quotient • denominator • divide • remainder • numerator

1. When changing an improper fraction to a mixed number, _____ the _____ by the _____.

2. The _____ is the whole number part of the mixed number.

3. The numerator of the fraction part of the mixed number is a _____.

Follow along with the two Guided Learning Video examples, and fill in the blanks on the left as you learn with the instructor. When the pencil icon appears, pause and try the Student Practice on your own.

Guided Learning Video ▶ GUIDED LEARNING VIDEO	Pause: Student Practice ✏️
1. Write $\dfrac{15}{4}$ as a mixed number.	2. Write $\dfrac{23}{6}$ as a mixed number.
Divide the numerator by the denominator and identify the remainder to write the mixed number.	

$$\Box\,\overline{)\Box}$$

$$\dfrac{15}{4} = \Box \dfrac{\Box}{\Box}$$

3. Write $\dfrac{61}{8}$ as a mixed number.

Divide the numerator by the denominator and identify the remainder to write the mixed number.

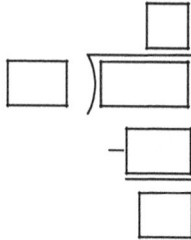

$$\boxed{}\,\overline{)\boxed{}^{\;\boxed{}}}$$
$$-\boxed{}$$
$$\boxed{}$$

$$\dfrac{61}{8} = \boxed{}\dfrac{\boxed{}}{\boxed{}}$$

4. Write $\dfrac{73}{9}$ as a mixed number.

Helpful Hint: Pause the video and write the helpful hint in your own words.

For each Active Video Lesson, when the pencil icon appears, pause and work the problem. Then press play to check your work.

Active Video Lesson 1 ✏️ | **Active Video Lesson 2** ✏️

5. Write $\dfrac{58}{9}$ as a mixed number.

6. Write $\dfrac{26}{3}$ as a mixed number.

Name: _____   Date: _____

Instructor: _____   Section: _____

Guided Learning Video Worksheet: Equivalent Fractions

Text: *Beginning Algebra: Early Graphing*
Student Learning Objective 0.1.4: Change a fraction to an equivalent fraction with a given denominator.

Follow along with *Guided Learning Video 0.1.4, Equivalent Fractions.*

Understanding the Big Picture: As you listen to the first part of the video, when the pencil icon appears, pause and fill in the blanks, choosing from the words listed.

value • building • comparing • equivalent

1. Fractions that look different but have the same value are _____ fractions.

2. Whenever a fraction is multiplied by a fraction that is a form of one, the _____ of the original fraction does not change.

3. The process of multiplying a fraction by a form of one to create a new, equivalent fraction is called _____ a fraction.

Follow along with the two Guided Learning Video examples, and fill in the blanks on the left as you learn with the instructor. When the pencil icon appears, pause and try the Student Practice on your own.

Guided Learning Video ▶ GUIDED LEARNING VIDEO	Pause: Student Practice
1. Create a fraction equivalent to $\frac{5}{9}$ with a denominator of 36.	2. Create a fraction equivalent to $\frac{3}{5}$ with a denominator of 40.
Determine the number that is multiplied times the old denominator to give the new denominator. Then, build the equivalent fraction.	
$\frac{5}{9} \times \dfrac{\boxed{}}{\boxed{}} = \dfrac{\boxed{}}{36}$	

7

3. Change $\dfrac{4}{7}$ to an equivalent fraction with a denominator of 56.

Determine the number that is multiplied times the old denominator to give the new denominator. Then, build the equivalent fraction.

$$\dfrac{4}{7} \times \dfrac{\boxed{}}{\boxed{}} = \dfrac{\boxed{}}{56}$$

4. Change $\dfrac{7}{8}$ to an equivalent fraction with a denominator of 72.

Helpful Hint: Pause the video and write the helpful hint in your own words.

For each Active Video Lesson, when the pencil icon appears, pause and work the problem. Then press play to check your work.

Active Video Lesson 1 ✏   **Active Video Lesson 2** ✏

5. Build an equivalent fraction with a denominator of 45 for the fraction $\dfrac{7}{15}$.

6. Build an equivalent fraction with a denominator of 60 for the fraction $\dfrac{11}{12}$.

Name: _____   Date: _____

Instructor: _____   Section: _____

Guided Learning Video Worksheet: Least Common Denominator

Text: *Beginning Algebra: Early Graphing*
Student Learning Objective 0.2.2: Use prime factors to find the least common denominator of two or more fractions.

Follow along with *Guided Learning Video 0.2.2, Least Common Denominator.*

Understanding the Big Picture: As you listen to the first part of the video, when the pencil icon appears, pause and fill in the blanks, choosing from the words listed.

one • prime • compare • greatest

1. A least common denominator is the smallest denominator that allows us to _____ fractions directly.

2. To form an LCD, write each denominator as the product of _____ factors.

3. The LCD is composed of each factor the _____ number of times it appears in any _____ denominator.

Follow along with the two Guided Learning Video examples, and fill in the blanks on the left as you learn with the instructor. When the pencil icon appears, pause and try the Student Practice on your own.

Guided Learning Video ▶ GUIDED LEARNING VIDEO	Pause: Student Practice
1. Find the LCD of $\frac{1}{12}$ and $\frac{5}{9}$.	2. Find the LCD of $\frac{12}{25}$ and $\frac{7}{15}$.

Write each denominator as the
product of primes, then form the
LCD.

$9 = \boxed{} \times \boxed{}$

$12 = \boxed{} \times \boxed{} \times \boxed{}$

$LCD = \boxed{} \times \boxed{} \times \boxed{} \times \boxed{} = \boxed{}$

9

3. Find the LCD of $\dfrac{10}{27}$ and $\dfrac{5}{18}$.

4. Find the LCD of $\dfrac{3}{8}$ and $\dfrac{7}{18}$.

Write each denominator as the product of primes, then form the LCD.

$27 = \boxed{} \times \boxed{} \times \boxed{}$

$18 = \boxed{} \times \boxed{} \times \boxed{}$

$\text{LCD} = \boxed{} \times \boxed{} \times \boxed{} \times \boxed{} = \boxed{}$

Helpful Hint: Pause the video and write the helpful hint in your own words.

For each Active Video Lesson, when the pencil icon appears, pause and work the problem. Then press play to check your work.

Active Video Lesson 1   **Active Video Lesson 2**

5. Find the LCD of $\dfrac{9}{10}$ and $\dfrac{7}{15}$.

6. Find the LCD of $\dfrac{11}{24}$ and $\dfrac{7}{9}$.

Name: _____   Date: _____

Instructor: _____   Section: _____

Guided Learning Video Worksheet: Subtract Mixed Numbers

Text: *Beginning Algebra: Early Graphing*
Student Learning Objective 0.2.4: Add or subtract mixed numbers.

Follow along with *Guided Learning Video 0.2.4, Subtract Mixed Numbers.*

Understanding the Big Picture: As you listen to the first part of the video, when the pencil icon appears, pause and fill in the blanks, choosing from the words listed.

whole numbers • fractions • least • equivalent

1. When adding or subtracting fractions that do not have a common denominator, using the _____ common denominator makes the work easier.

2. When adding or subtracting mixed numbers, add or subtract the _____ first, then the _____.

Follow along with the two Guided Learning Video examples, and fill in the blanks on the left as you learn with the instructor. When the pencil icon appears, pause and try the Student Practice on your own.

Guided Learning Video	Pause: Student Practice
1. Subtract. $8\dfrac{3}{5} - 3\dfrac{2}{3}$	2. Subtract. $6\dfrac{1}{5} - 3\dfrac{3}{4}$

Find the least common denominator and convert each fraction to an equivalent fraction. Subtract, borrowing if necessary, and express the final answer as a mixed number.

LCD = ☐

$$8\dfrac{3\times\square}{5\times\square} = 8\dfrac{\square}{\square} = \square\dfrac{\square}{\square}$$

$$-3\dfrac{2\times\square}{3\times\square} = -3\dfrac{\square}{\square} = -3\dfrac{\square}{\square}$$

$$\square\dfrac{\square}{\square} - \square\dfrac{\square}{\square} = \square\dfrac{\square}{\square}$$

3. Subtract. $4\frac{1}{2} - 2\frac{2}{3}$

Change the mixed numbers to improper fractions, and then find the least common denominator. Convert each fraction to an equivalent fraction and subtract. Express the final answer as a mixed number.

LCD = ☐

$4\dfrac{1}{2} = \dfrac{\boxed{} \times \boxed{}}{\boxed{} \times \boxed{}} = \dfrac{\boxed{}}{\boxed{}}$

$2\dfrac{2}{3} = \dfrac{\boxed{} \times \boxed{}}{\boxed{} \times \boxed{}} = \dfrac{\boxed{}}{\boxed{}}$

$\dfrac{\boxed{}}{\boxed{}} - \dfrac{\boxed{}}{\boxed{}} = \dfrac{\boxed{}}{\boxed{}} = \boxed{}\,\dfrac{\boxed{}}{\boxed{}}$

4. Subtract. $7\frac{1}{3} - 1\frac{3}{5}$

Helpful Hint: Pause the video and write the helpful hint in your own words.

For each Active Video Lesson, when the pencil icon appears, pause and work the problem. Then press play to check your work.

Active Video Lesson 1   **Active Video Lesson 2**

5. Subtract. $3\frac{2}{5} - 1\frac{3}{4}$

6. Subtract. $7\frac{1}{4} - 5\frac{1}{2}$

Name: _____   Date: _____

Instructor: _____ .   Section: _____

Guided Learning Video Worksheet: Multiply Fractions

Text: *Beginning Algebra: Early Graphing*
Student Learning Objective 0.3.1: Multiply fractions, whole numbers, and mixed numbers.

Follow along with *Guided Learning Video 0.3.1, Multiply Fractions.*

Understanding the Big Picture: As you listen to the first part of the video, when the pencil icon appears, pause and fill in the blanks, choosing from the words listed.

division • denominators • numerators • multiplication • terms

1. The word "of" indicates the operation of _____.

2. To multiply fractions, multiply numerators times _____.

3. Multiplication of fractions also requires _____ to be multiplied.

Follow along with the two Guided Learning Video examples, and fill in the blanks on the left as you learn with the instructor. When the pencil icon appears, pause and try the Student Practice on your own.

Guided Learning Video (▶) GUIDED LEARNING VIDEO	**Pause: Student Practice**
1. Multiply $\dfrac{4}{7} \times \dfrac{2}{9}$.	**2.** Multiply $\dfrac{5}{9} \times \dfrac{2}{11}$.
Multiply the numerators, and then multiply the denominators.	
$\dfrac{4}{7} \times \dfrac{2}{9} = \dfrac{\boxed{} \times \boxed{}}{\boxed{} \times \boxed{}} = \dfrac{\boxed{}}{\boxed{}}$	

3. Simplify first, then multiply $\dfrac{7}{15} \times \dfrac{9}{14}$.

Write the product as a single fraction. Find the prime factors of the values in the numerator and denominator, and factor out the common factors. Then, multiply the remaining factors.

$$\frac{7}{15} \times \frac{9}{14} = \frac{\boxed{} \times \boxed{}}{\boxed{} \times \boxed{}}$$

$$= \frac{\boxed{} \times \boxed{} \times \boxed{}}{\boxed{} \times \boxed{} \times \boxed{} \times \boxed{}}$$

$$= \frac{\boxed{} \times \boxed{} \times \boxed{}}{\boxed{} \times \boxed{} \times \boxed{} \times \boxed{}}$$

$$\frac{\boxed{}}{\boxed{}}$$

4. Simplify first, then multiply $\dfrac{5}{28} \times \dfrac{21}{25}$.

Helpful Hint: Pause the video and write the helpful hint in your own words.

For each Active Video Lesson, when the pencil icon appears, pause and work the problem. Then press play to check your work.

Active Video Lesson 1 ✏ | **Active Video Lesson 2** ✏

5. Simplify first, and then multiply $\dfrac{15}{33} \times \dfrac{22}{35}$.

6. Simplify first, and then multiply $\dfrac{9}{14} \times \dfrac{7}{12}$.

Guided Learning Video Worksheet: Multiplying Mixed Numbers

Text: *Beginning Algebra: Early Graphing*
Student Learning Objective 0.3.1: Multiply fractions, whole numbers, and mixed numbers.

Follow along with *Guided Learning Video 0.3.1, Multiplying Mixed Numbers*.

Understanding the Big Picture: As you listen to the first part of the video, when the pencil icon appears, pause and fill in the blanks, choosing from the words listed.

mixed number • improper • divide

1. To multiply two mixed numbers, first transform the mixed numbers into _____ fractions.

2. Whenever it's possible, _____ out common factors before multiplying.

3. When multiplying mixed numbers, a final step is to change the answer to a _____ if possible.

Follow along with the two Guided Learning Video examples, and fill in the blanks on the left as you learn with the instructor. When the pencil icon appears, pause and try the Student Practice on your own.

Guided Learning Video GUIDED LEARNING VIDEO	**Pause: Student Practice**
1. Multiply $2\frac{3}{4}\times 5\frac{1}{3}$.	2. Multiply $4\frac{2}{3}\times 3\frac{3}{5}$.
Change the mixed numbers to improper fractions and multiply. Be sure to divide out common factors before multiplying. $$2\frac{3}{4}\times 5\frac{1}{3}=\frac{\Box}{\Box}\times\frac{\Box}{\Box}$$ $$=\frac{\Box}{\Box}\times\frac{\Box}{\Box}$$ $$=\frac{\Box}{\Box}=\Box\frac{\Box}{\Box}$$	

3. Multiply $3\frac{5}{12}\times 8$.

Change the mixed number to an improper fraction and the whole number to a fraction with a denominator of one. Then, multiply. Be sure to divide out common factors before multiplying.

$$3\frac{5}{12}\times 8 = \frac{\square}{\square}\times\frac{\square}{\square}$$

$$= \frac{\square}{\square}\times\frac{\square}{\square}$$

$$= \frac{\square}{\square} = \square\frac{\square}{\square}$$

4. Multiply $5\frac{2}{9}\times 6$.

Helpful Hint: Pause the video and write the helpful hint in your own words.

For each Active Video Lesson, when the pencil icon appears, pause and work the problem. Then press play to check your work.

Active Video Lesson 1 ✏ | **Active Video Lesson 2** ✏

5. Find the area, in square inches, of the rectangle with a length of $4\frac{2}{3}$ inches and a width of $2\frac{1}{2}$ inches.

Remember that Area = Length × Width.

6. Multiply $2\frac{5}{6}\times 3$.

Name: _____ Date: _____
Instructor: _____ Section: _____

Guided Learning Video Worksheet: Dividing Fractions

Text: *Beginning Algebra: Early Graphing*
Student Learning Objective 0.3.2: Divide fractions, whole numbers, and mixed numbers.

Follow along with *Guided Learning Video 0.3.2, Dividing Fractions.*

Understanding the Big Picture: As you listen to the first part of the video, when the pencil icon appears, pause and fill in the blanks, choosing from the words listed.

second • inverted • reciprocal • first

1. When a fraction is multiplied times its _____, the result is always one.

2. When a fraction is _____, its numerator and denominator are interchanged.

3. To divide two fractions, invert the _____ fraction and multiply.

Follow along with the two Guided Learning Video examples, and fill in the blanks on the left as you learn with the instructor. When the pencil icon appears, pause and try the Student Practice on your own.

Guided Learning Video ▶ GUIDED LEARNING VIDEO	Pause: Student Practice
1. Divide the following. $\dfrac{7}{10} \div \dfrac{8}{9}$ Rewrite the expression as multiplication by the reciprocal. Then, calculate the answer. $\dfrac{7}{10} \div \dfrac{8}{9} = \dfrac{\boxed{}}{\boxed{}}\boxed{}\dfrac{\boxed{}}{\boxed{}}$ $= \dfrac{\boxed{}}{\boxed{}}$	2. Divide the following. $\dfrac{4}{11} \div \dfrac{5}{7}$

Guided Learning Video ▶ GUIDED LEARNING VIDEO	Pause: Student Practice
3. Evaluate the following. $\dfrac{5}{9} \div \dfrac{2}{3}$	4. Evaluate the following. $\dfrac{8}{13} \div \dfrac{4}{5}$

3. Rewrite the expression as multiplication by the reciprocal. Then, calculate the answer. Be sure to divide out common factors before multiplying.

$$\frac{5}{9} \div \frac{2}{3} = \frac{\square}{\square} \cdot \frac{\square}{\square}$$

$$= \frac{\square}{\square} \cdot \frac{\square}{\square}$$

$$= \frac{\square}{\square}$$

Helpful Hint: Pause the video and write the helpful hint in your own words.

For each Active Video Lesson, when the pencil icon appears, pause and work the problem. Then press play to check your work.

Active Video Lesson 1	Active Video Lesson 2
5. Simplify. $\dfrac{3}{7} \div \dfrac{12}{21}$	6. Divide the following fractions. $\dfrac{5}{8} \div \dfrac{25}{36}$

Name: _____   Date: _____

Instructor: _____   Section: _____

Guided Learning Video Worksheet: Divide Mixed Numbers

Text: *Beginning Algebra: Early Graphing*
Student Learning Objective 0.3.2: Divide fractions, whole numbers, and mixed numbers.

Follow along with *Guided Learning Video 0.3.2, Divide Mixed Numbers.*

Understanding the Big Picture: As you listen to the first part of the video, when the pencil icon appears, pause and fill in the blanks, choosing from the words listed.

addition • multiplication • improper

1. If mixed numbers are being divided, they should be converted to _____ fractions before dividing.

2. After inverting the second fraction, change the division symbol to a _____ symbol.

Follow along with the two Guided Learning Video examples, and fill in the blanks on the left as you learn with the instructor. When the pencil icon appears, pause and try the Student Practice on your own.

Guided Learning Video ▶ GUIDED LEARNING VIDEO	Pause: Student Practice
1. Divide: $4\dfrac{1}{2} \div 3\dfrac{3}{8}$	2. Divide: $3\dfrac{3}{4} \div 5\dfrac{1}{2}$

Convert the mixed numbers to improper fractions. Then follow the rules for dividing fractions to calculate the answer. If necessary write the final answer as a mixed number.

$$4\frac{1}{20} \div 3\frac{3}{8} = \frac{\square}{\square} \div \frac{\square}{\square}$$

$$= \frac{\square}{\square} \cdot \frac{\square}{\square}$$

$$= \frac{\square}{\square} = \frac{\square}{\square}$$

$$= \square\frac{\square}{\square}$$

Guided Learning Video ▶ GUIDED LEARNING VIDEO	Pause: Student Practice
3. Divide: $5\frac{1}{3} \div 8$ Convert the mixed number to an improper fraction and the whole number to a fraction. Then follow the rules for dividing fractions to calculate the answer. Be sure to divide out common factors before multiplying. $5\frac{1}{3} \div 8 = \dfrac{\square}{\square} \div \dfrac{\square}{\square}$ $= \dfrac{\square}{\square} \cdot \dfrac{\square}{\square}$ $= \dfrac{\square}{\square} \cdot \dfrac{\square}{\square} = \dfrac{\square}{\square}$	**4.** Divide: $7\frac{3}{5} \div 19$

Helpful Hint: Pause the video and write the helpful hint in your own words.

For each Active Video Lesson, when the pencil icon appears, pause and work the problem. Then press play to check your work.

Active Video Lesson 1	Active Video Lesson 2
5. Divide: $7\frac{1}{2} \div 5$	**6.** Divide: $5\frac{1}{2} \div 6\frac{1}{4}$

Guided Learning Video Worksheet: Change a Fraction with a Denominator That Is a Power of 10 to a Decimal

Text: *Beginning Algebra: Early Graphing*

Student Learning Objective 0.4.2: Change a fraction to a decimal.

Follow along with *Guided Learning Video 0.4.2, Change a Fraction with a Denominator that is a Power of 10 to a Decimal.*

Understanding the Big Picture: As you listen to the first part of the video, when the pencil icon appears, pause and fill in the blanks, choosing from the words listed.

delete • equal • decimal • numerator • zeros

1. Fractions with denominators that are powers of ten are called _____ fractions.

2. When changing fractions with denominators that are powers of ten to decimals, move the decimal point in the _____ a number of places _____ to the number of _____ in the denominator.

3. After moving the decimal point in the numerator, _____ the denominator.

Follow along with the two Guided Learning Video examples, and fill in the blanks on the left as you learn with the instructor. When the pencil icon appears, pause and try the Student Practice on your own.

Guided Learning Video ⏵ GUIDED LEARNING VIDEO	Pause: Student Practice
1. Change $6\dfrac{59}{100}$ to a decimal.	2. Change $4\dfrac{6}{100}$ to a decimal.
Write the whole number part, and count the number of zeros in the denominator. Then, shift the decimal point in the numerator the necessary number of places.	
$6\dfrac{59}{100} = \boxed{}$	

Guided Learning Video (▶) GUIDED LEARNING VIDEO	**Pause: Student Practice** ✏️
3. Change $\dfrac{78}{1000}$ to a decimal.	**4.** Change $\dfrac{45}{1000}$ to a decimal.

Count the number of zeros in the denominator. Then, shift the decimal point in the numerator the necessary number of places.

$$\frac{78}{1000} = \boxed{}$$

Helpful Hint: Pause the video and write the helpful hint in your own words.

For each Active Video Lesson, when the pencil icon appears, pause and work the problem. Then press play to check your work.

Active Video Lesson 1 ✏️	**Active Video Lesson 2** ✏️
5. Change $8\dfrac{3}{1000}$ to a decimal.	**6.** Change $\dfrac{7}{100}$ to a decimal.

Name: _____   Date: _____

Instructor: _____   Section: _____

Guided Learning Video Worksheet: Convert a Fraction to a Decimal

Text: *Beginning Algebra: Early Graphing*
Student Learning Objective 0.4.2: Change a fraction to a decimal.

Follow along with *Guided Learning Video 0.4.2, Convert a Fraction to a Decimal.*

Understanding the Big Picture: As you listen to the first part of the video, when the pencil icon appears, pause and fill in the blanks, choosing from the words listed.

denominator • terminating • repeating • numerator • repeat

1. To convert a fraction to a decimal, divide the _____ by the _____.

2. When converting fractions to decimals, if the division yields a remainder of zero, the decimal is called a _____ decimal.

3. Decimals having a digit or group of digits that _____ are called _____ decimals.

Follow along with the two Guided Learning Video examples, and fill in the blanks on the left as you learn with the instructor. When the pencil icon appears, pause and try the Student Practice on your own.

Guided Learning Video ▶ GUIDED LEARNING VIDEO	**Pause: Student Practice** ✏️
1. Write $\frac{7}{8}$ as an equivalent decimal.	2. Write $\frac{3}{8}$ as an equivalent decimal.
Perform the division operation until a remainder of zero occurs or the remainder repeats itself.	

3. Write $\dfrac{4}{9}$ as an equivalent decimal.

Write the expression in long division form. Then, follow the procedure for dividing a decimal by a decimal to calculate the answer.

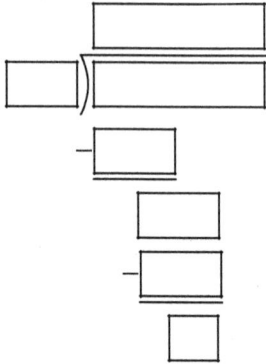

4. Write $\dfrac{5}{6}$ as an equivalent decimal.

Helpful Hint: Pause the video and write the helpful hint in your own words.

For each Active Video Lesson, when the pencil icon appears, pause and work the problem. Then press play to check your work.

Active Video Lesson 1 ✏ | **Active Video Lesson 2** ✏

5. Write $\dfrac{5}{12}$ as an equivalent decimal.

6. Write $\dfrac{43}{80}$ as an equivalent decimal.

Name: _____   Date: _____

Instructor: _____   Section: _____

Guided Learning Video Worksheet: Changing a Decimal to a Fraction

Text: *Beginning Algebra: Early Graphing*
Student Learning Objective 0.4.3: Change a decimal to a fraction.

Follow along with *Guided Learning Video 0.4.3, Changing a Decimal to a Fraction.*

Understanding the Big Picture: As you listen to the first part of the video, when the pencil icon appears, pause and fill in the blanks, choosing from the words listed.

denominator • last • point • numerator • and • first • whole

1. Numbers written in decimal notation are composed of three parts; the _____ number part, the decimal part, and the decimal _____.

2. The word _____ represents the decimal _____.

3. When changing a decimal number to a fraction, the fraction's denominator is the same as the _____ place value used in the decimal part of the decimal number.

4. The decimal part of the decimal number becomes the _____ of the fraction to which the decimal number is changed.

Follow along with the two Guided Learning Video examples, and fill in the blanks on the left as you learn with the instructor. When the pencil icon appears, pause and try the Student Practice on your own.

Guided Learning Video ⏵ GUIDED LEARNING VIDEO	Pause: Student Practice ✏
1. Write 6.317 as a fraction.	2. Write 4.629 as a fraction.
To write the fraction, identify the whole number part, the decimal part, and the last place value used in the decimal part.	
$6.317 = \boxed{}\dfrac{\boxed{}}{\boxed{}}$	

Guided Learning Video ▶ GUIDED LEARNING VIDEO	**Pause: Student Practice** ✏
3. Write 0.75 as a fraction.	**4.** Write 0.32 as a fraction.
To write the fraction, identify the whole number part, the decimal part, and the last place value used in the decimal part. Remember to reduce the fraction if possible.	

$$0.75 = \frac{\boxed{}}{\boxed{}}$$

Helpful Hint: Pause the video and write the helpful hint in your own words.

For each Active Video Lesson, when the pencil icon appears, pause and work the problem. Then press play to check your work.

Active Video Lesson 1 ✏	**Active Video Lesson 2** ✏
5. Convert 0.135 to a fraction. Remember to reduce the fraction if possible.	**6.** Change 4.29 to a fraction.

Name: _____ Date: _____
Instructor: _____ Section: _____

Guided Learning Video Worksheet: Adding Decimals

Text: *Beginning Algebra: Early Graphing*
Student Learning Objective 0.4.4: Add and subtract decimals.

Follow along with *Guided Learning Video 0.4.4, Adding Decimals*.

Understanding the Big Picture: As you listen to the first part of the video, when the pencil icon appears, pause and fill in the blanks, choosing from the words listed.

difference • sum • lining • fractions • same

1. The addition of decimals can be related to the addition of _____.

2. The procedure for adding decimals begins with writing the numbers vertically and _____ up the decimal points.

3. Add all digits with the _____ place value moving from the right column to the left, then place the decimal point in the _____ in line with the decimal points of the numbers added.

Follow along with the two Guided Learning Video examples, and fill in the blanks on the left as you learn with the instructor. When the pencil icon appears, pause and try the Student Practice on your own.

Guided Learning Video	Pause: Student Practice
1. Add $1.2 + 3.7 + 6.9$. Align the numbers vertically, lining up the decimal points. Then, add the numbers.	2. Add $4.3 + 1.6 + 7.4$.

27

Guided Learning Video ▶ GUIDED LEARNING VIDEO	Pause: Student Practice
3. Add $8.73 + 25.1 + 0.679$.	**4.** Add $3.25 + 46.8 + 0.356$.

Align the numbers vertically, lining up the decimal points. Fill in any empty place values with zeros. Then, add the numbers.

```
  ┌──────────┐
  │          │
  ├──────────┤
  │          │
+ ├──────────┤
  │          │
  └──────────┘
  ┌──────────┐
  │          │
  └──────────┘
```

Helpful Hint: Pause the video and write the helpful hint in your own words.

For each Active Video Lesson, when the pencil icon appears, pause and work the problem. Then press play to check your work.

Active Video Lesson 1	Active Video Lesson 2
5. Add $0.528 + 36.9 + 7.41$.	**6.** Victoria has $739.52 in her savings account. If she deposits $214.86 into the account, how much does she have in savings?

Name: _____   Date: _____

Instructor: _____   Section: _____

Guided Learning Video Worksheet: Subtracting Decimals

Text: *Beginning Algebra: Early Graphing*
Student Learning Objective 0.4.4: Add and subtract decimals.

Follow along with *Guided Learning Video 0.4.4, Subtracting Decimals.*

Understanding the Big Picture: As you listen to the first part of the video, when the pencil icon appears, pause and fill in the blanks, choosing from the words listed.

right • left • difference • borrow • zeros • same

1. Just like subtraction of mixed numbers with the same denominators, when subtracting decimals, we must sometimes _____ from the whole number.

2. When subtracting decimals, additional _____ may be placed to the _____ of the decimal point if not all numbers have the _____ number of decimal places.

3. Place the decimal point in the _____ so it is in line with the decimal points of the two numbers being subtracted.

Follow along with the two Guided Learning Video examples, and fill in the blanks on the left as you learn with the instructor. When the pencil icon appears, pause and try the Student Practice on your own.

Guided Learning Video ▶ GUIDED LEARNING VIDEO	Pause: Student Practice
1. a. Subtract $26.9 - 15.3$. **b.** Subtract $24.39 - 5.8$. For both examples, align the numbers vertically, lining up the decimal points and adding zeros where needed. Then, subtract the numbers, borrowing when necessary. a. ☐ − ☐ ☐ b. ☐ − ☐ ☐	**2. a.** Subtract $47.5 - 26.3$. **b.** Subtract $61.27 - 9.3$.

3. Subtract $17 - 8.769$.

Align the numbers vertically, lining up the decimal points and adding zeros where needed. Then, subtract the numbers, borrowing when necessary.

```
 ┌──────┐
 │      │
 └──────┘
 ┌──────┐
−│      │
 └──────┘
 ┌──────────┐
 │          │
 └──────────┘
```

4. Subtract $23 - 6.592$.

Helpful Hint: Pause the video and write the helpful hint in your own words.

For each Active Video Lesson, when the pencil icon appears, pause and work the problem. Then press play to check your work.

Active Video Lesson 1   **Active Video Lesson 2**

5. Subtract $26 - 11.045$.

6. Subtract $96.35 - 48.29$.

Guided Learning Video Worksheet: Multiplying Decimals

Text: *Beginning Algebra: Early Graphing*
Student Learning Objective 0.4.5: Multiply decimals.

Follow along with *Guided Learning Video 0.4.5, Multiplying Decimals.*

Understanding the Big Picture: As you listen to the first part of the video, when the pencil icon appears, pause and fill in the blanks, choosing from the words listed.

left • total • equal • zeros • whole

1. To multiply decimal numbers, multiply the numbers as if they were _____ numbers.

2. To position the decimal point in the product of decimal numbers, the _____ number of decimal places in the factors must be determined.

3. The number of decimal places in the product of decimal numbers must _____ the total number of decimal places in the factors.

4. Sometimes, _____ need to be inserted to the _____ of the product of decimal numbers.

Follow along with the two Guided Learning Video examples, and fill in the blanks on the left as you learn with the instructor. When the pencil icon appears, pause and try the Student Practice on your own.

Guided Learning Video ▶ GUIDED LEARNING VIDEO	**Pause: Student Practice** ✏️
1. Multiply 0.07×0.8. Write the factors in a vertical format. Then, multiply and position the decimal point in the answer. ☐ × ☐ ☐	**2.** Multiply 0.06×0.4.

3. Multiply 4.679×53.

Write the factors in a vertical format. Then, multiply and position the decimal point in the answer.

×

4. Multiply 3.275×46.

Helpful Hint: Pause the video and write the helpful hint in your own words.

For each Active Video Lesson, when the pencil icon appears, pause and work the problem. Then press play to check your work.

Active Video Lesson 1 ✏

5. Multiply 0.9×0.04.

Active Video Lesson 2 ✏

6. Multiply 0.758×96.

Guided Learning Video Worksheet: Dividing Decimals

Text: *Beginning Algebra: Early Graphing*
Student Learning Objective 0.4.6: Divide decimals.

Follow along with *Guided Learning Video 0.4.6, Dividing Decimals.*

Understanding the Big Picture: As you listen to the first part of the video, when the pencil icon appears, pause and fill in the blanks, choosing from the words listed.

left • quotient • right • divisor • dividend

1. When dividing a decimal by a decimal, make the _____ a whole number by moving the decimal point to the _____.

2. Move the decimal point in the _____ the same number of places it was moved in the _____.

3. Position the decimal point in the _____ directly above its position in the _____.

Follow along with the two Guided Learning Video examples, and fill in the blanks on the left as you learn with the instructor. When the pencil icon appears, pause and try the Student Practice on your own.

Guided Learning Video ⏵ GUIDED LEARNING VIDEO	Pause: Student Practice
1. Divide: $0.537 \div 0.3$ Write the expression in long division form. Then, follow the procedure for dividing a decimal by a decimal to calculate the answer.	**2.** Divide: $0.918 \div 0.6$

Guided Learning Video	**Pause: Student Practice** ✏
3. Divide: $0.756 \div 0.24$	**4.** Divide: $1.272 \div 0.48$

Write the expression in long division form. Then, follow the procedure for dividing a decimal by a decimal to calculate the answer.

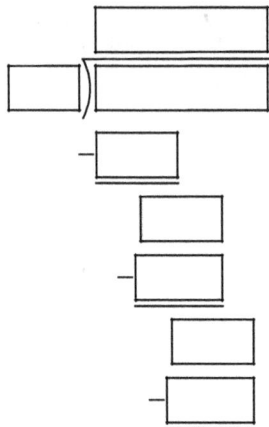

Helpful Hint: Pause the video and write the helpful hint in your own words.

For each Active Video Lesson, when the pencil icon appears, pause and work the problem. Then press play to check your work.

Active Video Lesson 1 ✏	**Active Video Lesson 2** ✏
5. Divide: $8.93 \div 3.8$	**6.** Divide: $15.75 \div 3.5$

Guided Learning Video Worksheet: Multiplying Decimals by a Power of Ten

Text: *Beginning Algebra: Early Graphing*
Student Learning Objective 0.4.7: Multiply and divide a decimal by a multiple of 10.

Follow along with *Guided Learning Video 0.4.7, Multiplying Decimals by a Power of Ten.*

Understanding the Big Picture: As you listen to the first part of the video, when the pencil icon appears, pause and fill in the blanks, choosing from the words listed.

left • right • multiplication • zeros • places • less

1. The operation of _____ is used to write a number with digits only when its value is expressed in word form.

2. When multiplying by a power of ten, the number of _____ in the power of ten is the number of _____ the decimal point is moved to the _____.

3. Zeros are sometimes needed when the number of digits to the right of the decimal point is _____ than the number of zeros in the power of ten.

Follow along with the two Guided Learning Video examples, and fill in the blanks on the left as you learn with the instructor. When the pencil icon appears, pause and try the Student Practice on your own.

Guided Learning Video	Pause: Student Practice
1. Multiply 81.74×10^5. To multiply, move the decimal point the same number of places as there are zeros in the power of ten. If necessary, add zeros. $81.74 \times 10^5 = \boxed{}$	2. Multiply 66.73×10^4.

Guided Learning Video	Pause: Student Practice
3. Multiply $9.16 \times 10{,}000$. To multiply, move the decimal point the same number of places as there are zeros in the power of ten. If necessary, add zeros. $9.16 \times 10{,}000 = \boxed{}$	**4.** Multiply 4.95×1000.

Helpful Hint: Pause the video and write the helpful hint in your own words.

For each Active Video Lesson, when the pencil icon appears, pause and work the problem. Then press play to check your work.

Active Video Lesson 1	Active Video Lesson 2
5. Multiply 1.7×10^{3}.	**6.** Change $62.8\,\text{km}$ to m. $1\,\text{km} = 1000\,\text{m}$.

Guided Learning Video Worksheet: Changing a Decimal to a Percent

Text: *Beginning Algebra: Early Graphing*
Student Learning Objective 0.5.1: Change a decimal to a percent.

Follow along with *Guided Learning Video 0.5.1, Changing a Decimal to a Percent.*

Understanding the Big Picture: As you listen to the first part of the video, when the pencil icon appears, pause and fill in the blanks, choosing from the words listed.

parts • ratios • numerators • percent • denominators

1. The word _____ means per hundred.

2. Percents can be described as _____ with _____ of one hundred.

3. The percent symbol means _____ per one hundred.

Follow along with the two Guided Learning Video examples, and fill in the blanks on the left as you learn with the instructor. When the pencil icon appears, pause and try the Student Practice on your own.

Guided Learning Video	Pause: Student Practice
1. Change 0.726 to a percent. Move the decimal point the necessary number of places, and write the percent symbol at the end of the number. $0.726 \Rightarrow \boxed{}$	2. Change 0.539 to a percent.

Guided Learning Video ▶ GUIDED LEARNING VIDEO	**Pause: Student Practice** ✏
3. Change 5.8 to a percent.	**4.** Change 3.1 to a percent.
Move the decimal point the necessary number of places, and write the percent symbol at the end of the number.	
$5.8 \Rightarrow$ ☐	

Helpful Hint: Pause the video and write the helpful hint in your own words.

For each Active Video Lesson, when the pencil icon appears, pause and work the problem. Then press play to check your work.

Active Video Lesson 1 ✏	**Active Video Lesson 2** ✏
5. Convert 2.9 to a percent.	**6.** Convert 0.165 to a percent.

Guided Learning Video Worksheet: Solve Percent Equations

Text: *Beginning Algebra: Early Graphing*
Student Learning Objective 0.5.3: Find the percent of a given number.

Follow along with *Guided Learning Video 0.5.3, Solve Percent Equations.*

Understanding the Big Picture: As you listen to the first part of the video, when the pencil icon appears, pause and fill in the blanks, choosing from the words listed.

multiplication • base • decimal • percent • amount

1. The percent equation is _____ equals _____ times
    _____.

2. When translating percent statements into equations, the word "of" means
    _____.

3. Before performing any calculations with percents, change the _____ to an
    equivalent _____.

Follow along with the two Guided Learning Video examples, and fill in the blanks on the left as you learn with the instructor. When the pencil icon appears, pause and try the Student Practice on your own.

Guided Learning Video ▶ GUIDED LEARNING VIDEO	**Pause: Student Practice**
1. What is 32% of 260 ? Use the procedure for solving applied percent problems to find the answer. ☐ = ☐ × ☐ ☐ = ☐ × ☐ ☐ = ☐	2. What is 45% of 730 ?

Guided Learning Video ▶ GUIDED LEARNING VIDEO	Pause: Student Practice
3. 76% of college freshman receive financial aid. If 1862 freshman receive financial aid, how many freshmen are there in total? Use the procedure for solving applied percent problems to find the answer.	**4.** 62% of the total number of ballots mailed to voting members of a union are returned and considered valid. If 3565 ballots were returned and considered valid, what was the total number of ballots mailed?

☐ = ☐ × ☐

☐ = ☐ × ☐

$\dfrac{☐}{☐} = \dfrac{☐}{☐}$

☐ = ☐

Helpful Hint: Pause the video and write the helpful hint in your own words.

For each Active Video Lesson, when the pencil icon appears, pause and work the problem. Then press play to check your work.

Active Video Lesson 1	Active Video Lesson 2
5. 15% of the students at West Lake College are currently taking math courses. There are 3400 students at West Lake College. How many students are currently taking math courses?	**6.** 12% of the residents of Clinton live in apartment buildings. 1524 of the residents of Clinton live in apartment buildings. How many people live in Clinton?

Name: _____   Date: _____

Instructor: _____   Section: _____

Guided Learning Video Worksheet: Solve for a Missing Percent

Text: *Beginning Algebra: Early Graphing*
Student Learning Objective 0.5.4: Find the missing percent when given two numbers.

Follow along with *Guided Learning Video 0.5.4, Solve for a Missing Percent.*

Understanding the Big Picture: As you listen to the first part of the video, when the pencil icon appears, pause and fill in the blanks, choosing from the words listed.

right • base • divide • two • percent • amount

1. The percent equation is _____ equals _____ times
   _____.

2. When solving the percent equation for the percent quantity, _____ the
   _____ by the _____.

3. To convert a decimal to a percent, shift the decimal point _____ places to the
   _____.

Follow along with the two Guided Learning Video examples, and fill in the blanks on the left as you learn with the instructor. When the pencil icon appears, pause and try the Student Practice on your own.

Guided Learning Video ▶ GUIDED LEARNING VIDEO	Pause: Student Practice ✏
1. 75 is what percent of 300? Use the procedure for solving applied percent problems to find the answer. $\boxed{} = \boxed{} \times \boxed{}$ $\dfrac{\boxed{}}{\boxed{}} = \dfrac{\boxed{}}{\boxed{}}$ $\boxed{} = \boxed{}$	2. 21 is what percent of 35?

Guided Learning Video ▶ GUIDED LEARNING VIDEO	Pause: Student Practice ✏
3. Adelina purchases a new laptop for $950 and paid $76 in sales tax. What percent is the sales tax in her area?	**4.** A new set of golf clubs costs $1250 plus $75 in sales tax. What percent is the sales tax rate?

3. Adelina purchases a new laptop for $950 and paid $76 in sales tax. What percent is the sales tax in her area?

Use the procedure for solving applied percent problems to find the answer.

$$\boxed{} = \boxed{} \times \boxed{}$$

$$\frac{\boxed{}}{\boxed{}} = \frac{\boxed{}}{\boxed{}}$$

$$\boxed{} = \boxed{}$$

Helpful Hint: Pause the video and write the helpful hint in your own words.

For each Active Video Lesson, when the pencil icon appears, pause and work the problem. Then press play to check your work.

Active Video Lesson 1 ✏	Active Video Lesson 2 ✏
5. Tammy correctly answered 28 problems out of 35 problems on her Chemistry quiz. What percent did she get correct?	**6.** What percent of 75 is 30?

Guided Learning Video Worksheet: Rounding Decimals

Text: *Beginning Algebra: Early Graphing*
Student Learning Objective 0.5.5: Use rounding to estimate.

Follow along with *Guided Learning Video 0.5.5, Rounding Decimals.*

Understanding the Big Picture: As you listen to the first part of the video, when the pencil icon appears, pause and fill in the blanks, choosing from the words listed.

decrease • less • increase • place • one • number • greater

1. To round a decimal, first identify the given _____ value to which rounding is required.

2. When the digit to the right of the given place value is _____ than five, drop it and all the digits to its right.

3. If the first digit to the right of the given place value is five or more, _____ the number in the given place value by _____.

Follow along with the two Guided Learning Video examples, and fill in the blanks on the left as you learn with the instructor. When the pencil icon appears, pause and try the Student Practice on your own.

Guided Learning Video ▶ GUIDED LEARNING VIDEO	Pause: Student Practice
1. Round 14.259 to the nearest hundredth. Follow the rules for rounding decimals to round this number to the nearest hundredth. The number in the given place value is ☐ , and the first number to its right is ☐ . 14.259 rounded to the nearest hundredth is ☐ .	2. Round 63.536 to the nearest hundredth.

Guided Learning Video ▶ GUIDED LEARNING VIDEO	Pause: Student Practice ✏️
3. Round 0.70328 to the nearest thousandth.	**4.** Round 0.09545 to the nearest thousandth.

Follow the rules for rounding decimals to round this number to the nearest thousandth.

The number in the given place value is ☐, and the first number to its right is ☐.

0.70328 rounded to the nearest thousandth is ☐.

Helpful Hint: Pause the video and write the helpful hint in your own words.

For each Active Video Lesson, when the pencil icon appears, pause and work the problem. Then press play to check your work.

Active Video Lesson 1 ✏️	Active Video Lesson 2 ✏️
5. A student purchases a single cup coffee brewer. The sales tax is calculated to be $9.8736. Round this sales tax amount to the nearest cent.	**6.** Round 3.195 to the nearest tenth.

Guided Learning Video Worksheet: Solving Problems Using a Mathematics Blueprint

Text: *Beginning Algebra: Early Graphing*
Student Learning Objective 0.6.1: Use the Mathematics Blueprint to solve real-life problems.

Follow along with *Guided Learning Video 0.6.1, Solving Problems Using a Mathematics Blueprint*.

Understanding the Big Picture: As you listen to the first part of the video, when the pencil icon appears, pause and fill in the blanks, choosing from the words listed.

calculate • plan • understand • check • blueprint • estimate • organize

1. To solve word problems with many facts and steps, it's helpful to _____ the information and _____ the process in a _____.

2. The first two steps of the problem solving process are to _____ the problem and _____ the answer.

3. After solving a problem, it's important to _____ the answer.

Follow along with the two Guided Learning Video examples, and fill in the blanks on the left as you learn with the instructor. When the pencil icon appears, pause and try the Student Practice on your own.

Guided Learning Video ▶ GUIDED LEARNING VIDEO	Pause: Student Practice
1. Trang works 40 hours a week and earns $15 an hour as a payroll clerk. She is considering accepting a job offer to work as an assistant office manager earning a salary of $2400 per month. Which job pays more per year? Organize the information and plan the problem solving process in a mathematics blueprint. Then, perform the calculations to solve the problem. Current weekly pay: □ × □ = □ Current yearly pay: □ × □ = □ Offered yearly pay: □ × □ = □ □ is greater than □	2. Bob is evaluating two job offers. The first requires a 40-hour work week and pays $22 per hour. The second offer pays $3900 per month. Which job pays more per year?

Guided Learning Video (▶) GUIDED LEARNING VIDEO	**Pause: Student Practice** ✏️

3. The Hernandez Insurance Agency restocked its office supplies recently. The agency bought 15 reams of paper at $3 each, 2 ink cartridges at $32 each, and 4 boxes of folders at $7 each. How much did the agency pay for the office supplies?

Organize the information and plan the problem solving process in a mathematics blueprint. Then, perform the calculations to solve the problem.

Paper: ☐ × ☐ = ☐

Ink: ☐ × ☐ = ☐

Folders: ☐ × ☐ = ☐

Total:

☐ + ☐ + ☐ = ☐

4. Clarissa's checking account has an initial balance of $1850. This month she made 3 deposits of $140. She also wrote 2 checks for $74 and 3 checks for $45. What is her final balance?

Helpful Hint: Pause the video and write the helpful hint in your own words.

For each Active Video Lesson, when the pencil icon appears, pause and work the problem. Then press play to check your work.

Active Video Lesson 1 ✏️	**Active Video Lesson 2** ✏️

5. Juan has investments in the stock market. Last month, his stocks were worth a total of $2347. When he checked his investments this month, 2 stocks had increased in value by $146 and $135. Three stocks had decreased in value by $48, $86, and $93. What is the total value of his stocks this month?

6. Alec drives a taxi. He began his day with a full tank of gas and his odometer read 103,276. At the end of the day, the odometer read 103,591. Alec filled his tank with 12 gallons of gas at noon and filled it again at the end of the day with 9 gallons. How many miles per gallon did the taxi get that day?

Name: _____   Date: _____

Instructor: _____   Section: _____

Guided Learning Video Worksheet: Absolute Value

Text: *Beginning Algebra: Early Graphing*
Student Learning Objective 1.1.3: Add real numbers with the same sign.

Follow along with *Guided Learning Video 1.1.3, Absolute Value.*

Understanding the Big Picture: As you listen to the first part of the video, when the pencil icon appears, pause and fill in the blanks, choosing from the words listed.

zero • absolute • vertical • positive • negative • direction • distance

1. Distance is always a _____ number on a number line regardless of the _____ moved.

2. The _____ value of a number is the _____ between the number and _____ on a number line.

3. The symbol for absolute value is two _____ lines on either side of the value.

4. The absolute value of any nonzero number is always _____.

Follow along with the two Guided Learning Video examples, and fill in the blanks on the left as you learn with the instructor. When the pencil icon appears, pause and try the Student Practice on your own.

Guided Learning Video ▶ GUIDED LEARNING VIDEO	Pause: Student Practice
1. Simplify. $\left\|-21\right\|$ Use the concept of absolute value to determine the answer. $\left\|-21\right\| = \boxed{}$	2. Simplify. $\left\|-14\right\|$

Guided Learning Video ▶ GUIDED LEARNING VIDEO	**Pause: Student Practice** ✎
3. Simplify. $-\lvert-16\rvert$ Use the concept of absolute value to determine the answer. $-\lvert-16\rvert = \boxed{}$	**4.** Simplify. $-\lvert-33\rvert$ Use the concept of absolute value to determine the answer. $-\lvert-33\rvert = \boxed{}$

Helpful Hint: Pause the video and write the helpful hint in your own words.

For each Active Video Lesson, when the pencil icon appears, pause and work the problem. Then press play to check your work.

Active Video Lesson 1 ✎	**Active Video Lesson 2** ✎
5. Find the absolute value. $\lvert-53\rvert$	**6.** Find the absolute value. $\lvert 0\rvert$

Name: _____ Date: _____

Instructor: _____ Section: _____

Guided Learning Video Worksheet: Adding with Same Sign

Text: *Beginning Algebra: Early Graphing*
Student Learning Objective 1.1.3: Add real numbers with the same sign.

Follow along with *Guided Learning Video 1.1.3, Adding with Same Sign.*

Understanding the Big Picture: As you listen to the first part of the video, when the pencil icon appears, pause and fill in the blanks, choosing from the words listed.

negative • adding • positive • common • absolute values • same

1. To add two numbers with the same sign, first add the _____ of the numbers.

2. When _____ two numbers with the _____ sign, use the _____ sign in the answer.

3. If all the numbers being added are negative, the answer will be _____.

Follow along with the two Guided Learning Video examples, and fill in the blanks on the left as you learn with the instructor. When the pencil icon appears, pause and try the Student Practice on your own.

Guided Learning Video GUIDED LEARNING VIDEO	**Pause: Student Practice**
1. Add. $-76+(-32)$ Apply the addition rule for adding two numbers with the same sign. $\left|\boxed{}\right|+\left|\boxed{}\right|$ $=\boxed{}+\boxed{}=\boxed{}$	2. Add. $-37+(-64)$

3. Find the total value of surplus or deficit for the two years, 2010 and 2014.

The Federal Budget

Approximate value of surplus or deficit in the federal budget (in billions)

Year	2010	2011	2012	2013	2014
	−1290	−1300	−1090	−680	−480

Use the graph to identify the numbers to be added, then apply the addition rule for adding numbers with the same sign.

$$\boxed{} + \boxed{}$$

$$= \boxed{} + \boxed{}$$

$$= \boxed{} = \boxed{}$$

4. Find the total value of the deficit for the two years, 2013 and 2014.

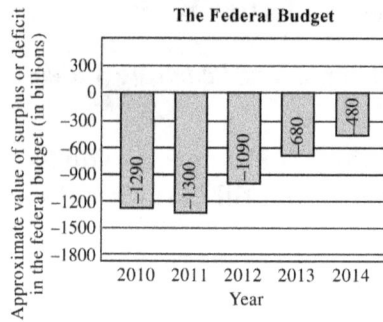

The Federal Budget

Approximate value of surplus or deficit in the federal budget (in billions)

Year	2010	2011	2012	2013	2014
	−1290	−1300	−1090	−680	−480

Helpful Hint: Pause the video and write the helpful hint in your own words.

For each Active Video Lesson, when the pencil icon appears, pause and work the problem. Then press play to check your work.

Active Video Lesson 1 | **Active Video Lesson 2**

5. Find the total value of surplus or deficit for the two years, 2011 and 2013.

The Federal Budget

Approximate value of surplus or deficit in the federal budget (in billions)

Year	2010	2011	2012	2013	2014
	−1290	−1300	−1090	−680	−480

6. Add. $-479 + (-365)$

Guided Learning Video Worksheet: Adding Integers with Different Signs

Text: *Beginning Algebra: Early Graphing*
Student Learning Objective 1.1.4: Add real numbers with different signs.

Follow along with *Guided Learning Video 1.1.4, Adding Integers with Different Signs.*

Understanding the Big Picture: As you listen to the first part of the video, when the pencil icon appears, pause and fill in the blanks, choosing from the words listed.

difference • positive • smaller • negative • larger • different

1. A temperature below zero can be represented by a _____ number.

2. When adding two numbers with _____ signs, use the sign of the number with the _____ absolute value in the answer.

3. To add two numbers with different signs, find the _____ between the _____ absolute value and the _____ absolute value.

Follow along with the two Guided Learning Video examples, and fill in the blanks on the left as you learn with the instructor. When the pencil icon appears, pause and try the Student Practice on your own.

Guided Learning Video ⏵ GUIDED LEARNING VIDEO	Pause: Student Practice ✎
1. Suppose we spend $52 and we earn $38. Calculate the sum. Apply the addition rule for adding two numbers with different signs. ☐ + ☐ = ☐ ☐ + ☐ = ☐	2. Suppose we spend $73 and we earn $25. Calculate the sum.

3. If you were to gain $7 and lose $15, how much money do you have?

   Apply the addition rule for adding two numbers with different signs.

   $\boxed{} + \boxed{} = \boxed{}$

   $\boxed{} + \boxed{} = \boxed{}$

4. If there is a gain of $150 and a loss of $184, calculate the final balance.

Helpful Hint: Pause the video and write the helpful hint in your own words.

For each Active Video Lesson, when the pencil icon appears, pause and work the problem. Then press play to check your work.

Active Video Lesson 1   **Active Video Lesson 2**

5. Add. $8+(-14)$

6. Add. $-21+16$

Guided Learning Video Worksheet: Adding Two Signed Numbers with Different Signs

Text: *Beginning Algebra: Early Graphing*
Student Learning Objective 1.1.4: Add real numbers with different signs.

Follow along with *Guided Learning Video 1.1.4, Adding Two Signed Numbers with Different Signs*.

Understanding the Big Picture: As you listen to the first part of the video, when the pencil icon appears, pause and fill in the blanks, choosing from the words listed.

difference • positive • smaller • negative • larger • different

1. A temperature below zero can be represented by a _____ number.

2. When adding two numbers with _____ signs, use the sign of the number with the _____ absolute value in the answer.

3. To add two numbers with different signs, find the _____ between the _____ absolute value and the _____ absolute value.

Follow along with the two Guided Learning Video examples, and fill in the blanks on the left as you learn with the instructor. When the pencil icon appears, pause and try the Student Practice on your own.

Guided Learning Video ▶ GUIDED LEARNING VIDEO	**Pause: Student Practice**
1. Add each of the following.	2. Add each of the following.
a. $10+(-12)$	**a.** $6+(-13)$
b. $-15+2$	**b.** $-23+4$
c. $-5+12$	**c.** $-8+17$
Apply the addition rule for adding two numbers with different signs.	
a. $10+(-12)=\boxed{}$	
b. $-15+2=\boxed{}$	
c. $-5+12=\boxed{}$	

Guided Learning Video ▶ GUIDED LEARNING VIDEO	Pause: Student Practice
3. Add. $-10.7+12.8$ Apply the addition rule for adding two numbers with different signs. $\boxed{} + \boxed{} = \boxed{}$	**4.** Add. $-13.9+11.5$

Helpful Hint: Pause the video and write the helpful hint in your own words.

For each Active Video Lesson, when the pencil icon appears, pause and work the problem. Then press play to check your work.

Active Video Lesson 1	Active Video Lesson 2
5. Add. $6+(-13)$	**6.** Add. $-20.8+15.2$

Name: _____   Date: _____

Instructor: _____   Section: _____

Guided Learning Video Worksheet: Adding and Subtracting Fractions

Text: *Beginning Algebra: Early Graphing*
Student Learning Objective 1.1.5: Use the addition properties for real numbers.

Follow along with *Guided Learning Video 1.1.5, Adding and Subtracting Fractions*.

Understanding the Big Picture: As you listen to the first part of the video, when the pencil icon appears, pause and fill in the blanks, choosing from the words listed.

denominators • simplify • numerators • common

1. In order to add or subtract fractions, _____ denominators are needed.

2. To add or subtract fractions, add or subtract only the _____ .

3. The last step in the process for adding or subtracting fractions is to _____ the answer if possible.

Follow along with the two Guided Learning Video examples, and fill in the blanks on the left as you learn with the instructor. When the pencil icon appears, pause and try the Student Practice on your own.

Guided Learning Video ⏵ GUIDED LEARNING VIDEO	Pause: Student Practice ✏
1. Subtract. $\dfrac{8}{14} - \dfrac{2}{21}$ Use the process for adding/subtracting fractions with different denominators to calculate the answer. LCD = □ × □ × □ = □ 14 = □ × □ 21 = □ × □ $\dfrac{□}{□} \times \dfrac{□}{□} = \dfrac{□}{□}$ $\dfrac{□}{□} \times \dfrac{□}{□} = \dfrac{□}{□}$ $\dfrac{□}{□} - \dfrac{□}{□} = \dfrac{□}{□} = \dfrac{□}{□}$	2. Subtract. $\dfrac{15}{25} - \dfrac{3}{10}$

3. Add. $\dfrac{11}{30} + \dfrac{7}{45}$

Use the process for adding/subtracting fractions with different denominators to calculate the answer.

LCD

= □ × □ × □ × □ = □

30 = □ × □ × □

45 = □ × □ × □

$\dfrac{\square}{\square} \times \dfrac{\square}{\square} = \dfrac{\square}{\square}$

$\dfrac{\square}{\square} \times \dfrac{\square}{\square} = \dfrac{\square}{\square}$

$\dfrac{\square}{\square} - \dfrac{\square}{\square} = \dfrac{\square}{\square}$

4. Add. $\dfrac{7}{20} + \dfrac{9}{70}$

Helpful Hint: Pause the video and write the helpful hint in your own words.

For each Active Video Lesson, when the pencil icon appears, pause and work the problem. Then press play to check your work.

Active Video Lesson 1 ✎	**Active Video Lesson 2** ✎
5. Add. $\dfrac{7}{25} + \dfrac{12}{45}$	**6.** Subtract. $\dfrac{5}{12} - \dfrac{15}{42}$

Guided Learning Video Worksheet: Subtracting Integers

Text: *Beginning Algebra: Early Graphing*
Student Learning Objective 1.2.1: Subtract real numbers with the same or different signs.

Follow along with *Guided Learning Video 1.2.1, Subtracting Integers.*

Understanding the Big Picture: As you listen to the first part of the video, when the pencil icon appears, pause and fill in the blanks, choosing from the words listed.

zero • opposite • same • negative • add • equal

1. The opposite of a positive number is a _____ number with the
   _____ absolute value.

2. If a number is the opposite of another number, the numbers are an _____ distance
   from _____ on the number line.

3. To subtract signed numbers, _____ the _____ of the second number
   to the first number.

Follow along with the two Guided Learning Video examples, and fill in the blanks on the left as you learn with the instructor. When the pencil icon appears, pause and try the Student Practice on your own.

Guided Learning Video ▶ GUIDED LEARNING VIDEO	Pause: Student Practice ✎
1. Subtract. $24 - $32 Apply the rule for subtracting signed numbers. $24 - $32 $= \boxed{}\boxed{}\boxed{} = \boxed{}$	2. Subtract. $65 - $80

Guided Learning Video ▶ GUIDED LEARNING VIDEO	**Pause: Student Practice** ✎
3. Subtract. $-21-5$	**4.** Subtract. $-37-14$

Apply the rule for subtracting signed numbers.

$-21-5$

$= \boxed{} \boxed{}\boxed{} = \boxed{}$

Helpful Hint: Pause the video and write the helpful hint in your own words.

For each Active Video Lesson, when the pencil icon appears, pause and work the problem. Then press play to check your work.

Active Video Lesson 1 ✎	**Active Video Lesson 2** ✎
5. Subtract. $-19-(-4)$	**6.** Subtract. $28-(-15)$

Guided Learning Video Worksheet: Subtracting Signed Numbers

Text: *Beginning Algebra: Early Graphing*
Student Learning Objective 1.2.1: Subtract real numbers with the same or different signs.

Follow along with *Guided Learning Video 1.2.1, Subtracting Signed Numbers.*

Understanding the Big Picture: As you listen to the first part of the video, when the pencil icon appears, pause and fill in the blanks, choosing from the words listed.

zero • opposite • same • negative • add • equal

1. The opposite of a positive number is a(n) _____ number with the _____ absolute value.

2. If a number is the opposite of another number, the numbers are a(n) _____ distance from _____ on the number line.

3. To subtract signed numbers, _____ the _____ of the second number to the first number.

Follow along with the two Guided Learning Video examples, and fill in the blanks on the left as you learn with the instructor. When the pencil icon appears, pause and try the Student Practice on your own.

Guided Learning Video ▶ GUIDED LEARNING VIDEO	Pause: Student Practice
1. **a.** The opposite of 7 is [ ] .	2. **a.** The opposite of 20 is [ ] .
b. The opposite of −9 is [ ] .	**b.** The opposite of −17 is [ ] .

3. Subtract. $-13-(-5)$

Apply the rule for subtracting signed numbers.

$-13-(-5)$

$=$ ☐ ☐ ☐ $=$ ☐

4. Subtract. $-23-(-12)$

Helpful Hint: Pause the video and write the helpful hint in your own words.

For each Active Video Lesson, when the pencil icon appears, pause and work the problem. Then press play to check your work.

Active Video Lesson 1 ✏ | **Active Video Lesson 2** ✏

5. Subtract. $-9.4-(-5.1)$

6. Subtract. $57-92$

Name: _____   Date: _____
Instructor: _____   Section: _____

Guided Learning Video Worksheet: Multiply Integers

Text: *Beginning Algebra: Early Graphing*
Student Learning Objective 1.3.1: Multiply real numbers.

Follow along with *Guided Learning Video 1.3.1, Multiply Integers.*

Understanding the Big Picture: As you listen to the first part of the video, when the pencil icon appears, pause and fill in the blanks, choosing from the words listed.

dot • odd • negative • parentheses • even

1. A positive number times a negative number equals a(n) _____ number.

2. Multiplication can be indicated by placing a raised a(n) _____ between numbers or by placing _____ on either side of each number.

3. When multiplying nonzero numbers, the product is positive if there are a(n) _____ number of negative signs and negative if there are a(n) _____ number of negative signs.

Follow along with the two Guided Learning Video examples, and fill in the blanks on the left as you learn with the instructor. When the pencil icon appears, pause and try the Student Practice on your own.

Guided Learning Video ▶ GUIDED LEARNING VIDEO	Pause: Student Practice
1. Multiply. $9(-6)$ Determine the sign of the product, then multiply. $9(-6) = \boxed{}$	2. Multiply. $5(-4)$

3. Multiply. $(-7)(-6)$

Determine the sign of the product, then multiply.

$(-7)(-6) = \boxed{}$

4. Multiply. $(-3)(-11)$

Helpful Hint: Pause the video and write the helpful hint in your own words.

For each Active Video Lesson, when the pencil icon appears, pause and work the problem. Then press play to check your work.

Active Video Lesson 1 ✏ | **Active Video Lesson 2** ✏

5. Multiply. $-8 \cdot 7$

6. Multiply. $-4 \times (-9)$

Guided Learning Video Worksheet: Divide Integers

Text: *Beginning Algebra: Early Graphing*
Student Learning Objective 1.3.3: Divide real numbers.

Follow along with *Guided Learning Video 1.3.3, Divide Integers.*

Understanding the Big Picture: As you listen to the first part of the video, when the pencil icon appears, pause and fill in the blanks, choosing from the words listed.

same • multiplication • odd • negative • division • even

1. Division statements can be rewritten as _____ statements.

2. The signed number rules for multiplication and _____ are the _____.

3. When dividing nonzero numbers, the answer will be positive if there are a(n) _____ number of negative signs and negative if there are a(n) _____ number of negative signs.

Follow along with the two Guided Learning Video examples, and fill in the blanks on the left as you learn with the instructor. When the pencil icon appears, pause and try the Student Practice on your own.

Guided Learning Video	Pause: Student Practice
1. Divide. $56 \div (-7)$ Follow the rules for dividing integers to calculate the answer. $56 \div (-7) = \boxed{}$	2. Divide. $-32 \div 8$

Guided Learning Video ▶ GUIDED LEARNING VIDEO	**Pause: Student Practice**
3. Divide. $-27 \div (-3)$	**4.** Divide. $-63 \div (-9)$
Follow the rules for dividing integers to calculate the answer.	
$-27 \div (-3) = \boxed{}$	

Helpful Hint: Pause the video and write the helpful hint in your own words.

For each Active Video Lesson, when the pencil icon appears, pause and work the problem. Then press play to check your work.

Active Video Lesson 1	**Active Video Lesson 2**
5. Divide. $-48 \div 6$	**6.** Divide. $-54 \div (-9)$

Guided Learning Video Worksheet: Exponents

Text: *Beginning Algebra: Early Graphing*
Student Learning Objective 1.4.1: Write numbers in exponent form.

Follow along with *Guided Learning Video 1.4.1, Exponents.*

Understanding the Big Picture: As you listen to the first part of the video, when the pencil icon appears, pause and fill in the blanks, choosing from the words listed.

superscript • exponent • shorthand • base

1. An exponent is a _____ number for expressing multiplication of the same number.

2. The _____ is the number being multiplied.

3. The _____ names the number of times the base is multiplied.

4. An exponent is sometimes called a(n) _____.

Follow along with the two Guided Learning Video examples, and fill in the blanks on the left as you learn with the instructor. When the pencil icon appears, pause and try the Student Practice on your own.

Guided Learning Video ▶ GUIDED LEARNING VIDEO	**Pause: Student Practice** ✏️
1. Find the value of the expression: "Four cubed" Identify the base and the exponent. Then, rewrite the expression and find its value. The exponent is ☐ . The base is ☐ . ☐^☐ = ☐ = ☐ ☐^☐ = ☐	2. Find the value of the expression: "Eight cubed"

Guided Learning Video ▶ GUIDED LEARNING VIDEO	**Pause: Student Practice** ✏️
3. Write the product in exponent form: "five times five times five times five" Identify the base. To determine the exponent, identify the number of times the base is repeated. Then, write the exponent form. The base is ☐. The exponent is ☐ The exponent form is ☐[☐]	4. Write the product in exponent form:"six times six times six times six"

Helpful Hint: Pause the video and write the helpful hint in your own words.

For each Active Video Lesson, when the pencil icon appears, pause and work the problem. Then press play to check your work.

Active Video Lesson 1 ✏️	**Active Video Lesson 2** ✏️
5. Divide 5523 by 46.	6. Divide 7490 by 214.

Guided Learning Video Worksheet: Evaluate Exponents

Text: *Beginning Algebra: Early Graphing*
Student Learning Objective 1.4.2: Evaluate numerical expressions that contain exponents.

Follow along with *Guided Learning Video 1.4.2, Evaluate Exponents*.

Understanding the Big Picture: As you listen to the first part of the video, when the pencil icon appears, pause and fill in the blanks, choosing from the words listed.

negative • factor • powers • positive • multiplication • base

1. Exponents or _____ are used to indicate repeated _____.

2. The exponent indicates how many times the _____ occurs as a
    _____.

3. When a number is written in exponential notation and the base is a negative, the result is
    _____ if the exponent is even and _____ if the exponent is odd.

Follow along with the two Guided Learning Video examples, and fill in the blanks on the left as you learn with the instructor. When the pencil icon appears, pause and try the Student Practice on your own.

Guided Learning Video	Pause: Student Practice
1. Evaluate. $(-5)^3$ To evaluate, express the exponential notation as repeated multiplication. $\boxed{} = \boxed{}$	2. Evaluate. $(-6)^3$

Guided Learning Video	**Pause: Student Practice**
3. Evaluate. $\left(\dfrac{2}{3}\right)^4$	4. Evaluate. $\left(\dfrac{3}{5}\right)^4$
To evaluate, express the exponential notation as repeated multiplication.	
$\boxed{} = \boxed{}$	

Helpful Hint: Pause the video and write the helpful hint in your own words.

For each Active Video Lesson, when the pencil icon appears, pause and work the problem. Then press play to check your work.

Active Video Lesson 1	**Active Video Lesson 2**
5. Evaluate. $\left(-\dfrac{4}{7}\right)^2$	6. Evaluate. $(-2)^4$

Guided Learning Video Worksheet: Order of Operations with Signed Numbers

Text: *Beginning Algebra: Early Graphing*
Student Learning Objective 1.5.1: Use the order of operations to simplify numerical expressions.

Follow along with *Guided Learning Video 1.5.1, Order of Operations with Signed Numbers.*

Understanding the Big Picture: As you listen to the first part of the video, when the pencil icon appears, pause and fill in the blanks, choosing from the words listed.

exponents • right • parentheses • left • priorities

1. The order of operations is a list of _____ for working with the numbers in computation problems.

2. When there is more than one operation in a problem, first perform any operations inside _____ .

3. Evaluating _____ is the second priority in the order of operations.

4. The operations of multiplication and division are performed in the order they appear from _____ to _____ .

Follow along with the two Guided Learning Video examples, and fill in the blanks on the left as you learn with the instructor. When the pencil icon appears, pause and try the Student Practice on your own.

Guided Learning Video	Pause: Student Practice
1. Simplify. $-6(-3)-4(3-7)^2$	2. Simplify. $-5(-2)-6(5-8)^2$
Use the order of operations to simplify the expression.	
$-6(-3)-4(3-7)^2$	
$= \boxed{}$	
$= \boxed{}$	
$= \boxed{}$	
$= \boxed{} = \boxed{}$	

3. Simplify. $\left(\dfrac{1}{3}\right)^3 + 2\left(\dfrac{5}{3} - \dfrac{1}{6}\right) \div \left(-\dfrac{9}{4}\right)$

4. Simplify. $\left(\dfrac{1}{2}\right)^2 - 4\left(\dfrac{1}{4} - \dfrac{3}{2}\right) \div \left(-\dfrac{3}{2}\right)$

Use the order of operations to simplify the expression.

$$\left(\dfrac{1}{3}\right)^3 + 2\left(\dfrac{5}{3} - \dfrac{1}{6}\right) \div \left(-\dfrac{9}{4}\right)$$

$$= \boxed{}$$

$$= \boxed{}$$

$$= \boxed{}$$

$$= \boxed{} = \boxed{}$$

Helpful Hint: Pause the video and write the helpful hint in your own words.

For each Active Video Lesson, when the pencil icon appears, pause and work the problem. Then press play to check your work.

Active Video Lesson 1

Active Video Lesson 2

5. Multiply. $3(4-6)^3 + 12 \div (-4) + 2$

6. Multiply. $\left(\dfrac{1}{5}\right)^2 + 4\left(\dfrac{1}{5} - \dfrac{3}{10}\right) \div \dfrac{2}{3}$

Guided Learning Video Worksheet: Order of Operations: Signed Decimals

Text: *Beginning Algebra: Early Graphing*
Student Learning Objective 1.5.1: Use the order of operations to simplify numerical expressions.

Follow along with *Guided Learning Video 1.5.1, Order of Operations: Signed Decimals*.

Understanding the Big Picture: As you listen to the first part of the video, when the pencil icon appears, pause and fill in the blanks, choosing from the words listed.

powers • right • parentheses • equal • priorities • left

1. The order of operations is a list of _____ for working with the numbers in computation problems.

2. Performing operations with _____ is the second step in the order of operations.

3. Multiplication and division have _____ priority in the order of operations.

4. The operations of multiplication and division are performed in the order they appear from _____ to _____.

Follow along with the two Guided Learning Video examples, and fill in the blanks on the left as you learn with the instructor. When the pencil icon appears, pause and try the Student Practice on your own.

Guided Learning Video ▶ GUIDED LEARNING VIDEO	**Pause: Student Practice** ✏️
1. Simplify. $(0.8)^2 - 2.3(-4.5)$	2. Simplify. $(0.4)^2 - 5.1(-3.2)$

Use the order of operations to simplify the expression.

$(0.8)^2 - 2.3(-4.5)$

= [   ]

= [   ]

= [   ]

= [   ] = [   ]

3. Simplify.

$$1.7 + 0.54 \div 0.6 - (-0.2)^3 - (1.1 + 0.3)$$

Use the order of operations to simplify the expression.

$$1.7 + 0.54 \div 0.6 - (-0.2)^3 - (1.1 + 0.3)$$

= ☐

= ☐

= ☐

= ☐ = ☐

4. Simplify.

$$3.9 + 0.81 \div 0.9 - (5.3 + 0.6) - (0.3)^3$$

Helpful Hint: Pause the video and write the helpful hint in your own words.

For each Active Video Lesson, when the pencil icon appears, pause and work the problem. Then press play to check your work.

Active Video Lesson 1   **Active Video Lesson 2**

5. Simplify.

$$-1.3 + 0.8 \times 0.07 - (0.4)^2 + (3.5 - 0.9)$$

6. Simplify. $0.3 + 2.5 - (0.7 - 0.9)^3$

Name: _____   Date: _____

Instructor: _____   Section: _____

Guided Learning Video Worksheet: The Distributive Property

Text: *Beginning Algebra: Early Graphing*
Student Learning Objective 1.6.1: Use the distributive property to simplify algebraic expressions.

Follow along with *Guided Learning Video 1.6.1, The Distributive Property*.

Understanding the Big Picture: As you listen to the first part of the video, when the pencil icon appears, pause and fill in the blanks, choosing from the words listed.

multiplied • parentheses • subtraction • distributive • addition

1. The _____ property is used to remove _____.

2. The distributive property can be used over _____ and _____.

3. Distributing a number to all terms inside a set of parentheses means that all terms inside the parentheses are _____ by the number.

Follow along with the two Guided Learning Video examples, and fill in the blanks on the left as you learn with the instructor. When the pencil icon appears, pause and try the Student Practice on your own.

Guided Learning Video ▶ GUIDED LEARNING VIDEO	**Pause: Student Practice**
1. Simplify. $4(x+3)$ Use the distributive property to simplify the expression. $4(x+3)$ $= \square(\square) + \square(\square)$ $= \square + \square$	2. Simplify. $-5(x-6y)$

3. Simplify. $-3(5x-2y-8z)$

Use the distributive property to simplify the expression.

$-3(5x-2y-8z)$

$= \square(\square) - \square(\square) - \square(\square)$

$= \square + \square + \square$

4. Simplify. $-4(6x-3y+2z)$

Helpful Hint: Pause the video and write the helpful hint in your own words.

For each Active Video Lesson, when the pencil icon appears, pause and work the problem. Then press play to check your work.

Active Video Lesson 1 ✏ | **Active Video Lesson 2** ✏

5. Simplify. $5(9x-y)$

6. Simplify. $-7(8x+6)$

Name: _____   Date: _____

Instructor: _____   Section: _____

Guided Learning Video Worksheet: Identify Like Terms

Text: *Beginning Algebra: Early Graphing*
Student Learning Objective 1.7.1: Identify like terms.

Follow along with *Guided Learning Video 1.7.1, Identify Like Terms.*

Understanding the Big Picture: As you listen to the first part of the video, when the pencil icon appears, pause and fill in the blanks, choosing from the words listed.

multiplication • minus • exponents • term • plus • product • variables

1. A _____ is a number, a variable, or the _____ of numbers and variables.

2. Terms are parts of an algebraic expression separated by _____ or _____ signs.

3. Terms having identical _____ raised to identical _____ are like terms.

Follow along with the two Guided Learning Video examples, and fill in the blanks on the left as you learn with the instructor. When the pencil icon appears, pause and try the Student Practice on your own.

Guided Learning Video ▶ GUIDED LEARNING VIDEO	Pause: Student Practice ✏️
1. List the like terms of $8x + 3y - 7x$. Identify the like terms using the definition. Like terms: [＿＿＿＿]	2. List the like terms of $5x - 3y - 4z + 5y$.

Guided Learning Video	Pause: Student Practice
3. List the like terms of $4a^2 + 6a + a^2 - 2a$. Identify the like terms using the definition. Like terms: ☐ Like terms: ☐	4. List the like terms of $3x^2 + 2y^4 - 5x^2 - y^4$.

Helpful Hint: Pause the video and write the helpful hint in your own words.

For each Active Video Lesson, when the pencil icon appears, pause and work the problem. Then press play to check your work.

Active Video Lesson 1	Active Video Lesson 2
5. List the like terms of $9y^2 - 5y - 4y^2 + 12y$	6. List the like terms of $15a + 14b + 5b$

Name: _____   Date: _____

Instructor: _____   Section: _____

Guided Learning Video Worksheet: Combining Like Terms

Text: *Beginning Algebra: Early Graphing*
Student Learning Objective 1.7.2: Combine like terms.

Follow along with *Guided Learning Video 1.7.2, Combining Like Terms.*

Understanding the Big Picture: As you listen to the first part of the video, when the pencil icon appears, pause and fill in the blanks, choosing from the words listed.

subtraction • one • coefficients • term • like • addition

1. A _____ is a number, a variable, or a product of a number and one or more variables.

2. The terms of an expressions are separated by _____ and _____ symbols.

3. Terms with identical variables with identical exponents are called _____ terms.

4. To combine _____ terms, combine the numerical _____ using the rules for adding signed numbers.

5. The numerical coefficient of a variable term is understood to be equal to _____ when it is not written.

Follow along with the two Guided Learning Video examples, and fill in the blanks on the left as you learn with the instructor. When the pencil icon appears, pause and try the Student Practice on your own.

Guided Learning Video ⓖ GUIDED LEARNING VIDEO	**Pause: Student Practice** ✏️
1. Simplify by combining like terms. $-2a + 4b - 5a$	2. Simplify by combining like terms. $-5x + 8y - 6x$
Use the rules for combining like terms to simplify the expression.	
$-2a + 4b - 5a$	
$= \boxed{} + \boxed{} + (\boxed{})$	
$= \boxed{} + (\boxed{}) + \boxed{}$	
$= \boxed{} + \boxed{}$	

3. Simplify by combining like terms.
$$9x + y + (-6y) + 7xy$$

Use the rules for combining like terms to simplify the expression.

$$9x + y + (-6y) + 7xy$$

$$= \boxed{} + \left(\boxed{} \right) + \boxed{}$$

4. Simplify by combining like terms.
$$5a + b + (-3b) + 2ab$$

Helpful Hint: Pause the video and write the helpful hint in your own words.

For each Active Video Lesson, when the pencil icon appears, pause and work the problem. Then press play to check your work.

Active Video Lesson 1 ✎

5. Simplify by combining like terms.
$$-8x + 2xy - 4x$$

Active Video Lesson 2 ✎

6. Simplify by combining like terms.
$$4x + 9y - 7x - 3y$$

Name: _____   Date: _____
Instructor: _____   Section: _____

Guided Learning Video Worksheet: Evaluate Algebraic Expressions with Integers

Text: *Beginning Algebra: Early Graphing*
Student Learning Objective 1.8.1: Evaluate an algebraic expression for a specified value.

Follow along with *Guided Learning Video 1.8.1, Evaluate Algebraic Expressions with Integers*.

Understanding the Big Picture: As you listen to the first part of the video, when the pencil icon appears, pause and fill in the blanks, choosing from the words listed.

operation • value • parentheses • variable • expressions

1. Use the order of operations when evaluating algebraic _____.

2. When evaluating an algebraic expression, first replace each _____ by its numerical _____.

3. If a negative integer is substituted for a variable, put _____ around the integer.

Follow along with the two Guided Learning Video examples, and fill in the blanks on the left as you learn with the instructor. When the pencil icon appears, pause and try the Student Practice on your own.

Guided Learning Video ▶ GUIDED LEARNING VIDEO	Pause: Student Practice ✏
1. Evaluate $\dfrac{(x^2 - y)}{5}$ for $x = -7$ and $y = -1$.	2. Evaluate $\dfrac{(x^3 + y)}{6}$ for $x = -2$ and $y = -4$.
Replace each variable by its numerical value, and then use the order of operations to evaluate the expression. $$\frac{(x^2 - y)}{5} = \frac{((\;)^2 - (\;))}{5}$$ $$= \frac{\boxed{}}{5} = \frac{\boxed{}}{5} = \boxed{}$$	

Guided Learning Video 🔘 GUIDED LEARNING VIDEO	**Pause: Student Practice** ✏

3. Evaluate $m^2 - 8m - 6$ for $m = -2$.	**4.** Evaluate $x^2 - 3x + 5$ for $x = -4$.

Replace the variable by its numerical value, and then use the order of operations to evaluate the expression.

$m^2 - 8m - 6 = (\quad)^2 - 8(\quad) - 6$

$= \boxed{} = \boxed{}$

$= \boxed{} = \boxed{}$

Helpful Hint: Pause the video and write the helpful hint in your own words.

For each Active Video Lesson, when the pencil icon appears, pause and work the problem. Then press play to check your work.

Active Video Lesson 1 ✏	**Active Video Lesson 2** ✏
5. Evaluate $a^2 - 4a + 9$ for $a = 3$.	**6.** Evaluate $\dfrac{s^2 + t}{7}$ for $s = 5$ and $t = -4$.

Name: _____   Date: _____

Instructor: _____   Section: _____

Guided Learning Video Worksheet: Evaluate a Formula by Substituting Values

Text: *Beginning Algebra: Early Graphing*
Student Learning Objective 1.8.2: Evaluate a formula by substituting values.

Follow along with *Guided Learning Video 1.8.2, Evaluate a Formula by Substituting Values.*

Understanding the Big Picture: As you listen to the first part of the video, when the pencil icon appears, pause and fill in the blanks, choosing from the words listed.

center • square • circumference • triangle • circle

1. A _____ is a closed plane figure with three sides.

2. The units of measurement for area are _____ units.

3. A _____ is a plane curve consisting of all points equidistant from a fixed point called the _____.

4. The distance around a circle is called its _____.

Follow along with the two Guided Learning Video examples, and fill in the blanks on the left as you learn with the instructor. When the pencil icon appears, pause and try the Student Practice on your own.

Guided Learning Video ▶ GUIDED LEARNING VIDEO	Pause: Student Practice
1. Find the area (A) of a triangle with a base (b) of 9 inches and an altitude (a) of 20 inches.	2. Find the area (A) of a triangle with a base (b) of 13 inches and an altitude (a) of 16 inches.

Use the formula $A = \dfrac{1}{2}ab$ to calculate the area.

$A = \dfrac{1}{2}ab$

$A = \left(\dfrac{1}{2}\right)\left(\boxed{}\right)\left(\boxed{}\right)$

$A = \left(\boxed{}\right)\left(\boxed{}\right)$

$A = \boxed{}$

Guided Learning Video (▶ GUIDED LEARNING VIDEO)	**Pause: Student Practice**
3. Find the area (A) of a circle if the radius (r) is 5 cm.	**4.** Find the area (A) of a circle if the radius (r) is 9 cm.

Use the formula $A = \pi r^2$ and let $\pi \approx 3.14$ to calculate the area.

$A \approx \left(\boxed{}\right)\left(\boxed{}\right)^2$

$A \approx \left(\boxed{}\right)\left(\boxed{}\right)$

$A \approx \left(\boxed{}\right)$

Helpful Hint: Pause the video and write the helpful hint in your own words.

For each Active Video Lesson, when the pencil icon appears, pause and work the problem. Then press play to check your work.

Active Video Lesson 1	**Active Video Lesson 2**
5. Find the circumference (C) of a circle when the radius (r) is 7 inches.	**6.** Find the area (A) of a triangle with a base (b) of 10 cm and an altitude (a) of 11 cm.

Guided Learning Video Worksheet: The Distributive Property with Combining Like Terms

Text: *Beginning Algebra: Early Graphing*
Student Learning Objective 1.9.1: Simplify algebraic expressions by removing grouping symbols.

Follow along with *Guided Learning Video 1.9.1, The Distributive Property with Combining Like Terms.*

Understanding the Big Picture: As you listen to the first part of the video, when the pencil icon appears, pause and fill in the blanks, choosing from the words listed.

opposite • one • multiplied • distributive • equal

1. A negative sign in front of a set of parentheses is _____ to a numerical coefficient of negative _____.

2. If a set of parentheses is preceded by a negative sign, all terms inside the parentheses are changed to their signed _____.

3. When a set of parentheses is preceded by a positive sign, all terms inside the parentheses are _____ by positive _____.

Follow along with the two Guided Learning Video examples, and fill in the blanks on the left as you learn with the instructor. When the pencil icon appears, pause and try the Student Practice on your own.

Guided Learning Video	Pause: Student Practice
1. Simplify. $4(2x-y)+5(x+3y)$	2. Simplify. $3(x+4y)-2(5x+y)$
Use the distributive property and combine like terms.	
$4(2x-y)+5(x+3y)$	
$=$ [_____]	
$=$ [_____]	

Guided Learning Video ⏵ GUIDED LEARNING VIDEO	**Pause: Student Practice** ✎
3. Simplify. $9(7s+4t)-8(6s-3)$	**4.** Simplify. $-6(3x-y+2)+4(4x+3y)$

Use the distributive property and combine like terms.

$9(7s+4t)-8(6s-3)$

$= \boxed{}$

$= \boxed{}$

Helpful Hint: Pause the video and write the helpful hint in your own words.

For each Active Video Lesson, when the pencil icon appears, pause and work the problem. Then press play to check your work.

Active Video Lesson 1 ✎	**Active Video Lesson 2** ✎
5. Simplify. $6(5x-7y+3z)-2(9x-4z)$	**6.** Simplify. $2(-3a+2b)-5(a-2b)$

Guided Learning Video Worksheet: The Distributive Property with Combining Like Terms

Text: *Beginning Algebra: Early Graphing*
Student Learning Objective 1.9.1: Simplify algebraic expressions by removing grouping symbols.

Follow along with *Guided Learning Video 1.9.1, The Distributive Property with Combining Like Terms.*

Understanding the Big Picture: As you listen to the first part of the video, when the pencil icon appears, pause and fill in the blanks, choosing from the words listed.

opposite • one • multiplied • distributive • equal

1. A negative sign in front of a set of parentheses is _____ to a numerical coefficient of negative _____.

2. If a set of parentheses is preceded by a negative sign, all terms inside the parentheses are changed to their signed _____.

3. When a set of parentheses is preceded by a positive sign, all terms inside the parentheses are _____ by positive _____.

Follow along with the two Guided Learning Video examples, and fill in the blanks on the left as you learn with the instructor. When the pencil icon appears, pause and try the Student Practice on your own.

Guided Learning Video ▶ GUIDED LEARNING VIDEO	Pause: Student Practice
1. Simplify. $4(2x - y) + 5(x + 3y)$ Use the distributive property and combine like terms. $4(2x - y) + 5(x + 3y)$ = [box] = [box]	2. Simplify. $3(x + 4y) - 2(5x + y)$

3. Simplify. $9(7s+4t)-8(6s-3)$

Use the distributive property and combine like terms.

$9(7s+4t)-8(6s-3)$

$=$ [   ]

$=$ [   ]

4. Simplify. $-6(3x-y+2)+4(4x+3y)$

Helpful Hint: Pause the video and write the helpful hint in your own words.

For each Active Video Lesson, when the pencil icon appears, pause and work the problem. Then press play to check your work.

Active Video Lesson 1

Active Video Lesson 2

5. Simplify.
$6(5x-7y+3z)-2(9x-4z)$

6. Simplify. $2(-3a+2b)-5(a-2b)$

Name: _____   Date: _____

Instructor: _____   Section: _____

Guided Learning Video Worksheet: Simplifying Algebraic Expressions with Nested Grouping Symbols

Text: *Beginning Algebra: Early Graphing*
Student Learning Objective 1.9.1: Simplify algebraic expressions by removing grouping symbols.

Follow along with *Guided Learning Video 1.9.1, Simplifying Algebraic Expressions With Nested Grouping Symbols.*

Understanding the Big Picture: As you listen to the first part of the video, when the pencil icon appears, pause and fill in the blanks, choosing from the words listed.

distributive • braces • parentheses • innermost • brackets

1. Algebraic expressions are sometimes enclosed in grouping symbols like parentheses, _____, and _____.

2. When expressions are inside a set of grouping symbols within another set of grouping symbols, simplify by starting with the _____ grouping symbols.

3. Remove grouping symbols by using the _____ property.

Follow along with the two Guided Learning Video examples, and fill in the blanks on the left as you learn with the instructor. When the pencil icon appears, pause and try the Student Practice on your own.

Guided Learning Video ▶ GUIDED LEARNING VIDEO	**Pause: Student Practice** ✏
1. Simplify the following expression:	2. Simplify the following expression:
$2\left[5y - 3(y-4)\right]$	$4\left[3x - 2(y+6)\right]$
Begin with the innermost set of grouping symbols and work outward to simplify the expression.	
$2\left[5y - 3(y-4)\right]$ $= \boxed{}$ $= \boxed{}$ $= \boxed{}$	

Guided Learning Video ▶ GUIDED LEARNING VIDEO	**Pause: Student Practice** ✏️
3. Simplify the following expression:	**4.** Simplify the following expression:

3. Simplify the following expression:

$$3\left[4a-(2b+c)\right]-2\left[5a+2(3a-4b)\right]$$

Begin with the innermost set of grouping symbols and work outward to simplify the expression.

$$3\left[4a-(2b+c)\right]-2\left[5a+2(3a-4b)\right]$$

$$=\rule{6cm}{0.4pt}$$

$$=\rule{6cm}{0.4pt}$$

$$=\rule{6cm}{0.4pt}$$

$$=\rule{5cm}{0.4pt}$$

4. Simplify the following expression:

$$2\left[4x-3(x+1)\right]-5\left[x-(2y+4)\right]$$

Helpful Hint: Pause the video and write the helpful hint in your own words.

For each Active Video Lesson, when the pencil icon appears, pause and work the problem. Then press play to check your work.

Active Video Lesson 1 ✏️	**Active Video Lesson 2** ✏️
5. Simplify the following expression:	**6.** Simplify the following expression:
$$-3\left\{6x-2\left[x-(5x-3)\right]\right\}$$	$$2\left[7a+4(3b-5c)\right]-3\left[4a-(2b-6c)\right]$$

Name: _____   Date: _____

Instructor: _____   Section: _____

Guided Learning Video Worksheet: Addition Principle of Equality

Text: *Beginning Algebra: Early Graphing*
Student Learning Objective 2.1.1: Use the addition principle to solve equations of the form
$$x + b = c.$$

Follow along with *Guided Learning Video 2.1.1, Addition Principle of Equality*.

Understanding the Big Picture: As you listen to the first part of the video, when the pencil icon appears, pause and fill in the blanks, choosing from the words listed.

solution • addition • inverse • same • equal • expression

1. An equation is a statement indicating two _____ are _____.

2. A number's signed opposite is its additive _____.

3. The numerical _____ of an equation is the number(s) that makes the equation a true statement.

4. The _____ principle of equality states that if the _____ number is added to both sides of an equation, the results on both sides are _____ in value.

Follow along with the two Guided Learning Video examples, and fill in the blanks on the left as you learn with the instructor. When the pencil icon appears, pause and try the Student Practice on your own.

Guided Learning Video	Pause: Student Practice
1. Solve. $x - 9 = 4$ Use the addition principle of equality and the additive inverse to solve the equation. $x - 9 = 4$ $\boxed{} = \boxed{}$ $\boxed{} = \boxed{}$ $\boxed{} = \boxed{}$	2. Solve. $x - 12 = 3$

3. Solve. $x + 7 = 12$

Use the addition principle of equality and the additive inverse to solve the equation.

$x + 7 = 12$

$$\boxed{} = \boxed{}$$

$$\boxed{} = \boxed{}$$

$$\boxed{} = \boxed{}$$

4. Solve. $x - 6 = 5$

Helpful Hint: Pause the video and write the helpful hint in your own words.

For each Active Video Lesson, when the pencil icon appears, pause and work the problem. Then press play to check your work.

Active Video Lesson 1

5. Solve. $15 = x + 8$

Active Video Lesson 2

6. Solve. $6 = x - 11$

Guided Learning Video Worksheet: Multiplication Principle of Equality

Text: *Beginning Algebra: Early Graphing*

Student Learning Objective 2.2.1: Solve equations of the form $\frac{1}{a}x = b$.

Follow along with *Guided Learning Video 2.2.1, Multiplication Principle of Equality*.

Understanding the Big Picture: As you listen to the first part of the video, when the pencil icon appears, pause and fill in the blanks, choosing from the words listed.

numerical • reciprocal • equal • coefficient • multiplication • division

1. The _____ principle of equality states if both sides of an equation are multiplied by the same nonzero number, the results on both sides are _____ in value.

2. When solving an equation of the form $ax = b$, if the coefficient of the variable is a fraction, multiply both sides by the _____ of the _____.

3. In an equation of the form $ax = b$, where a and b are real numbers, a is called the _____ coefficient of the variable.

Follow along with the two Guided Learning Video examples, and fill in the blanks on the left as you learn with the instructor. When the pencil icon appears, pause and try the Student Practice on your own.

Guided Learning Video ⏵ GUIDED LEARNING VIDEO	Pause: Student Practice ✏
1. Solve. $\frac{1}{9}x = -6$	2. Solve. $\frac{1}{5}x = -8$
Use the multiplication principle of equality to solve the equation. $\frac{1}{9}x = -6$ $\square\dfrac{\square}{\square} = \square(\square)$ $\square = \square$ $\square = \square$	

3. Solve. $\dfrac{x}{-8} = 7$

Use the multiplication principle of equality to solve the equation.

$\dfrac{x}{-8} = 7$

$$\boxed{}\dfrac{\boxed{}}{\boxed{}} = \boxed{}\left(\boxed{}\right)$$

$$\boxed{} = \boxed{}$$

$$\boxed{} = \boxed{}$$

4. Solve. $\dfrac{x}{-6} = 3$

Helpful Hint: Pause the video and write the helpful hint in your own words.

For each Active Video Lesson, when the pencil icon appears, pause and work the problem. Then press play to check your work.

Active Video Lesson 1 | **Active Video Lesson 2**

5. Solve. $-\dfrac{1}{7}x = 11$

6. Solve. $\dfrac{x}{12} = 8$

Name: _____   Date: _____

Instructor: _____   Section: _____

Guided Learning Video Worksheet: The Division Principle of Equality

Text: *Beginning Algebra: Early Graphing*
Student Learning Objective 2.2.2: Solve equations of the form $ax = b$.

Follow along with *Guided Learning Video 2.2.2, The Division Principle of Equality*.

Understanding the Big Picture: As you listen to the first part of the video, when the pencil icon appears, pause and fill in the blanks, choosing from the words listed.

nonzero • divide • equal • coefficient • undefined • division

1. The _____ principle of equality states if both sides of an equation are divided by the same nonzero number, the results on both sides are _____ in value.

2. Division by zero is _____; therefore, when applying the division principle, the divisor must be a _____ number.

3. To solve an equation of the form $ax = b$, _____ both sides by the _____ of x.

Follow along with the two Guided Learning Video examples, and fill in the blanks on the left as you learn with the instructor. When the pencil icon appears, pause and try the Student Practice on your own.

Guided Learning Video ▶ GUIDED LEARNING VIDEO	**Pause: Student Practice** ✏️
1. Solve. $7x = 63$	2. Solve. $5x = 65$

Use the division principle of equality to solve the equation.

$7x = 63$

$$\frac{\boxed{}}{\boxed{}} = \frac{\boxed{}}{\boxed{}}$$

$$\boxed{} = \boxed{}$$

Guided Learning Video ⏵ GUIDED LEARNING VIDEO	Pause: Student Practice ✏
3. Solve. $-8x = 56$	**4.** Solve. $-4x = 24$

Use the division principle of equality to solve the equation.

$-8x = 56$

$$\dfrac{\boxed{}}{\boxed{}} = \dfrac{\boxed{}}{\boxed{}}$$

$$\boxed{} = \boxed{}$$

Helpful Hint: Pause the video and write the helpful hint in your own words.

For each Active Video Lesson, when the pencil icon appears, pause and work the problem. Then press play to check your work.

Active Video Lesson 1 ✏	Active Video Lesson 2 ✏
5. Solve. $-6x = -72$	**6.** Solve. $-3x = 51$

Guided Learning Video Worksheet: Solving Equations $ax + b = c$

Text: *Beginning Algebra: Early Graphing*
Student Learning Objective 2.3.1: Solve equations of the form $ax + b = c$.

Follow along with *Guided Learning Video 2.3.1, Solving Equations* $ax + b = c$.

Understanding the Big Picture: As you listen to the first part of the video, when the pencil icon appears, pause and fill in the blanks, choosing from the words listed.

value • reciprocal • addition • true • variable • division

1. When solving equations, sometimes both the _____ principle of equality and the _____ principle of equality will be used to get the _____ alone on one side of the equation.

2. To solve equations, determine the _____ of the variable that will make the equation _____.

3. Solving an equation is oftentimes made easier if the _____ principle of equality is used before the _____ principle of equality.

Follow along with the two Guided Learning Video examples, and fill in the blanks on the left as you learn with the instructor. When the pencil icon appears, pause and try the Student Practice on your own.

Guided Learning Video	**Pause: Student Practice**
1. Solve. $3x - 7 = 8$	2. Solve. $8x - 3 = 37$

Use the addition and division principles of equality to solve the equation.

$3x - 7 = 8$

$\boxed{} = \boxed{}$

$\boxed{} = \boxed{}$

$\dfrac{\boxed{}}{\boxed{}} = \dfrac{\boxed{}}{\boxed{}}$

$\boxed{} = \boxed{}$

Guided Learning Video ▶ GUIDED LEARNING VIDEO	**Pause: Student Practice**
3. Solve. $-27 = 6x + 9$	**4.** Solve. $-54 = 5x + 6$

Use the addition and division principles of equality to solve the equation.

$-27 = 6x + 9$

$$\boxed{} = \boxed{}$$

$$\boxed{} = \boxed{}$$

$$\frac{\boxed{}}{\boxed{}} = \frac{\boxed{}}{\boxed{}}$$

$$\boxed{} = \boxed{}$$

Helpful Hint: Pause the video and write the helpful hint in your own words.

For each Active Video Lesson, when the pencil icon appears, pause and work the problem. Then press play to check your work.

Active Video Lesson 1	**Active Video Lesson 2**
5. Solve. $4y + 5 = 29$	**6.** Solve. $-8x - 12 = 20$

Name: _____   Date: _____

Instructor: _____   Section: _____

Guided Learning Video Worksheet: Solving Equations of the Form $ax + b = cx + d$

Text: *Beginning Algebra: Early Graphing*

Student Learning Objective 2.3.2: Solve equations with the variable on both sides of the equation.

Follow along with *Guided Learning Video 2.3.2, Solving Equations of the Form* $ax + b = cx + d$.

Understanding the Big Picture: As you listen to the first part of the video, when the pencil icon appears, pause and fill in the blanks, choosing from the words listed.

opposite • variable • simplified • addition • terms • numerical

1. The process of solving an equation is easier if each side of the equation is _____ first.

2. To simplify each side of an equation, combine any like _____ and simplify _____ work.

3. The _____ principle of equality should be used to get all _____ terms on one side of the equation and all numerical terms on the _____ side.

Follow along with the two Guided Learning Video examples, and fill in the blanks on the left as you learn with the instructor. When the pencil icon appears, pause and try the Student Practice on your own.

Guided Learning Video ▶ GUIDED LEARNING VIDEO	Pause: Student Practice
1. Solve. $6x + 5 = 4x - 15$ Use the procedure for solving equations to determine the solution. $6x + 5 = 4x - 15$ □ □ □ = □ □/□ = □/□ □ = □	2. Solve. $8x + 3 = 4x - 21$

Guided Learning Video ▶ GUIDED LEARNING VIDEO	**Pause: Student Practice** ✎
3. Solve. $2y - 9 + 4y + 5 = 11y - 7$	**4.** Solve. $6x - 5 + 3x + 2 = 4x + 12$

Use the procedure for solving equations to determine the solution.

$$2y - 9 + 4y + 5 = 11y - 7$$

$$\boxed{} = \boxed{}$$

$$\boxed{} = \boxed{}$$

$$\boxed{} = \boxed{}$$

$$\frac{\boxed{}}{\boxed{}} = \frac{\boxed{}}{\boxed{}}$$

$$\boxed{} = \boxed{}$$

Helpful Hint: Pause the video and write the helpful hint in your own words.

For each Active Video Lesson, when the pencil icon appears, pause and work the problem. Then press play to check your work.

Active Video Lesson 1 ✎	**Active Video Lesson 2** ✎
5. Solve. $7z - 6 - 4z + 11 = 2z - 8$	**6.** Solve. $10x - 7 = 13x + 5$

Guided Learning Video Worksheet: Solve Equations with Parentheses

Text: *Beginning Algebra: Early Graphing*
Student Learning Objective 2.3.3: Solve equations with parentheses.

Follow along with *Guided Learning Video 2.3.3, Solve Equations with Parentheses.*

Understanding the Big Picture: As you listen to the first part of the video, when the pencil icon appears, pause and fill in the blanks, choosing from the words listed.

opposite • multiplication • variable • like • addition • combine • division

1. To simplify each side of an equation, _____ any _____ terms and simplify any numerical work.

2. At the final step in the procedure to solve equations, the _____ or _____ principle of equality is used to solve for the variable.

3. The _____ principle of equality is used to isolate all numerical values on one side of the equation and all _____ terms on the _____ side.

Follow along with the two Guided Learning Video examples, and fill in the blanks on the left as you learn with the instructor. When the pencil icon appears, pause and try the Student Practice on your own.

Guided Learning Video ▶ GUIDED LEARNING VIDEO	**Pause: Student Practice**
1. Solve. $7(2x-1)-11x=5$ Use the procedure for solving equations to determine the solution. $7(2x-1)-11x=5$ $\boxed{} = \boxed{}$ $\boxed{} = \boxed{}$ $\boxed{} = \boxed{}$ $\dfrac{\boxed{}}{\boxed{}} = \dfrac{\boxed{}}{\boxed{}}$ $\boxed{} = \boxed{}$	2. Solve. $3(4x-1)-8x=17$

3. Solve. $5(2-y)=8-(3+6y)$	**4.** Solve. $2(3-y)=9-(5+3y)$

Use the procedure for solving equations to determine the solution.

$5(2-y)=8-(3+6y)$

$$\boxed{}=\boxed{}$$

$$\boxed{}=\boxed{}$$

$$\boxed{}=\boxed{}$$

$$\boxed{}=\boxed{}$$

$$\boxed{}=\boxed{}$$

Helpful Hint: Pause the video and write the helpful hint in your own words.

For each Active Video Lesson, when the pencil icon appears, pause and work the problem. Then press play to check your work.

Active Video Lesson 1 ✏	**Active Video Lesson 2** ✏
5. Solve. $4(x-1)=-6(x+2)+48$	**6.** Solve. $13y-20=3(5y+2)-8$

Guided Learning Video Worksheet: Solve Equations with Parentheses and Decimals

Text: *Beginning Algebra: Early Graphing*
Student Learning Objective 2.3.3: Solve equations with parentheses.

Follow along with *Guided Learning Video 2.3.3, Solve Equations with Parentheses and Decimals*.

Understanding the Big Picture: As you listen to the first part of the video, when the pencil icon appears, pause and fill in the blanks, choosing from the words listed.

divisor • multiply • total • subtract • product • add

1. To _____ or _____ decimals, align the decimal points vertically, and to _____ decimals, align the last digits.

2. In the _____ of two decimals, position the decimal point so the answer has the same number of decimal places as the _____ number of decimal places in the numbers being multiplied.

3. When dividing decimals, the _____ should be a whole number.

Follow along with the two Guided Learning Video examples, and fill in the blanks on the left as you learn with the instructor. When the pencil icon appears, pause and try the Student Practice on your own.

Guided Learning Video ▶ GUIDED LEARNING VIDEO	Pause: Student Practice
1. Solve for x. $2.5(3x - 0.2) + 1.3 = 4x + 7.8$ Remove parentheses, combine like terms, and isolate the variable term to solve for x. $2.5(3x - 0.2) + 1.3 = 4x + 7.8$ [ ] = [ ] [ ] = [ ] [ ] = [ ] [ ] = [ ] [ ] = [ ] [ ] = [ ]	2. Solve for x. $0.2(x + 3.2) = 0.1x + 9.9$

3. Solve for x.

$$0.15 - 0.2x - 0.11 = 0.2(0.1 - 0.8x)$$

Remove parentheses, combine like terms, and isolate the variable term to solve for x.

$$0.15 - 0.2x - 0.11 = 0.2(0.1 - 0.8x)$$

☐	=	☐
☐	=	☐
☐	=	☐
☐	=	☐
☐	=	☐
☐	=	☐

4. Solve for x.

$$0.3(2x - 0.5) + 1.2 = 0.8x + 4.2$$

Helpful Hint: Pause the video and write the helpful hint in your own words.

For each Active Video Lesson, when the pencil icon appears, pause and work the problem. Then press play to check your work.

Active Video Lesson 1 | **Active Video Lesson 2**

5. Solve for x.

$$0.8x + 0.18 - 0.4x = 0.3(x + 0.2)$$

6. Solve for x.

$$0.2(1.3x + 4.2) = 0.51x + 0.09$$

Guided Learning Video Worksheet: Solving Equations with More Than One Set of Parentheses

Text: *Beginning Algebra: Early Graphing*
Student Learning Objective 2.3.3: Solve equations with parentheses.

Follow along with *Guided Learning Video 2.3.3, Solving Equations with More Than One Set of Parentheses*.

Understanding the Big Picture: As you listen to the first part of the video, when the pencil icon appears, pause and fill in the blanks, choosing from the words listed.

right • remove • multiple • first • left • parentheses

1. The order of operations states that any and all operations within _____ must be performed _____.

2. If _____ sets of parentheses are present in an equation, simplify the equation by working from _____ to _____.

3. Use the distributive property to _____ parentheses and create a simpler equation.

Follow along with the two Guided Learning Video examples, and fill in the blanks on the left as you learn with the instructor. When the pencil icon appears, pause and try the Student Practice on your own.

Guided Learning Video ▶ GUIDED LEARNING VIDEO	**Pause: Student Practice** ✏️
1. Solve for x. $11-(6x+1)=5x-2(x+4)$ Use the distributive property to remove parentheses, then combine like terms, and isolate the variable term to solve for x. $11-(6x+1)=5x-2(x+4)$ $\boxed{}=\boxed{}$ $\boxed{}=\boxed{}$ $\boxed{}=\boxed{}$ $\boxed{}\ \boxed{}=\boxed{}$	2. Solve for x. $8-(3x+5)=12-6(x+2)$

Guided Learning Video ▶ GUIDED LEARNING VIDEO	**Pause: Student Practice**
3. Solve for x. $5(2-x)=8-(3+6x)$	**4.** Solve for x. $12(x+1)=30-(2+4x)$

Use the distributive property to remove parentheses, then combine like terms, and isolate the variable term to solve for x.

$5(2-x)=8-(3+6x)$

$$\boxed{} = \boxed{}$$

$$\boxed{} = \boxed{}$$

$$\boxed{} = \boxed{}$$

$$\boxed{} = \boxed{}$$

Helpful Hint: Pause the video and write the helpful hint in your own words.

For each Active Video Lesson, when the pencil icon appears, pause and work the problem. Then press play to check your work.

Active Video Lesson 1	**Active Video Lesson 2**
5. Solve for x. $20-(2x+6)=5(2-x)+2x$	**6.** Solve for z. $5(2z-1)+7=7z-4(z+3)$

Name: _____   Date: _____

Instructor: _____   Section: _____

Guided Learning Video Worksheet: Solve Equations Using the LCD Method

Text: *Beginning Algebra: Early Graphing*
Student Learning Objective 2.4.1: Solve equations with fractions.

Follow along with *Guided Learning Video 2.4.1, Solve Equations Using the LCD Method.*

Understanding the Big Picture: As you listen to the first part of the video, when the pencil icon appears, pause and fill in the blanks, choosing from the words listed.

multiplying • every • clearing • LCD • equivalent

1. Transforming an equation containing fractions to an _____ equation without fractions is done using a process referred to as "_____ the fractions."

2. The process of _____ the fractions is accomplished by _____ all the terms of the equation by the _____ of all the fractions.

3. It's important to remember that when using the procedure for solving an equation with fractions, _____ term must be multiplied by the _____.

Follow along with the two Guided Learning Video examples, and fill in the blanks on the left as you learn with the instructor. When the pencil icon appears, pause and try the Student Practice on your own.

Guided Learning Video ▶ GUIDED LEARNING VIDEO	Pause: Student Practice ✏
1. Solve. $\dfrac{x}{2}+\dfrac{x}{3}=10$ Follow the procedure for solving an equation with fractions using the LCD to solve the equation. $\Box\cdot\dfrac{\Box}{\Box}+\Box\cdot\dfrac{\Box}{\Box}=\Box\cdot\Box$ $\Box\Box+\Box=\Box$ $\Box=\Box$ $\dfrac{\Box}{\Box}=\dfrac{\Box}{\Box}$ $\Box=\Box$	2. Solve. $\dfrac{x}{4}+\dfrac{x}{6}=10$

3. Solve. $\dfrac{y}{4} + y = 25$

Follow the procedure for solving an equation with fractions using the LCD to solve the equation.

$$\square \cdot \dfrac{\square}{\square} + \square \cdot \square = \square \cdot \square$$

$$\boxed{} + \boxed{} = \boxed{}$$

$$\boxed{} = \boxed{}$$

$$\dfrac{\square}{\square} = \dfrac{\square}{\square}$$

$$\square = \square$$

4. Solve. $\dfrac{y}{6} + 2y = 26$

Helpful Hint: Pause the video and write the helpful hint in your own words.

For each Active Video Lesson, when the pencil icon appears, pause and work the problem. Then press play to check your work.

Active Video Lesson 1 ✏️ | **Active Video Lesson 2** ✏️

5. Solve. $-4x + \dfrac{1}{3} = \dfrac{5}{6}$

6. Solve. $\dfrac{x}{4} + \dfrac{x}{5} = 9$

Name: _____   Date: _____

Instructor: _____   Section: _____

Guided Learning Video Worksheet: Solve Equations with Fractions

Text: *Beginning Algebra: Early Graphing*
Student Learning Objective 2.4.1: Solve equations with fractions.

Follow along with *Guided Learning Video 2.4.1, Solve Equations with Fractions.*

Understanding the Big Picture: As you listen to the first part of the video, when the pencil icon appears, pause and fill in the blanks, choosing from the words listed.

denominator • distributive • without • multiply • equivalent

1. To avoid working with fractions, transform the given equation with fractions into a(n) _____ equation _____ fractions.

2. To clear an equation of fractions, _____ each side by the lowest common _____ of all the fractions in the equation.

3. Ensure that each term of an equation is multiplied by the LCD by using the _____ property.

Follow along with the two Guided Learning Video examples, and fill in the blanks on the left as you learn with the instructor. When the pencil icon appears, pause and try the Student Practice on your own.

Guided Learning Video	Pause: Student Practice
1. Solve for x. $$\frac{1}{2}(2x-27)+\frac{3x}{4}=\frac{1}{2}(x+3)$$ Use the procedure to solve equations to find the value of the variable x. $$\frac{1}{2}(2x-27)+\frac{3x}{4}=\frac{1}{2}(x+3)$$	2. Solve for x. $$\frac{1}{5}(x+1)-\frac{3}{5}=\frac{3}{4}(x-2)$$

Placeholder boxes:

[ ] = [ ]
[ ] = [ ]
[ ] = [ ]
[ ] = [ ]
[ ] = [ ]

Guided Learning Video ▶ GUIDED LEARNING VIDEO	**Pause: Student Practice**

3. Solve for x.

$$\frac{1}{4}(3x-2)=\frac{1}{3}(2x-1)-\frac{1}{4}$$

Use the procedure to solve equations to find the value of the variable x.

$$\frac{1}{4}(3x-2)=\frac{1}{3}(2x-1)-\frac{1}{4}$$

☐ = ☐

☐ = ☐

☐ = ☐

☐ = ☐

☐ = ☐

4. Solve for x.

$$\frac{3}{8}(3x-4)=\frac{1}{2}(x-2)+2$$

Helpful Hint: Pause the video and write the helpful hint in your own words.

For each Active Video Lesson, when the pencil icon appears, pause and work the problem. Then press play to check your work.

Active Video Lesson 1	**Active Video Lesson 2**

5. Solve for x.

$$\frac{2}{3}(x+8)+\frac{3}{5}=\frac{1}{5}(11-6x)$$

6. Solve for x.

$$\frac{1}{2}(x+5)=\frac{1}{5}(x-2)+\frac{1}{2}$$

Name: _____   Date: _____

Instructor: _____   Section: _____

Guided Learning Video Worksheet: Translate English Statements into Equations

Text: *Beginning Algebra: Early Graphing*
Student Learning Objective 2.5.1: Translate English phrases into algebraic expressions.

Follow along with *Guided Learning Video 2.5.1, Translate English Statements into Equations.*

Understanding the Big Picture: As you listen to the first part of the video, when the pencil icon appears, pause and fill in the blanks, choosing from the words listed.

multiplication • subtraction • division • translated • addition • minus • equivalents

1. English words can be _____ into their math _____ and then used to solve math problems.

2. English phrases like "increased by", "more than", and "sum of" can all be represented by a(n) _____ symbol.

3. Phrases like "less than", "decreased by", and "difference between" indicate the operation of _____ and are represented by a(n) _____ sign.

4. The word "of" indicates _____ and the word "ratio" indicates _____.

Follow along with the two Guided Learning Video examples, and fill in the blanks on the left as you learn with the instructor. When the pencil icon appears, pause and try the Student Practice on your own.

Guided Learning Video ▶ GUIDED LEARNING VIDEO	Pause: Student Practice
1. Translate the English statement into an equation.	2. Translate the English statement into an equation.
"The sum of 15 and what number is 35?"	"The sum of 8 and what number is 23?"
Identify the math operation and use the variable x to represent the number.	
[ ]	

Guided Learning Video (▶) GUIDED LEARNING VIDEO	**Pause: Student Practice** ✏️
3. Translate the English statement into an equation. "The quotient of 56 and what number is 7 ?" Identify the math operation and use the variable x to represent the number. [　　　　　　　　]	4. Translate the English statement into an equation. "The difference between 45 and what number is 17 ?"

Helpful Hint: Pause the video and write the helpful hint in your own words.

For each Active Video Lesson, when the pencil icon appears, pause and work the problem. Then press play to check your work.

Active Video Lesson 1 ✏️	**Active Video Lesson 2** ✏️
5. Translate the English statement into an equation. "What number decreased by 4 is the same as 9 ?"	6. Translate the English statement into an equation. "8 times what number is equal to 72 ?"

Name: _____   Date: _____
Instructor: _____   Section: _____

Guided Learning Video Worksheet: Write an Algebraic Expression to Compare Quantities Using One Variable

Text: *Beginning Algebra: Early Graphing*
Student Learning Objective 2.5.2: Write an algebraic expression to compare two or more quantities.

Follow along with *Guided Learning Video 2.5.2, Write an Algebraic Expression to Compare Quantities Using One Variable.*

Understanding the Big Picture: As you listen to the first part of the video, when the pencil icon appears, pause and fill in the blanks, choosing from the words listed.

basis • same • subtracted • variable • before • commutative

1. The number appearing _____ the phrase "less than" would be the number being _____.

2. The operation of addition is _____; therefore, it's easier to write the values and mathematical symbols in the _____ order as they appear in the English statement.

3. When two or more quantities are described in terms of another different quantity, let the _____ represent the quantity that is the _____ of comparison.

Follow along with the two Guided Learning Video examples, and fill in the blanks on the left as you learn with the instructor. When the pencil icon appears, pause and try the Student Practice on your own.

Guided Learning Video ▶ GUIDED LEARNING VIDEO	**Pause: Student Practice** ✏️
1. Use a variable and algebraic expressions to describe the quantities in the English expression. "Olivia is 4 years older than Jake." Identify the quantities to be expressed algebraically, and then write each in terms of the same variable. ☐ = ☐ ☐ = ☐	2. Use a variable and algebraic expressions to describe the quantities in the English expression. "Sally's height is 5 inches less than Neal's height."

3. Use a variable and algebraic expressions to describe the quantities in the English expression. Use the variable x.

"The measure of the first angle is 15° less than the measure of the second angle. The measure of the third angle is 12° more than twice the measure of the second angle."

Identify the quantities to be expressed algebraically, and then write each in terms of the single variable x.

	=	
	=	
	=	

4. Use a variable and algebraic expressions to describe the quantities in the English expression. Use the variable x.

"The measure of the second side of a triangle is 3 inches less than the measure of the first side. The measure of the third side is 6 inches less than triple the first side.

Helpful Hint: Pause the video and write the helpful hint in your own words.

For each Active Video Lesson, when the pencil icon appears, pause and work the problem. Then press play to check your work.

Active Video Lesson 1 ✏️ | **Active Video Lesson 2** ✏️

5. Write algebraic expressions for the length and width of the rectangle.

"The length of the rectangle is 5 inches shorter than double the width."

6. "Savannah's salary is $750 more than Ryan's salary."

Write algebraic expressions for Savannah's salary and Ryan's salary.

Guided Learning Video Worksheet: Solve Word Problems: Number Problems

Text: *Beginning Algebra: Early Graphing*
Student Learning Objective 2.6.1: Solve number problems.

Follow along with *Guided Learning Video 2.6.1, Solve Word Problems: Number Problems.*

Understanding the Big Picture: As you listen to the first part of the video, when the pencil icon appears, pause and fill in the blanks, choosing from the words listed.

quantity • equation • same • calculations • variable

1. To solve a word problem, select a(n) _____ to represent one unknown _____, and then write algebraic expressions in terms of the _____ variable to represent the other quantities in the problem.

2. Write a(n) _____ composed of algebraic expressions that expresses the relationship between the quantities in the problem.

3. Solve the _____ and perform any additional _____ to fully answer the problem's question.

Follow along with the two Guided Learning Video examples, and fill in the blanks on the left as you learn with the instructor. When the pencil icon appears, pause and try the Student Practice on your own.

Guided Learning Video ▶ GUIDED LEARNING VIDEO	Pause: Student Practice ✏
1. $\frac{5}{6}$ of a number is 35. What is the number? Solve the problem using the 3-step outline for solving word problems. unknown number $=$ ☐ ☐ ☐ ☐ ☐	2. $\frac{3}{4}$ of a number is -6. What is the number?

3. One number is 15 less than 7 times a second number. The sum of the numbers is 73. Find each number.

   Solve the problem using the 3-step outline for solving word problems.

   first number ⎢= ⎢   ⎢

   second number ⎢= ⎢   ⎢

   ⎢   ⎢

   ⎢   ⎢

   ⎢   ⎢

   ⎢   ⎢

   first number ⎢= ⎢   ⎢

   second number ⎢= ⎢   ⎢

4. One number is 5 more than double a second number. The sum of the numbers is 23. Find each number.

Helpful Hint: Pause the video and write the helpful hint in your own words.

For each Active Video Lesson, when the pencil icon appears, pause and work the problem. Then press play to check your work.

Active Video Lesson 1 ✏️ | **Active Video Lesson 2** ✏️

5. The larger of two numbers is 9 more than twice the smaller number. The sum of the two numbers is 51. Find each number.

6. Two-fifths of a number is −18. What is the number?

Guided Learning Video Worksheet: Solve Investment Problems Involving Simple Interest

Text: *Beginning Algebra: Early Graphing*
Student Learning Objective 2.7.1: Solve investment problems involving simple interest.

Follow along with *Guided Learning Video 2.7.1, Solve Investment Problems Involving Simple Interest.*

Understanding the Big Picture: As you listen to the first part of the video, when the pencil icon appears, pause and fill in the blanks, choosing from the words listed.

rate • interest • years • time • principle

1. Simple interest is equal to the product of _____ times _____ times _____.

2. The cost of borrowing money or the income received from investing money is called _____.

3. The amount of money borrowed or invested is called the _____.

4. When computing simple interest, the amount of time over which money is borrowed or invested is usually measured in _____ unless otherwise stated.

Follow along with the two Guided Learning Video examples, and fill in the blanks on the left as you learn with the instructor. When the pencil icon appears, pause and try the Student Practice on your own.

Guided Learning Video ▶ GUIDED LEARNING VIDEO	**Pause: Student Practice**
1. Find the interest on $4000 borrowed at a simple interest rate of 12% for one year. Convert the percent to a decimal and use the simple interest formula to find the interest. ▭ ▭ ▭	2. Find the interest on $2000 borrowed at a simple interest rate of 3.5% for one year.

3. Last year Haing invested $1500 that she had saved. She invested part of her money in a certificate of deposit that earned 6% simple interest and the rest in a savings account that earned 4% simple interest. At the end of the year, she had earned $80 in interest. How much did she invest in the certificate of deposit, and how much did she invest in the savings account?

   Use a Mathematics Blueprint for Problem Solving to assist in finding the solution.

   amount invested at 6% = ☐

   amount invested at 4% = ☐

   ☐

   ☐

   ☐

   ☐

   amount invested at 6% = ☐

   amount invested at 4% = ☐

4. A portion of $15,000 is invested at a simple interest rate of 2%, and the remaining portion is invested at a simple interest rate of 4%. If $420 is earned in simple interest at the end of one year, how much was invested at each rate?

Helpful Hint: Pause the video and write the helpful hint in your own words.

For each Active Video Lesson, when the pencil icon appears, pause and work the problem. Then press play to check your work.

Active Video Lesson 1	Active Video Lesson 2
5. Franco invested $4000 in money market funds. Part was invested at 14% interest and the rest at 11% interest. At the end of each year, the fund company pays interest. After one year, he earned $482 in simple interest. How much was invested at each interest rate?	**6.** Find the interest on $3800 borrowed at a simple interest rate of 9% for one year.

Name: _____   Date: _____

Instructor: _____   Section: _____

Guided Learning Video Worksheet: Interpreting Inequality Statements

Text: *Beginning Algebra: Early Graphing*
Student Learning Objective 2.8.1: Interpret inequality statements.

Follow along with *Guided Learning Video 2.8.1, Interpreting Inequality Statements.*

Understanding the Big Picture: As you listen to the first part of the video, when the pencil icon appears, pause and fill in the blanks, choosing from the words listed.

equations • less • greater • plus • inequalities

1. In mathematics, the relationships of "less than" and "greater than" are called _____.

2. When read left to right, an expression with the inequality symbol pointing to the right means _____ than.

3. When read left to right, an expression with the inequality symbol pointing to the left means _____ than.

4. On a number line, if a first number is positioned to the right of a second number, the first number is _____ than the second number.

Follow along with the two Guided Learning Video examples, and fill in the blanks on the left as you learn with the instructor. When the pencil icon appears, pause and try the Student Practice on your own.

Guided Learning Video 🔘 GUIDED LEARNING VIDEO	Pause: Student Practice ✏️
1. Replace ? with < or >.	2. Replace ? with < or >.

a. −6 ? 3
    b. 5 ? −11
	a. −4 ? −3
	    b. −1 ? −6

Use a number line to determine which inequality symbol to use.	**a.** 
a. −6 3 **b.** 5 −11	**b.** 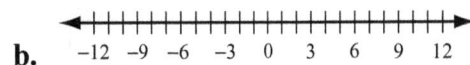
a. 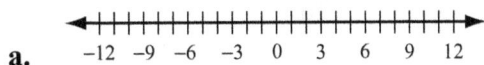	
b. 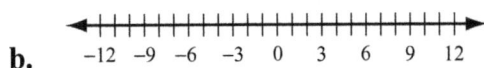	

119

Guided Learning Video ▶ GUIDED LEARNING VIDEO	**Pause: Student Practice**

3. Replace ? with < or >.

   a. −8 ? −9
   b. −7 ? 0

   Use a number line to determine which inequality symbol to use.

   a. −8   −9
   b. −7   0

   a.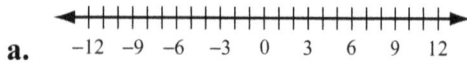

   b.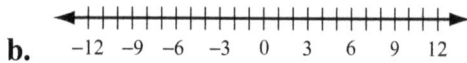

4. Replace ? with < or >.

   a. −4 ? 0
   b. 5 ? −5

   a.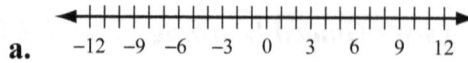

   b.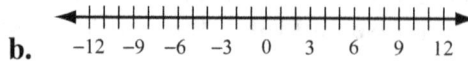

Helpful Hint: Pause the video and write the helpful hint in your own words.

For each Active Video Lesson, when the pencil icon appears, pause and work the problem. Then press play to check your work.

Active Video Lesson 1	**Active Video Lesson 2**

5. Insert the proper symbol between the numbers.

   a. 0 ? −5
   b. −3 ? −2

   a.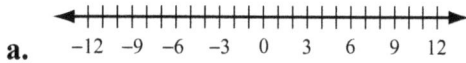

   b.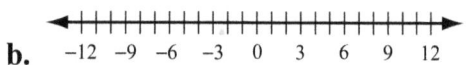

6. Insert the proper symbol between the numbers.

   a. −4 ? 1
   b. 7 ? −10

   a.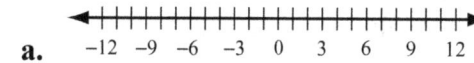

   b.

Name: _____   Date: _____

Instructor: _____   Section: _____

Guided Learning Video Worksheet: Graphing an Inequality on a Number Line

Text: *Beginning Algebra: Early Graphing*
Student Learning Objective 2.8.1: Graph an inequality on a number line.

Follow along with *Guided Learning Video 2.8.1, Graphing an Inequality on a Number Line*.

Understanding the Big Picture: As you listen to the first part of the video, when the pencil icon appears, pause and fill in the blanks, choosing from the words listed.

set • graph • inequality • solution • relationship

1. A(n) _____ is sometimes used to express the _____ between a number and a variable.

2. Any number that makes an inequality true is called a _____ of the inequality, and the _____ of all numbers that make the inequality true is called the solution _____.

3. A picture representing all of the solutions of an inequality is the _____ of the inequality.

Follow along with the two Guided Learning Video examples, and fill in the blanks on the left as you learn with the instructor. When the pencil icon appears, pause and try the Student Practice on your own.

Guided Learning Video ▶ GUIDED LEARNING VIDEO	Pause: Student Practice ✏
1. State the mathematical relationship in words, and then graph the inequality. $x \geq -1$	2. State the mathematical relationship in words, and then graph the inequality. $x \leq -3$
 ┌────────────────────┐ │ │ └────────────────────┘ ←─┼─┼─┼─┼─┼─┼─┼─┼─┼─┼─┼─┼─→	 ←─┼─┼─┼─┼─┼─┼─┼─┼─┼─┼─┼─┼─→

Guided Learning Video GUIDED LEARNING VIDEO	Pause: Student Practice

3. State the mathematical relationship in words, and then graph the inequality. $x < 1$

[box]

←+++++++++++++→

4. State the mathematical relationship in words, and then graph the inequality. $x > 0$

←+++++++++++++→

Helpful Hint: Pause the video and write the helpful hint in your own words.

For each Active Video Lesson, when the pencil icon appears, pause and work the problem. Then press play to check your work.

Active Video Lesson 1	Active Video Lesson 2

5. State the mathematical relationship in words, and then graph the inequality. $\frac{5}{2} < x$

←+++++++++++++→

6. State the mathematical relationship in words, and then graph the inequality. $x \leq 0$

←+++++++++++++→

Guided Learning Video Worksheet: Solve and Graph an Inequality

Text: *Beginning Algebra: Early Graphing*
Student Learning Objective 2.8.1: Solve and graph an inequality.

Follow along with *Guided Learning Video 2.8.1, Solve and Graph an Inequality.*

Understanding the Big Picture: As you listen to the first part of the video, when the pencil icon appears, pause and fill in the blanks, choosing from the words listed.

division • true • reversed • unchanged • solving • positive

1. When _____ an inequality, all of the values that make it _____ are being found.

2. When operations with a(n) _____ number are performed on an inequality, the inequality symbol is _____.

3. The direction of the inequality symbol is _____ when the operations of multiplication or _____ are performed by a negative number on both sides of the inequality.

Follow along with the two Guided Learning Video examples, and fill in the blanks on the left as you learn with the instructor. When the pencil icon appears, pause and try the Student Practice on your own.

Guided Learning Video	Pause: Student Practice
1. Solve and graph. $4x + 1 < -3$ Follow the procedure for solving inequalities, then graph the solution set. $4x + 1 < -3$ [ ] [ ] [ ]	2. Solve and graph. $7x - 2 > 19$

3. Solve and graph. $5 - 2x \geq 11$

Follow the procedure for solving inequalities, then graph the solution set.

$5 - 2x \geq 11$

4. Solve and graph. $4 - 8x \leq 12$

Helpful Hint: Pause the video and write the helpful hint in your own words.

For each Active Video Lesson, when the pencil icon appears, pause and work the problem. Then press play to check your work.

Active Video Lesson 1

Active Video Lesson 2

5. Solve and graph. $\frac{3}{4}(x - 6) > x - 5$

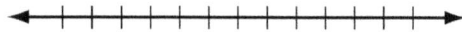

6. Solve and graph. $7 - 3x \leq 19$

Name: _____   Date: _____
Instructor: _____   Section: _____

Guided Learning Video Worksheet: Solve Word Problems: Plot a Point

Text: *Beginning Algebra: Early Graphing*
Student Learning Objective 3.1.1: Plot a point, given the coordinates.

Follow along with *Guided Learning Video 3.1.1, Plot a Point*.

Understanding the Big Picture: As you listen to the first part of the video, when the pencil icon appears, pause and fill in the blanks, choosing from the words listed.

coordinates • zero • *y*-coordinate • *x*-coordinate • ordered • origin

1. A rectangular coordinate system is composed of horizontal and vertical number lines intersecting at _____ on each number line which is a location called the _____.

2. Points on a rectangular coordinate system are represented by _____ pairs called the _____ of the point.

3. In an ordered pair, the first number is the _____, and the second number is the _____.

Follow along with the two Guided Learning Video examples, and fill in the blanks on the left as you learn with the instructor. When the pencil icon appears, pause and try the Student Practice on your own.

Guided Learning Video (▶) GUIDED LEARNING VIDEO	**Pause: Student Practice**
1. Plot and label the points. $L = (-1,\ 3)$ $M = (-3,\ -4)$ $N = (5,\ -1)$	2. Plot and label the points. $L = (4,\ -2)$ $M = (-3,\ 3)$ $N = (-2,\ -1)$

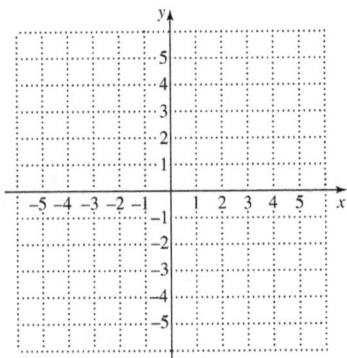

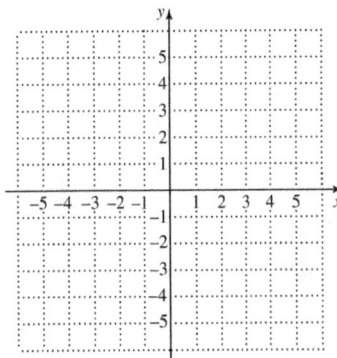

3. Plot and label the points. Use approximations when necessary.

$M = (1, \ 0)$

$N = (-2.2, \ -4.5)$

$O = \left(-5, \ -\dfrac{3}{2}\right)$

$P = (0, \ -5)$

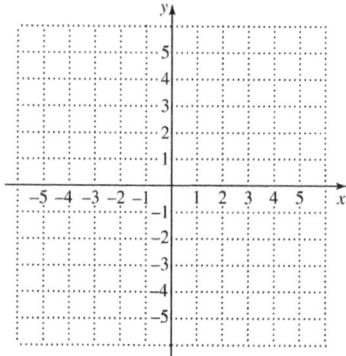

4. Plot and label the points. Use approximations when necessary.

$M = (0, \ -3)$

$N = (-1.5, \ -3.5)$

$O = \left(-\dfrac{5}{2}, \ 4\right)$

$P = (-2, \ 0)$

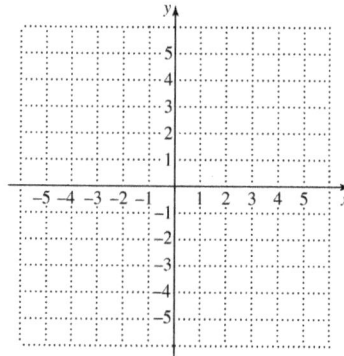

Helpful Hint: Pause the video and write the helpful hint in your own words.

For each Active Video Lesson, when the pencil icon appears, pause and work the problem. Then press play to check your work.

Active Video Lesson 1 | **Active Video Lesson 2**

5. Plot and label the points.

$A = (0, \ 4)$ $B = (-3.5, \ 2.5)$

$C = \left(\dfrac{3}{2}, \ -2\right)$ $D = (0, \ -3)$

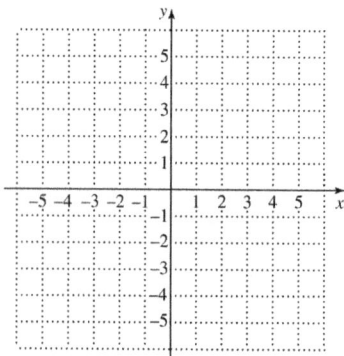

6. Plot and label the points.

$E = (2, \ 4)$ $F = (-3, \ -3)$

$G = (-4, \ 1)$ $H = (1, \ -4)$

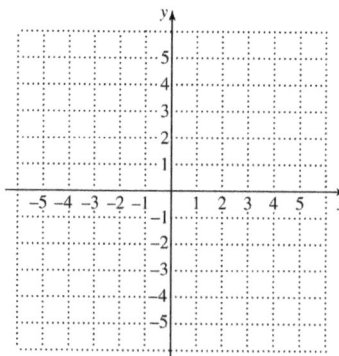

Guided Learning Video Worksheet: Determine the Coordinates of a Plotted Point

Text: *Beginning Algebra: Early Graphing*
Student Learning Objective 3.1.2: Determine the coordinates of a plotted point.

Follow along with *Guided Learning Video 3.1.2, Determine the Coordinates of a Plotted Point.*

Understanding the Big Picture: As you listen to the first part of the video, when the pencil icon appears, pause and fill in the blanks, choosing from the words listed.

down • *y*-axis • *x*-axis • up • origin • first

1. The coordinates of the _____ are $(0, 0)$.

2. The number of units moved left or right from the _____ along the _____ is the _____ number in the ordered pair.

3. The second number of the ordered pair identifies the number of units to move _____ or _____ from the *x*-axis.

Follow along with the two Guided Learning Video examples, and fill in the blanks on the left as you learn with the instructor. When the pencil icon appears, pause and try the Student Practice on your own.

Guided Learning Video ▶ GUIDED LEARNING VIDEO	Pause: Student Practice ✏️
1. What ordered pairs identify points *A*, *B*, and *C*?	2. What ordered pairs identify points *A*, *B*, and *C*?

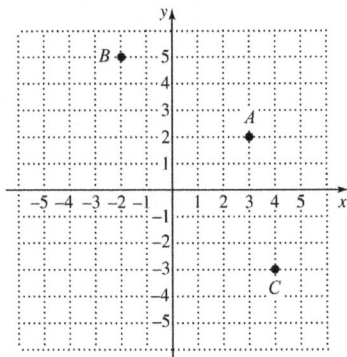

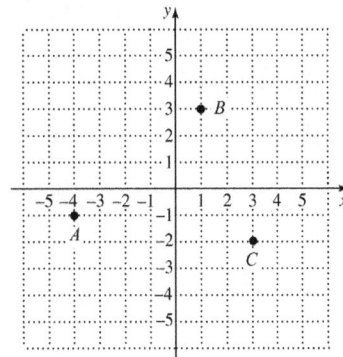

3. Plot the points representing the data in the table.

# of years since 1990	# of Landline Home Phones (in thousands)
0	550
5	428
10	172
15	135
20	96
25	50

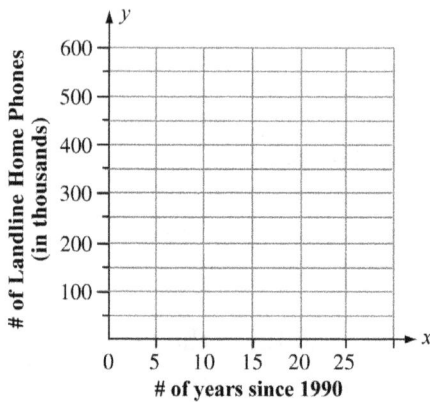

4. Plot the points representing the data in the table.

Number of years since 2004	Number of Cell Phones Sold (in millions)
0	6
2	8
4	14
6	16
8	22
10	30

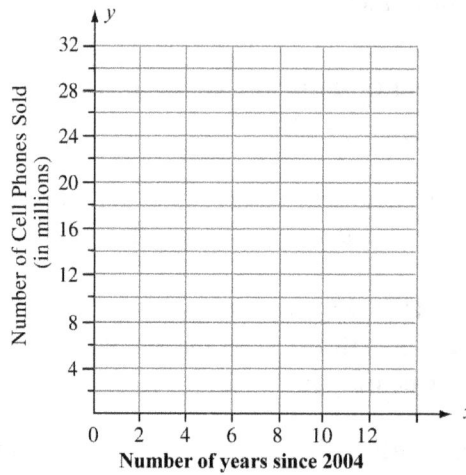

Helpful Hint: Pause the video and write the helpful hint in your own words.

128

For each Active Video Lesson, when the pencil icon appears, pause and work the problem. Then press play to check your work.

Active Video Lesson 1	**Active Video Lesson 2**

5. Plot the points representing the data in the table.

# of years since 1995	# of Condos Constructed in Essex County
0	100
5	400
10	600
15	800
20	1000

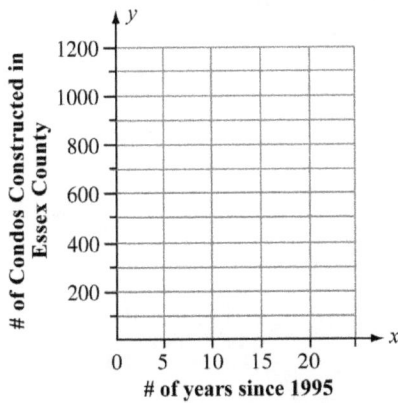

6. What ordered pairs identify points D, E, and F?

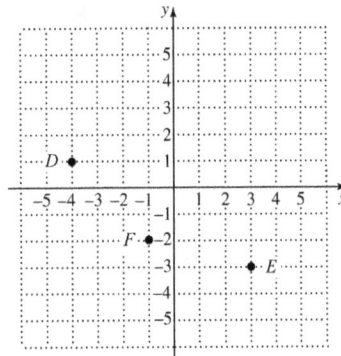

Name: _____   Date: _____
Instructor: _____   Section: _____

Guided Learning Video Worksheet: Finding Ordered Pair Solutions to Linear Equations

Text: *Beginning Algebra: Early Graphing*
Student Learning Objective 3.1.3: Find ordered pairs for a given linear equation.

Follow along with *Guided Learning Video 3.1.3, Finding Ordered Pair Solutions to Linear Equations*.

Understanding the Big Picture: As you listen to the first part of the video, when the pencil icon appears, pause and fill in the blanks, choosing from the words listed.

infinite • true • variables • linear • solution • zero

1. A _____ equation in two _____ is an equation that can be written in the form $Ax + By = C$ where A, B, and C are real numbers, but A and B are not both _____.

2. An ordered pair is a _____ of a linear equation if, when substituted for x and y, they make a _____ mathematical statement.

3. There are an _____ number of solutions to any given linear equation in two variables.

Follow along with the two Guided Learning Video examples, and fill in the blanks on the left as you learn with the instructor. When the pencil icon appears, pause and try the Student Practice on your own.

Guided Learning Video ▶ GUIDED LEARNING VIDEO	**Pause: Student Practice**
1. Complete the ordered pair solutions of the equation $x + 4y = 6$.	2. Complete the ordered pair solutions of the equation $2x + y = 8$.
a. (__, 0) **b.** (2, __)	**a.** (−4, __) **b.** (__, 2)
Replace the variable in the equation with the known value and solve for the remaining variable.	
a. $\boxed{} = \boxed{}$ $\boxed{} = \boxed{}$, so $\left(\boxed{},\boxed{}\right)$	
b. $\boxed{} = \boxed{}$ $\boxed{} = \boxed{}$, so $\left(\boxed{},\boxed{}\right)$	

3. Complete the ordered pair solutions of the equation $5x - 2y = 10$.

   a. $(\underline{},\ 0)$ **b.** $(-2,\ \underline{})$

   Replace the variable in the equation with the known value and solve for the remaining variable.

   a. $\boxed{} = \boxed{}$
   $\boxed{} = \boxed{}$, so $\left(\boxed{},\boxed{}\right)$

   b. $\boxed{} = \boxed{}$
   $\boxed{} = \boxed{}$, so $\left(\boxed{},\boxed{}\right)$

4. Complete the ordered pair solutions of the equation $3x + 2y = 12$.

   a. $(0,\ \underline{})$ **b.** $(\underline{},\ -3)$

Helpful Hint: Pause the video and write the helpful hint in your own words.

For each Active Video Lesson, when the pencil icon appears, pause and work the problem. Then press play to check your work.

Active Video Lesson 1 | **Active Video Lesson 2**

5. Complete the ordered pair solutions of the equation $4x - 3y = 12$.

   a. $(0,\ \underline{})$ **b.** $(\underline{},\ 4)$

6. Complete the ordered pair solutions of the equation $2x + y = 8$.

   a. $(2,\ \underline{})$ **b.** $(\underline{},\ -2)$

Name: _____   Date: _____
Instructor: _____   Section: _____

Guided Learning Video Worksheet: Graph Linear Equations

Text: *Beginning Algebra: Early Graphing*
Student Learning Objective 3.2.1: Graph linear equations by plotting three ordered pairs.

Follow along with *Guided Learning Video 3.2.1, Graph Linear Equations.*

Understanding the Big Picture: As you listen to the first part of the video, when the pencil icon appears, pause and fill in the blanks, choosing from the words listed.

straight • *y*-value • linear • solutions • *x*-value • infinite

1. The first value in an ordered pair is the _____, and the second value is the
   _____.

2. A linear equation in two variables has a(n) _____ number of ordered pair
   _____.

3. The graph of any _____ equation in two variables is a(n) _____
   line.

Follow along with the two Guided Learning Video examples, and fill in the blanks on the left as you learn with the instructor. When the pencil icon appears, pause and try the Student Practice on your own.

Guided Learning Video ▶ GUIDED LEARNING VIDEO	Pause: Student Practice

1. Determine three ordered pair solutions to $y = 3x - 1$.
 	
     a. Let $x = -1, 0, 1$.
 	
 $y = \boxed{} = \boxed{} = \boxed{}$, so $\left(\boxed{}, \boxed{}\right)$
 	
 $y = \boxed{} = \boxed{} = \boxed{}$, so $\left(\boxed{}, \boxed{}\right)$
 	
 $y = \boxed{} = \boxed{} = \boxed{}$, so $\left(\boxed{}, \boxed{}\right)$
 	
     b. Plot the ordered pairs on a rectangular coordinate system.
	2. Determine three ordered pair solutions to $y = 2x + 1$.
	
	     a. Let $x = -1, 0, 1$.
	
	     b. Plot the ordered pairs on a rectangular coordinate system.

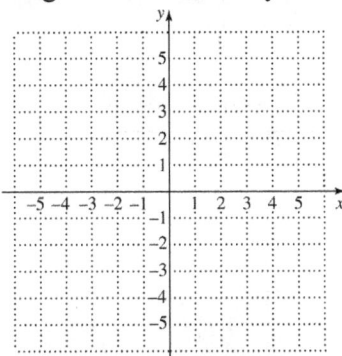

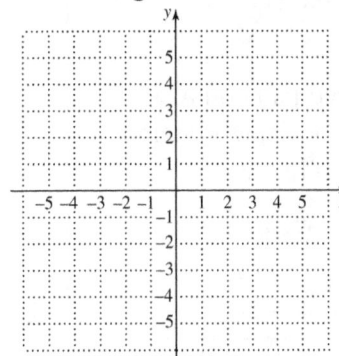

133
Copyright © 2017 Pearson Education, Inc.

3. Determine three ordered pair solutions to $y = -2x + 3$.

    a. Let $x = -1, 0, 1$.

$y = \boxed{} = \boxed{} = \boxed{}$, so $\left(\boxed{}, \boxed{}\right)$

$y = \boxed{} = \boxed{} = \boxed{}$, so $\left(\boxed{}, \boxed{}\right)$

$y = \boxed{} = \boxed{} = \boxed{}$, so $\left(\boxed{}, \boxed{}\right)$

    b. Plot the ordered pairs on a rectangular coordinate system.

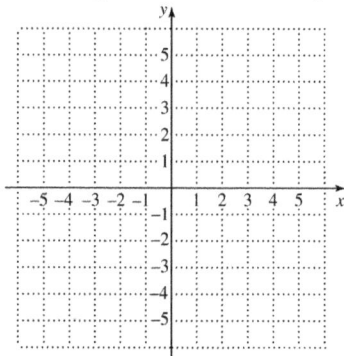

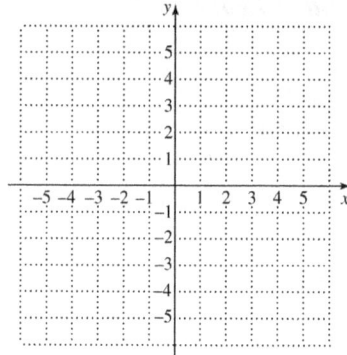

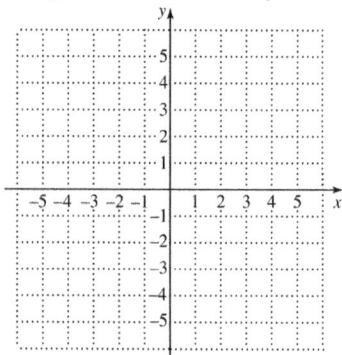

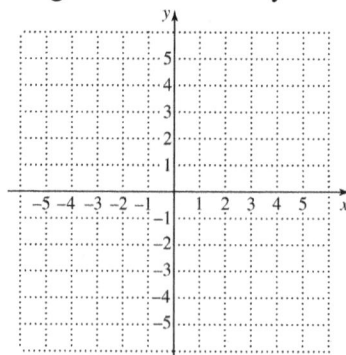

4. Determine three ordered pair solutions to $y = x - 3$.

    a. Let $x = -1, 0, 1$.

    b. Plot the ordered pairs on a rectangular coordinate system.

Helpful Hint: Pause the video and write the helpful hint in your own words.

For each Active Video Lesson, when the pencil icon appears, pause and work the problem. Then press play to check your work.

Active Video Lesson 1 | **Active Video Lesson 2**

5. Determine three ordered pair solutions to $y = -4x - 1$.

    a. Let $x = -1, 0, 1$.

    b. Plot the ordered pairs on a rectangular coordinate system.

6. Determine three ordered pair solutions to $y = 2x + 4$.

    a. Let $x = -1, 0, 1$.

    b. Plot the ordered pairs on a rectangular coordinate system.

Name: _____   Date: _____

Instructor: _____   Section: _____

Guided Learning Video Worksheet: Graph Lines by Plotting Intercepts

Text: *Beginning Algebra: Early Graphing*
Student Learning Objective 3.2.2: Graph a straight line by plotting its intercepts.

Follow along with *Guided Learning Video 3.2.2, Graph Lines by Plotting Intercepts.*

Understanding the Big Picture: As you listen to the first part of the video, when the pencil icon appears, pause and fill in the blanks, choosing from the words listed.

y-intercept • *y*-axis • horizontal • *x*-axis • *x*-intercept • vertical

1. The _____ line in the rectangular coordinate system is the *y*-axis.

2. The _____ line in the rectangular coordinate system is the *x*-axis.

3. Straight lines that aren't vertical or horizontal have both a(n) _____ and a(n) _____.

4. The *x*-intercept of a line is the point where the line crosses the _____, and the *y*-intercept is the point where the line crosses the _____.

Follow along with the two Guided Learning Video examples, and fill in the blanks on the left as you learn with the instructor. When the pencil icon appears, pause and try the Student Practice on your own.

Guided Learning Video ▶ GUIDED LEARNING VIDEO	**Pause: Student Practice**
1. Find the *x*-intercept, *y*-intercept, and an additional ordered pair. Graph the equation. $3x + 2y = 6$	2. Find the *x*-intercept, *y*-intercept, and an additional ordered pair. Graph the equation. $6x + 3y = 12$

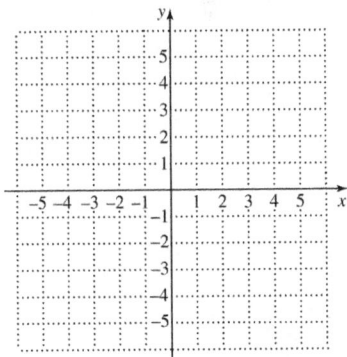

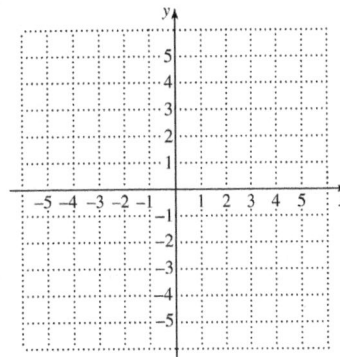

3. Find the x-intercept, y-intercept, and an additional ordered pair. Graph the equation. $5x - 2y = 10$

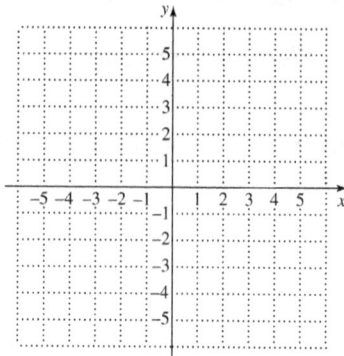

4. Find the x-intercept, y-intercept, and an additional ordered pair. Graph the equation. $2x - 4y = 8$

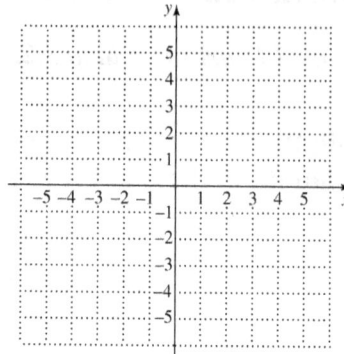

Helpful Hint: Pause the video and write the helpful hint in your own words.

For each Active Video Lesson, when the pencil icon appears, pause and work the problem. Then press play to check your work.

Active Video Lesson 1 | **Active Video Lesson 2**

5. Find the x-intercept, y-intercept, and an additional ordered pair. Graph the equation. $2x - y = 4$

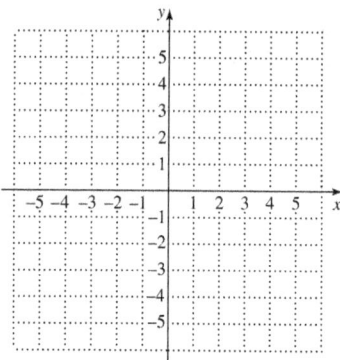

6. Find the x-intercept, y-intercept, and an additional ordered pair. Graph the equation. $x + 4y = -4$

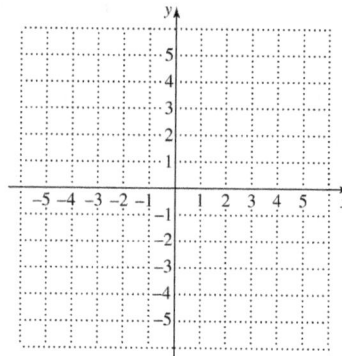

Name: _____   Date: _____

Instructor: _____   Section: _____

Guided Learning Video Worksheet: Graph Horizontal and Vertical Lines

Text: *Beginning Algebra: Early Graphinga*
Student Learning Objective 3.2.3: Graph horizontal and vertical lines.

Follow along with *Guided Learning Video 3.2.3, Graph Horizontal and Vertical Lines.*

Understanding the Big Picture: As you listen to the first part of the video, when the pencil icon appears, pause and fill in the blanks, choosing from the words listed.

straight • *y*-axis • vertical • solutions • *x*-axis • horizontal

1. The equation of the line $y = 0$ represents the _____ axis on the rectangular coordinate system.

2. The equation of the line $x = 0$ represents the _____ axis on the rectangular coordinate system.

3. Horizontal lines, when graphed, are parallel to the _____, and vertical lines, when graphed are parallel to the _____.

4. Linear equations of the form $x = a$ represent _____ lines, and linear equations of the form $y = b$ represent _____ lines.

Follow along with the two Guided Learning Video examples, and fill in the blanks on the left as you learn with the instructor. When the pencil icon appears, pause and try the Student Practice on your own.

Guided Learning Video ▶ GUIDED LEARNING VIDEO	Pause: Student Practice
1. Graph. $x = 4$	2. Graph. $x = -2$.

1. Graph. $x = 4$

   Use the fact that for every value of y, $x = 4$ to graph the line.

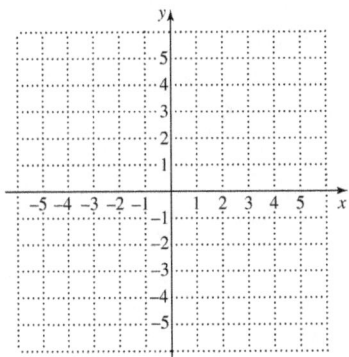

2. Graph. $x = -2$.

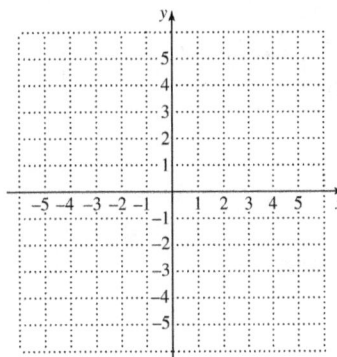

137
Copyright © 2017 Pearson Education, Inc.

3. Graph. $y = -5$

Use the fact that for every value of x, $y = -5$ to graph the line.

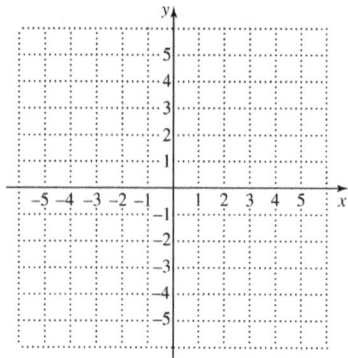

4. Graph. $y = -1$.

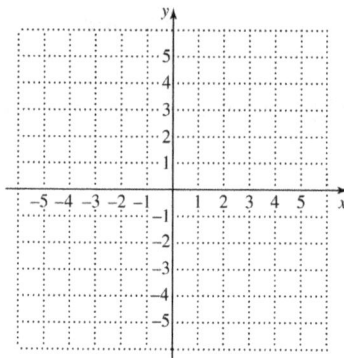

Helpful Hint: Pause the video and write the helpful hint in your own words.

For each Active Video Lesson, when the pencil icon appears, pause and work the problem. Then press play to check your work.

Active Video Lesson 1 ✏️ | **Active Video Lesson 2** ✏️

5. Graph. $x = -3$.

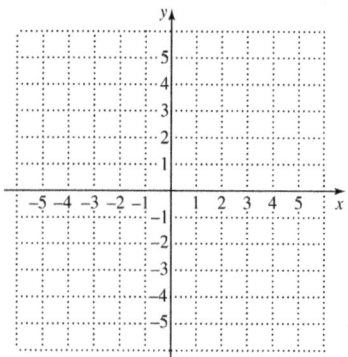

6. Graph. $y = 4$.

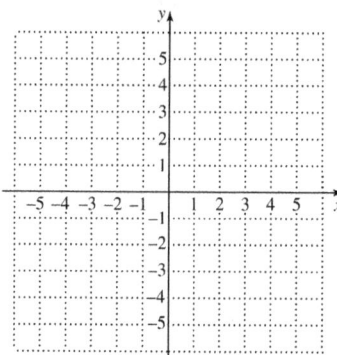

Guided Learning Video Worksheet: Finding the Slope if Two Points Are Known

Text: *Beginning Algebra: Early Graphing*
Student Learning Objective 3.3.1: Find the slope of a line given two points on the line.

Follow along with *Guided Learning Video 3.3.1, Finding the Slope if Two Points Are Known.*

Understanding the Big Picture: As you listen to the first part of the video, when the pencil icon appears, pause and fill in the blanks, choosing from the words listed.

vertical • run • horizontal • slope • rise • steepness

1. The word _____ is often used to describe a line's _____.

2. The change in the _____ distance compared to the change in the _____ distance is called the _____.

3. The change in vertical distance is called the _____; whereas, the change in the horizontal distance is called the _____.

Follow along with the two Guided Learning Video examples, and fill in the blanks on the left as you learn with the instructor. When the pencil icon appears, pause and try the Student Practice on your own.

Guided Learning Video ▶ GUIDED LEARNING VIDEO	**Pause: Student Practice** ✏
1. Find the slope of the line passing through $(2, 1)$ and $(5, 3)$. Use the formula for the slope of a line to calculate the slope. $$m = \frac{y_2 - y_1}{x_2 - x_1} = \frac{(\boxed{} - \boxed{})}{(\boxed{} - \boxed{})}$$ $$m = \frac{\boxed{}}{\boxed{}}$$	2. Find the slope of the line passing through $(6, 9)$ and $(12, 18)$.

3. Find the slope of the line passing through $(-4, 1)$ and $(-1, -2)$.

Use the formula for the slope of a line to calculate the slope.

$$m = \frac{y_2 - y_1}{x_2 - x_1} = \frac{(\boxed{} - \boxed{})}{(\boxed{} - \boxed{})}$$

$$m = \frac{(\boxed{} - \boxed{})}{(\boxed{} - \boxed{})}$$

$$m = \frac{\boxed{}}{\boxed{}} = \boxed{}$$

4. Find the slope of the line passing through $(-5, 8)$ and $(-3, -4)$.

Helpful Hint: Pause the video and write the helpful hint in your own words.

For each Active Video Lesson, when the pencil icon appears, pause and work the problem. Then press play to check your work.

Active Video Lesson 1 ✏ | **Active Video Lesson 2** ✏

5. Find the slope of the line passing through $(2, -3)$ and $(4, 1)$.

6. Find the slope of the line passing through $(-4, 5)$ and $(2, 2)$.

Guided Learning Video Worksheet: Slope-Intercept Form of the Equation of a Line

Text: *Beginning Algebra: Early Graphing*
Student Learning Objective 3.3.4: Graph a line using the slope and *y*-intercept.

Follow along with *Guided Learning Video 3.3.4, Slope-Intercept Form of the Equation of a Line.*

Understanding the Big Picture: As you listen to the first part of the video, when the pencil icon appears, pause and fill in the blanks, choosing from the words listed.

y-axis • slope • graph • *y*-intercept • standard • slope-intercept

1. The _____ form of the equation of a line is $Ax + By = C$.

2. The _____ form of the equation of the line reveals the line's _____ and where it intersects the _____.

3. The _____ form of the equation of a line, $y = mx + b$, can be used to _____ the line by identifying its _____ and using the slope to plot a second point.

Follow along with the two Guided Learning Video examples, and fill in the blanks on the left as you learn with the instructor. When the pencil icon appears, pause and try the Student Practice on your own.

Guided Learning Video ▶ GUIDED LEARNING VIDEO	**Pause: Student Practice**
1. Write the slope-intercept form of the equation of the line with a slope of $\frac{5}{6}$ and a *y*-intercept of $(0, -3)$. Substitute the values of m and b in the form $y = mx + b$ to write the equation of the line. $m = \dfrac{\boxed{}}{\boxed{}}$ and $b = \boxed{}$ $y = \boxed{}$ $y = \boxed{}$	2. Write the slope-intercept form of the equation of the line with a slope of $-\frac{4}{5}$ and a *y*-intercept of $(0, 6)$.

3. Find the slope and the y-intercept, and then sketch the graph of the equation $8x + 6y = -12$.

   Write the equation in slope-intercept form, then graph the line.
   $8x + 6y = -12$

   $y = $ [ ]

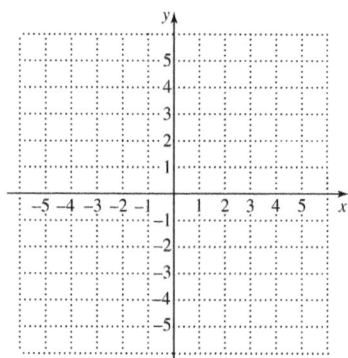

4. Find the slope and the y-intercept, and then sketch the graph of the equation $3x + 4y = -8$.

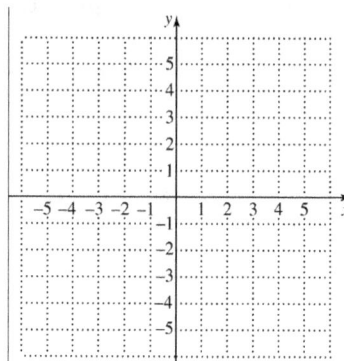

Helpful Hint: Pause the video and write the helpful hint in your own words.

For each Active Video Lesson, when the pencil icon appears, pause and work the problem. Then press play to check your work.

5. Find the slope and the y-intercept, and then sketch the graph of the equation $x - 2y = -2$.

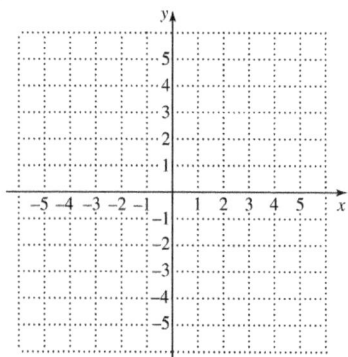

6. Write the slope-intercept form of the equation of the line with a slope of $\frac{2}{5}$ and a y-intercept of $(0, -1)$.

Name: _____   Date: _____

Instructor: _____   Section: _____

Guided Learning Video Worksheet: Finding the Slopes of Parallel and Perpendicular Lines

Text: *Beginning Algebra: Early Graphing*
Student Learning Objective 3.3.5: Find the slopes of parallel and perpendicular lines.

Follow along with *Guided Learning Video 3.3.5, Finding the Slopes of Parallel and Perpendicular Lines.*

Understanding the Big Picture: As you listen to the first part of the video, when the pencil icon appears, pause and fill in the blanks, choosing from the words listed.

perpendicular • slope • reciprocals • *y*-intercepts • parallel • product

1. Straight lines that never touch are _____ lines.

2. Parallel lines have the same _____ but different _____.

3. Straight lines intersecting at a 90° angle are _____ lines.

4. Perpendicular lines have slopes whose _____ equals −1, or are said to be negative _____ of each other.

Follow along with the two Guided Learning Video examples, and fill in the blanks on the left as you learn with the instructor. When the pencil icon appears, pause and try the Student Practice on your own.

Guided Learning Video ▶ GUIDED LEARNING VIDEO	**Pause: Student Practice** ✎
1. Line h has a slope of $-\dfrac{6}{11}$. If line f is parallel to line h, what is the slope of line f? If line g is perpendicular to line h, what is the slope of line g? Use the facts about the slopes of parallel and perpendicular lines to determine the slope of line f and line g. Slope of line $f =$ ▢ Slope of line $g =$ ▢	2. Line P has a slope of $-\dfrac{3}{4}$. If line N is parallel to line P, what is the slope of line N? If line R is perpendicular to line P, what is the slope of line R?

3. Find the slope of a line perpendicular to line l that passes through $(-4, -2)$ and $(0, 6)$.

   Calculate the slope of line l, then determine the slope of a line perpendicular to it.

   $m_1 = \dfrac{\boxed{}}{\boxed{}} = \boxed{}$

   $m_2 = \dfrac{\boxed{}}{\boxed{}}$

4. Find the slope of a line perpendicular to line l that passes through $(-3, 9)$ and $(0, 18)$.

Helpful Hint: Pause the video and write the helpful hint in your own words.

For each Active Video Lesson, when the pencil icon appears, pause and work the problem. Then press play to check your work.

Active Video Lesson 1 | **Active Video Lesson 2**

5. Find the slope of a line perpendicular to line h that passes through $(2, 1)$ and $(5, 3)$.

6. Line g has a slope of -5. If line f is parallel to line g, what is the slope of line f? If line h is perpendicular to line g, what is the slope of line h?

Guided Learning Video Worksheet: Write the Equation of a Line, Given a Point and the Slope

Text: *Beginning Algebra: Early Graphing*
Student Learning Objective 3.4.1: Write an equation of a line given a point and the slope.

Follow along with *Guided Learning Video 3.4.1, Write the Equation of a Line, Given a Point and the Slope.*

Understanding the Big Picture: As you listen to the first part of the video, when the pencil icon appears, pause and fill in the blanks, choosing from the words listed.

known • slope-intercept • *y*-intercept • point • slope

1. The equation $y = 4x + 2$ is written in _____ form.

2. Whenever the _____ and the _____ of a line are _____, the equation of the line can be written in slope-intercept form.

3. When given a line's slope and a _____ on the line other than the *y*-intercept, the form $y = mx + b$ can be used to solve for b to find the _____.

Follow along with the two Guided Learning Video examples, and fill in the blanks on the left as you learn with the instructor. When the pencil icon appears, pause and try the Student Practice on your own.

Guided Learning Video ▶ GUIDED LEARNING VIDEO	**Pause: Student Practice** ✏️
1. Find an equation for the line that passes through $(2, -3)$ with slope $-\dfrac{1}{2}$.	2. Find an equation for the line that passes through $(-4, 5)$ with slope $-\dfrac{1}{4}$.
Use the procedure for finding the equation of a line when given a point and the slope.	
$y = mx + b$	
[_____]	
[_____]	
[_____]	
$y = mx + b$	
[_____]	

3. Find an equation for the line that passes through $(6, 1)$ with slope $-\dfrac{1}{3}$.

   Use the procedure for finding the equation of a line when given a point and the slope.

   $y = mx + b$

   $y = mx + b$

4. Find an equation for the line that passes through $(-9, 1)$ with slope $\dfrac{2}{3}$.

Helpful Hint: Pause the video and write the helpful hint in your own words.

For each Active Video Lesson, when the pencil icon appears, pause and work the problem. Then press play to check your work.

Active Video Lesson 1 ✎ | **Active Video Lesson 2** ✎

5. Find an equation for the line that passes through $(4, 7)$ with slope $\dfrac{1}{2}$.

6. Find an equation for the line that passes through $(2, 0)$ with slope $\dfrac{3}{2}$.

Name: _____   Date: _____

Instructor: _____   Section: _____

Guided Learning Video Worksheet: Write the Equation of a Line, Given Two Points

Text: *Beginning Algebra: Early Graphing*
Student Learning Objective 3.4.2: Write an equation of a line given two points.

Follow along with *Guided Learning Video 3.4.2, Write the Equation of a Line, Given Two Points*.

Understanding the Big Picture: As you listen to the first part of the video, when the pencil icon appears, pause and fill in the blanks, choosing from the words listed.

coordinates • slope-intercept • two • *y*-intercept • slope

1. When given _____ points to find a line's equation, first find the line's
    _____ .

2. Determine the line's _____ by substituting the _____ and
    the _____ of one of the given points into $y = mx + b$, then solve for b.

3. To write the equation of the line in _____ form, substitute the values
    of m and b into the equation $y = mx + b$.

Follow along with the two Guided Learning Video examples, and fill in the blanks on the left as you learn with the instructor. When the pencil icon appears, pause and try the Student Practice on your own.

Guided Learning Video	**Pause: Student Practice**
1. Find the equation of the line that passes through $(1,\ 5)$ and $(2,\ 7)$. Determine the line's slope and *y*-intercept, then write the line's equation in slope-intercept form. $m = \dfrac{y_2 - y_1}{x_2 - x_1} = \dfrac{\boxed{} - \boxed{}}{\boxed{} - \boxed{}} = \dfrac{\boxed{}}{\boxed{}} = \boxed{}$ $y = mx + b$	2. Find the equation of the line that passes through $(3,\ 4)$ and $(2, 6)$.

3. Find the equation of the line that passes through $(-3, 2)$ and $(6, -1)$.

Determine the line's slope and y-intercept, then write the line's equation in slope-intercept form.

$$m = \frac{y_2 - y_1}{x_2 - x_1} = \frac{\boxed{} - \boxed{}}{\boxed{} - \boxed{}} = \frac{\boxed{}}{\boxed{}} = \boxed{}$$

$y = mx + b$

4. Find the equation of the line that passes through $(4, -4)$ and $(-1, 1)$.

Helpful Hint: Pause the video and write the helpful hint in your own words.

For each Active Video Lesson, when the pencil icon appears, pause and work the problem. Then press play to check your work.

Active Video Lesson 1 ✏	**Active Video Lesson 2** ✏
5. Find the equation of the line that passes through $(4, -7)$ and $(-2, -4)$.	6. Find the equation of the line that passes through $(1, 1)$ and $(2, 5)$.

Name: _____   Date: _____
Instructor: _____   Section: _____

Guided Learning Video Worksheet: Write the Equation of a Line, Given a Graph

Text: *Beginning Algebra: Early Graphing*
Student Learning Objective 3.4.3: Write an equation of a line given a graph of the line.

Follow along with *Guided Learning Video 3.4.3, Write the Equation of a Line, Given a Graph.*

Understanding the Big Picture: As you listen to the first part of the video, when the pencil icon appears, pause and fill in the blanks, choosing from the words listed.

coordinates • horizontal • slope • *y*-intercept • points • vertical

1. Previously, the equation of a line was found when given either two _____ or when given its _____ and a point on the line.

2. To find the equation of a line when given its graph, first identify the _____ of the _____ to determine *b* .

3. To determine the line's slope, count the number of _____ and _____ units from the *y*-intercept to a second point on the line.

Follow along with the two Guided Learning Video examples, and fill in the blanks on the left as you learn with the instructor. When the pencil icon appears, pause and try the Student Practice on your own.

Guided Learning Video ▶ GUIDED LEARNING VIDEO	**Pause: Student Practice**
1. What is the equation of the line in the following graph?	2. What is the equation of the line in the following graph?

Identify the *y*-intercept and slope of the line to write the equation of the line.

$$m = \dfrac{\boxed{}}{\boxed{}} = \boxed{}$$

$$b = \boxed{}$$

$$\boxed{}$$

3. What is the equation of the line in the following graph?

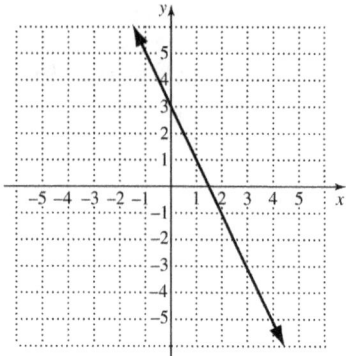

Identify the *y*-intercept and slope of the line to write the equation of the line.

$$m = \frac{\boxed{}}{\boxed{}} = \boxed{}$$

$$b = \boxed{}$$

$$\boxed{}$$

4. What is the equation of the line in the following graph?

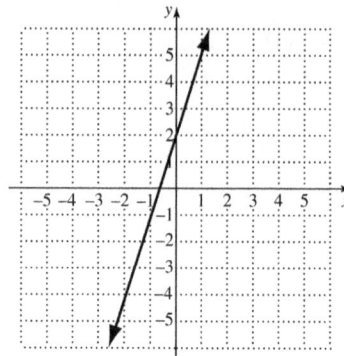

Helpful Hint: Pause the video and write the helpful hint in your own words.

For each Active Video Lesson, when the pencil icon appears, pause and work the problem. Then press play to check your work.

Active Video Lesson 1 | **Active Video Lesson 2**

5. What is the equation of the line in the following graph?

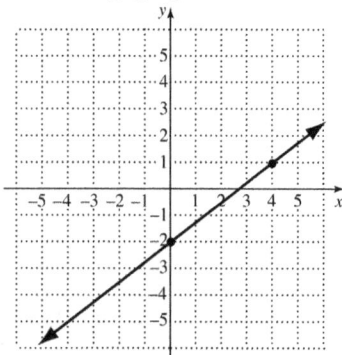

6. What is the equation of the line in the following graph?

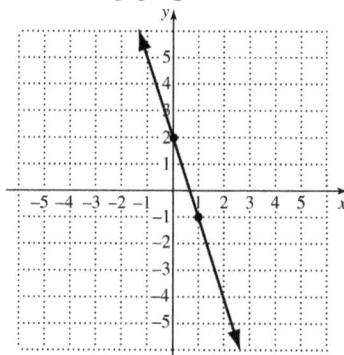

Name: _____   Date: _____

Instructor: _____   Section: _____

Guided Learning Video Worksheet: Graph Linear Inequalities in Two Variables

Text: *Beginning Algebra: Early Graphing*
Student Learning Objective 3.5.1: Graph linear inequalities in two variables.

Follow along with *Guided Learning Video 3.5.1, Graph Linear Inequalities in Two Variables.*

Understanding the Big Picture: As you listen to the first part of the video, when the pencil icon appears, pause and fill in the blanks, choosing from the words listed.

includes • dashed • set • true • false • inequality • solid

1. A solution to a linear _____ is the _____ of all possible ordered pairs which when substituted into the _____ lead to a true statement.

2. When graphing a linear _____, graph a _____ line when the inequality symbol is ≤ or ≥, and graph a _____ line when the symbol is < or >.

3. If a test point leads to an inequality that is _____, shade the region that _____ the test point.

4. If a test point leads to an inequality that is _____, shade the region that doesn't include the test point.

Follow along with the two Guided Learning Video examples, and fill in the blanks on the left as you learn with the instructor. When the pencil icon appears, pause and try the Student Practice on your own.

Guided Learning Video	Pause: Student Practice
1. Graph the inequality $3x + 2y > 6$.	2. Graph the inequality $4x + y < -4$.

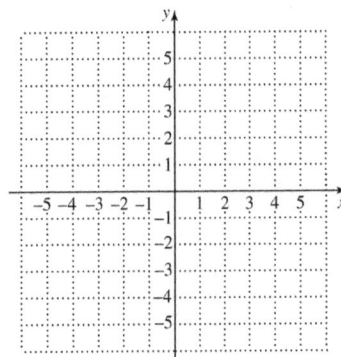

Use the procedure for graphing linear inequalities to graph the inequality and identify the solution region.

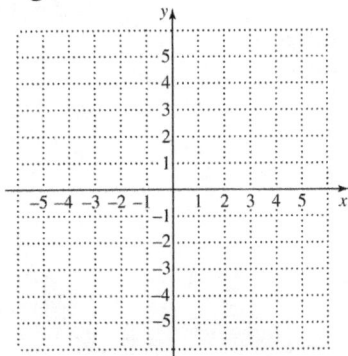

3. Graph the inequality $y \le -x + 2$.

Use the procedure for graphing linear inequalities to graph the inequality and identify the solution region.

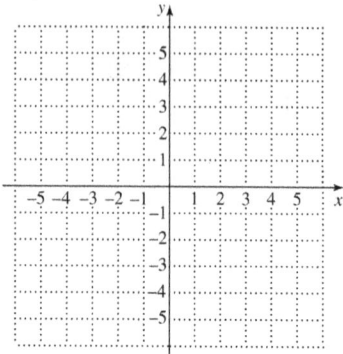

4. Graph the inequality $y \ge 2x + 1$.

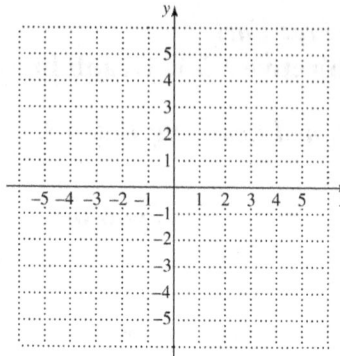

Helpful Hint: Pause the video and write the helpful hint in your own words.

For each Active Video Lesson, when the pencil icon appears, pause and work the problem. Then press play to check your work.

Active Video Lesson 1 ✏ | **Active Video Lesson 2** ✏

5. Graph the inequality $y < 3x - 1$.

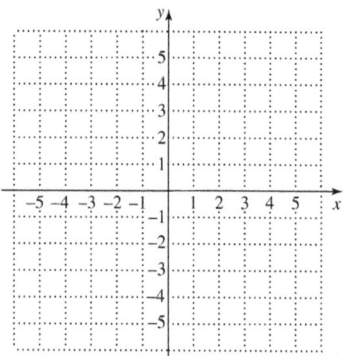

6. Graph the inequality $4x - 3y \ge 12$.

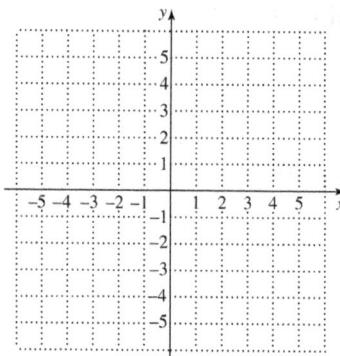

Name: _____   Date: _____

Instructor: _____   Section: _____

Guided Learning Video Worksheet: Relations, Domain, Range and Function

Text: *Beginning Algebra: Early Graphing*
Student Learning Objective 3.6.1: Understand the meanings of a relation and a function.

Follow along with *Guided Learning Video 3.6.1, Relations, Domain, Range and Function.*

Understanding the Big Picture: As you listen to the first part of the video, when the pencil icon appears, pause and fill in the blanks, choosing from the words listed.

range • set • input • relation • output • function • first • domain

1. A _____ is any _____ of ordered pairs.

2. The set of first items of the ordered pairs in a relation is called the _____ and is considered _____.

3. The set of second items of the ordered pairs in a relation is called the _____ and is considered _____.

4. A(n) _____ is a(n) _____ in which no two ordered pairs have the same _____ coordinate.

Follow along with the two Guided Learning Video examples, and fill in the blanks on the left as you learn with the instructor. When the pencil icon appears, pause and try the Student Practice on your own.

Guided Learning Video ▶ GUIDED LEARNING VIDEO	Pause: Student Practice
1. Give the domain and range of the following relation. Indicate whether the relation is a function or not.	2. Give the domain and range of the following relation. Indicate whether the relation is a function or not.
$h = \{(-1,8),(0,5),(2,8),(7.6,-5.1)\}$	$h = \{(6,3),(0,-8),(4,9),(5,-8),(2,12)\}$
Use the definitions of domain and range to identify each set and to determine if the relation is a function.	
Domain $= \{ \qquad \}$ Range $= \{ \qquad \}$	

3. Give the domain and range of the following relation. Indicate whether the relation is a function or not.

Hours Studied	1.5	1.8	2	2	2.4	2.75	3
Grade on Exam	76	78	83	79	88	92	85

Use the definitions of domain and range to identify each set and to determine if the relation is a function.

Domain $=\{$ $\}$

Range $=\{$ $\}$

4. Give the domain and range of the following relation. Indicate whether the relation is a function or not.

Distance in Miles	120	85	100	120	75
Time in Hours	1.75	1.5	2	2.25	1.25

Helpful Hint: Pause the video and write the helpful hint in your own words.

For each Active Video Lesson, when the pencil icon appears, pause and work the problem. Then press play to check your work.

Active Video Lesson 1 ✏️   **Active Video Lesson 2** ✏️

5. Give the domain and range of the following relation. Indicate whether the relation is a function or not.

Distance from Home to Job in Miles	3	3.8	5	7	15
Length of Commute in Minutes	15	15	18	21	35

6. Give the domain and range of the following relation. Indicate whether the relation is a function or not.

$$\{(1,2),(5,9),(8,6),(5,1)\}$$

Guided Learning Video Worksheet: The Vertical Line Test

Text: *Beginning Algebra: Early Graphing*
Student Learning Objective 3.6.3: Determine whether a graph represents a function.

Follow along with *Guided Learning Video 3.6.3, The Vertical Line Test.*

Understanding the Big Picture: As you listen to the first part of the video, when the pencil icon appears, pause and fill in the blanks, choosing from the words listed.

once • not • unique • different • function • same • intersect

1. A function cannot have two _____ ordered pairs with the _____ first coordinate.

2. In a(n) _____, each value of x has a(n) _____ value of y.

3. If a vertical line can _____ the graph of a relation more than _____, the relation is _____ a function.

Follow along with the two Guided Learning Video examples, and fill in the blanks on the left as you learn with the instructor. When the pencil icon appears, pause and try the Student Practice on your own.

Guided Learning Video (▶) GUIDED LEARNING VIDEO	Pause: Student Practice
1. Determine whether the graph represents a function.	2. Determine whether the graph represents a function.
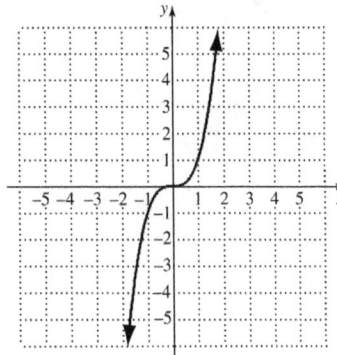	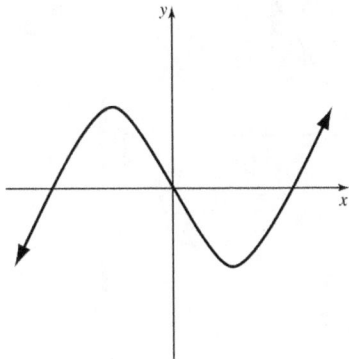
Use the vertical line test to determine if the graph represents a function.	

3. Determine whether the graph represents a function.

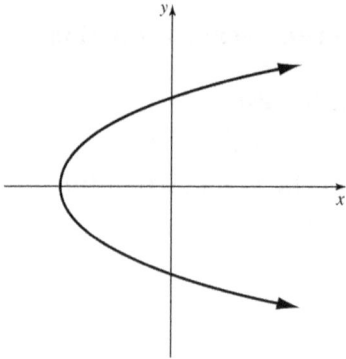

Use the vertical line test to determine if the graph represents a function.

4. Determine whether the graph represents a function.

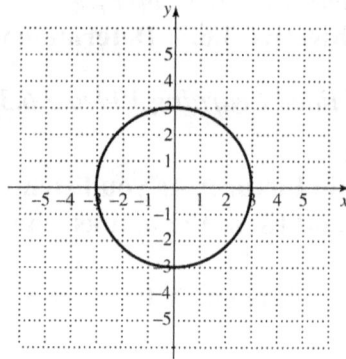

Helpful Hint: Pause the video and write the helpful hint in your own words.

For each Active Video Lesson, when the pencil icon appears, pause and work the problem. Then press play to check your work.

Active Video Lesson 1

5. Determine whether the graph represents a function.

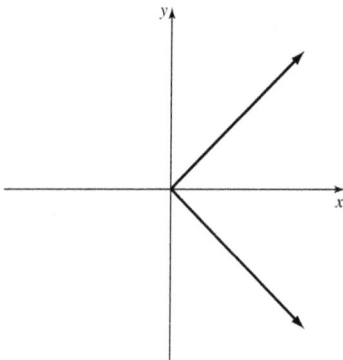

Active Video Lesson 2

6. Determine whether the graph represents a function.

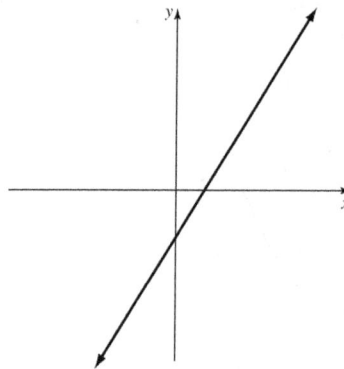

Guided Learning Video Worksheet: Function Notation

Text: *Beginning Algebra: Early Graphing*
Student Learning Objective 3.6.4: Use function notation.

Follow along with *Guided Learning Video 3.6.4, Function Notation.*

Understanding the Big Picture: As you listen to the first part of the video, when the pencil icon appears, pause and fill in the blanks, choosing from the words listed.

output • domain • multiplied • range • function • input

1. The notation $f(x)$ is called _____ notation.

2. The notation $f(x)$ does not mean f _____ by x.

3. When evaluating functions, values of x are considered _____ and are members of the function's _____.

4. When evaluating functions, values of y are considered _____ and are members of the function's _____.

Follow along with the two Guided Learning Video examples, and fill in the blanks on the left as you learn with the instructor. When the pencil icon appears, pause and try the Student Practice on your own.

Guided Learning Video ▶ GUIDED LEARNING VIDEO	Pause: Student Practice
1. Let $f(x) = x^3 - 4x + 2$. Find $f(-1)$ and $f(0)$. Substitute the value of x and evaluate the function for that value. $f(-1) = $ [____] $f(-1) = $ [____] $f(-1) = $ [____] $f(0) = $ [____] $f(0) = $ [____] $f(0) = $ [____]	2. Let $f(x) = 2x^3 + x - 4$. Find $f(2)$ and $f(-3)$.

3. Let $f(x) = -x^2 + 2x$.

Find $f(2)$ and $f\left(\dfrac{1}{2}\right)$.

Substitute the value of x and evaluate the function for that value.

$f(2) = $ [_____]

$f(2) = $ [_____]

$f(2) = $ [_____]

$f\left(\dfrac{1}{2}\right) = $ [_____]

$f\left(\dfrac{1}{2}\right) = $ [_____]

$f\left(\dfrac{1}{2}\right) = $ [_____]

4. Let $f(x) = -x^3 + x$.

Find $f(-2)$ and $f(1)$.

Helpful Hint: Pause the video and write the helpful hint in your own words.

For each Active Video Lesson, when the pencil icon appears, pause and work the problem. Then press play to check your work.

Active Video Lesson 1 | **Active Video Lesson 2**

5. Let $f(x) = -3x^2 - x + 4$.

Find $f(1)$ and $f(-2)$.

6. Let $f(x) = 2x^2 - 5x$.

Find $f(0)$ and $f(-1)$.

Name: _____   Date: _____

Instructor: _____   Section: _____

Guided Learning Video Worksheet: System of Linear Equations: Solve by Graphing

Text: *Beginning Algebra: Early Graphing*
Student Learning Objective 4.1.1: Use graphing to solve a system of linear equations with a unique solution.

Follow along with *Guided Learning Video 4.1.1, System of Linear Equations: Solve by Graphing.*

Understanding the Big Picture: As you listen to the first part of the video, when the pencil icon appears, pause and fill in the blanks, choosing from the words listed.

coordinates • intersect • solution • system • unique • equations

1. Two or more _____ in several variables that are considered simultaneously are called a(n) _____ of equations.

2. A system of linear equations has a(n) _____ solution if the lines _____.

3. The ordered pair of _____ of the point of intersection is the _____ to the system of equations.

Follow along with the two Guided Learning Video examples, and fill in the blanks on the left as you learn with the instructor. When the pencil icon appears, pause and try the Student Practice on your own.

Guided Learning Video	**Pause: Student Practice**
1. Solve by graphing.	2. Solve by graphing.
$x + 2y = 6$	$x - y = 3$
$x - y = 3$	$x + y = 5$
Graph each line on the same coordinate system, and identify the coordinates of the point of intersection.	
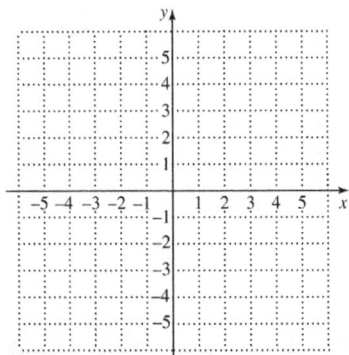	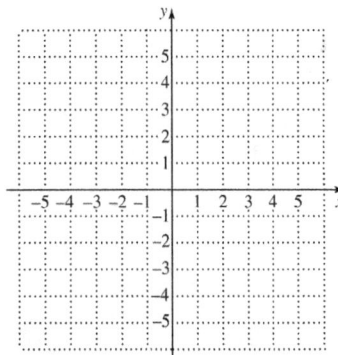

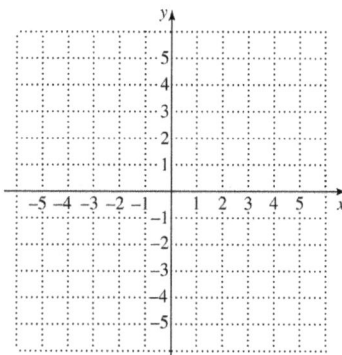

3. Solve by graphing.

$$2x - 3y = -12$$

$$2x + y = -4$$

Graph each line on the same coordinate system, and identify the coordinates of the point of intersection.

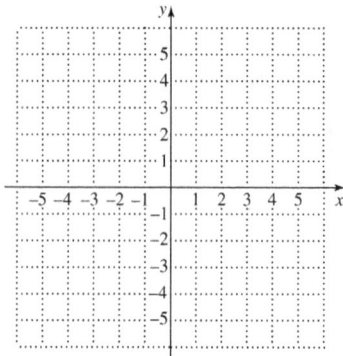

4. Solve by graphing.

$$-2x + y = 3$$

$$4x + y = -3$$

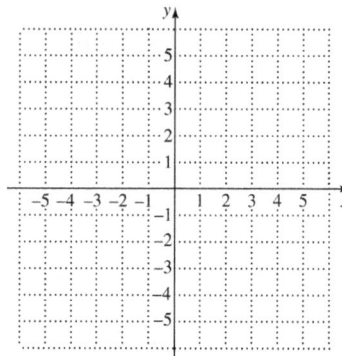

Helpful Hint: Pause the video and write the helpful hint in your own words.

For each Active Video Lesson, when the pencil icon appears, pause and work the problem. Then press play to check your work.

Active Video Lesson 1

Active Video Lesson 2

5. Solve by graphing.

$$4x - 3y = 15$$

$$2x + 3y = 3$$

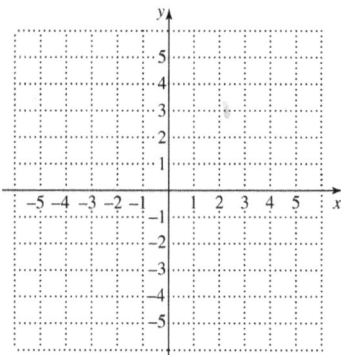

6. Solve by graphing.

$$x + y = -1$$

$$x + 2y = 0$$

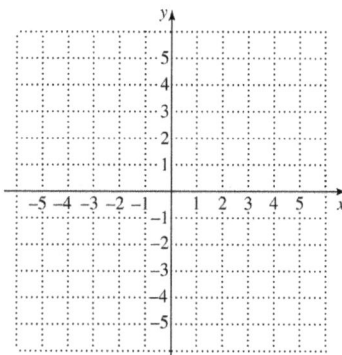

Guided Learning Video Worksheet: System of Linear Equations: Solve by Substitution

Text: *Beginning Algebra: Early Graphing*
Student Learning Objective 4.2.1: Solve a system of two linear equations with integer coefficients by the substitution method.

Follow along with *Guided Learning Video 4.2.1, System of Linear Equations: Solve by Substitution.*

Understanding the Big Picture: As you listen to the first part of the video, when the pencil icon appears, pause and fill in the blanks, choosing from the words listed.

original • substitute • second • solution • single • true • expression

1. The ordered pair _____ to a system of equations must lead to _____ statements in all equations of the system.

2. After solving one of the equations in the system for one of its variables, _____ the _____ into the other equation.

3. After solving the equation with the _____ variable, substitute the value into either of the _____ equations to find the solution for the _____ variable.

Follow along with the two Guided Learning Video examples, and fill in the blanks on the left as you learn with the instructor. When the pencil icon appears, pause and try the Student Practice on your own.

Guided Learning Video ▶ GUIDED LEARNING VIDEO	Pause: Student Practice ✏
1. Find the solution for the system of equations. $3x - 4y = -7$ $x + 2y = 1$ Solve by using the procedure for solving a system of equations by substitution. $3x - 4y = -7$ $x + 2y = 1$	2. Using the substitution method, find the solution for the system of equations. $x + 4y = 4$ $x - 2y = -2$

3. Find the solution for the system of equations.

$$4x + 3y = -6$$
$$2x - y = 12$$

Solve by using the procedure for solving a system of equations by substitution.

$$4x + 3y = -6$$
$$2x - y = 12$$

4. Using the substitution method, find the solution for the system of equations.

$$4x + 2y = 4$$
$$3x + y = 4$$

Helpful Hint: Pause the video and write the helpful hint in your own words.

For each Active Video Lesson, when the pencil icon appears, pause and work the problem. Then press play to check your work.

Active Video Lesson 1 | **Active Video Lesson 2**

5. Using the substitution method, find the solution for the system of equations.

$$3x - y = -5$$
$$-2x + 5y = -14$$

6. Using the substitution method, find the solution for the system of equations.

$$5x + 2y = 1$$
$$2x + y = 1$$

Guided Learning Video Worksheet: Solving Systems of Linear Equations with Fractional Coefficients Using the Addition Method

Text: *Beginning Algebra: Early Graphing*
Student Learning Objective 4.3.2: Use the addition method to solve a system of two linear equations with fractional coefficients.

Follow along with *Guided Learning Video 4.3.2, Solving Systems of Linear Equations with Fractional Coefficients Using the Addition Method.*

Understanding the Big Picture: As you listen to the first part of the video, when the pencil icon appears, pause and fill in the blanks, choosing from the words listed.

opposites • addition • fractional • coefficient • multiply • elimination

1. When the number 1 isn't a(n) _____ of any variable in the system, the _____ method is often used to solve the system.

2. The _____ method is also called the _____ method.

3. If necessary, _____ one or both equations by appropriate numbers so that coefficients of one of the variables are _____.

4. If _____ coefficients are present, clear each equation of the fractions by multiplying each term of the equation by the least common denominator of the fractions.

Follow along with the two Guided Learning Video examples, and fill in the blanks on the left as you learn with the instructor. When the pencil icon appears, pause and try the Student Practice on your own.

Guided Learning Video (▶) GUIDED LEARNING VIDEO	**Pause: Student Practice** ✏
1. Solve by the addition method. $x + \dfrac{2}{3}y = 5$ $\dfrac{2}{5}x + y = 2$ Clear each equation of fractions, and follow the procedure for solving a system of equations using the addition method. $x + \dfrac{2}{3}y = 5$ $\dfrac{2}{5}x + y = 2$	2. Solve by the addition method. $\dfrac{2}{3}x + y = 2$ $x + \dfrac{1}{2}y = 7$

3. Solve by the addition method.

$$\frac{x}{5} + \frac{y}{2} = \frac{7}{10}$$

$$\frac{x}{7} + \frac{y}{3} = \frac{11}{21}$$

Clear each equation of fractions, and follow the procedure for solving a system of equations using the addition method.

$$\frac{x}{5} + \frac{y}{2} = \frac{7}{10}$$

$$\frac{x}{7} + \frac{y}{3} = \frac{11}{21}$$

4. Solve by the addition method.

$$\frac{x}{5} + \frac{y}{2} = \frac{9}{10}$$

$$\frac{x}{3} + \frac{3y}{4} = \frac{5}{4}$$

Helpful Hint: Pause the video and write the helpful hint in your own words.

For each Active Video Lesson, when the pencil icon appears, pause and work the problem. Then press play to check your work.

Active Video Lesson 1   **Active Video Lesson 2**

5. Solve by the addition method.

$$\frac{2}{3}x - \frac{1}{5}y = 2$$

$$\frac{4}{3}x + 4y = 4$$

6. Solve by the addition method.

$$\frac{1}{3}x + \frac{5}{6}y = 2$$

$$\frac{3}{5}x - y = -\frac{7}{5}$$

Name: _____   Date: _____

Instructor: _____   Section: _____

Guided Learning Video Worksheet: System of Linear Equations: Solve by Addition Method with Decimal Coefficients

Text: *Beginning Algebra: Early Graphing*
Student Learning Objective 4.3.3: Use the addition method to solve a system of two linear equations with decimal coefficients.

Follow along with *Guided Learning Video 4.3.3, System of Linear Equations: Solve by Addition Method with Decimal Coefficients*.

Understanding the Big Picture: As you listen to the first part of the video, when the pencil icon appears, pause and fill in the blanks, choosing from the words listed.

true • addition • ten • integer • variables • multiplying • opposites

1. To eliminate one of the variables when using the _____ method, the coefficients must be _____.

2. The solution to a system of two equations in two _____ is the value of x and the value of y that make each equation _____.

3. If decimal coefficients are present in a system of equations, change them to _____ coefficients by _____ each term of the equation by the largest power of _____ necessary to clear the decimals in that equation.

Follow along with the two Guided Learning Video examples, and fill in the blanks on the left as you learn with the instructor. When the pencil icon appears, pause and try the Student Practice on your own.

Guided Learning Video ▶ GUIDED LEARNING VIDEO	Pause: Student Practice
1. Solve. $0.04x + 0.03y = 0.01$ $0.02x - 0.05y = -0.19$ Change the decimal coefficients to integer coefficients, and follow the procedure for solving a system of equations using the addition method. $0.04x + 0.03y = 0.01$ $0.02x - 0.05y = -0.19$	2. Solve by the addition method. $0.2x + 0.3y = 0.4$ $0.5x + 0.4y = 0.3$

Guided Learning Video (▶) GUIDED LEARNING VIDEO	**Pause: Student Practice**
3. Solve. $$0.2x + 0.3y = -0.6$$ $$0.3x + 0.5y = -1.1$$ Change the decimal coefficients to integer coefficients, and follow the procedure for solving a system of equations using the addition method. $$0.2x + 0.3y = -0.6$$ $$0.3x + 0.5y = -1.1$$	**4.** Solve by the addition method. $$0.2x + 0.3y = -0.1$$ $$0.5x - 0.1y = -1.1$$

Helpful Hint: Pause the video and write the helpful hint in your own words.

For each Active Video Lesson, when the pencil icon appears, pause and work the problem. Then press play to check your work.

Active Video Lesson 1	**Active Video Lesson 2**
5. Solve by the addition method. $$0.3x + 0.2y = 0.0$$ $$1.0x + 0.5y = -0.5$$	**6.** Solve by the addition method $$0.02x + 0.03y = 0.01$$ $$0.01x - 0.02y = 0.04$$

Name: _____   Date: _____
Instructor: _____   Section: _____

Guided Learning Video Worksheet: Using Algebraic Methods to Identify Inconsistent and Dependent Systems

Text: *Beginning Algebra: Early Graphing*
Student Learning Objective 4.4.2: Use algebraic methods to identify inconsistent and dependent systems.

Follow along with *Guided Learning Video 4.4.2 Using Algebraic Methods to Identify Inconsistent and Dependent Systems.*

Understanding the Big Picture: As you listen to the first part of the video, when the pencil icon appears, pause and fill in the blanks, choosing from the words listed.

coincide • identity • no • parallel • infinite • inconsistent • dependent • false

1. A(n) _____ system of linear equations is a system of
   _____ lines having no solution.

2. If a(n) _____ mathematical statement is obtained when solving a system of
   equations, the system has _____ solution.

3. A system where the lines _____ has a(n) _____ number of
   solutions.

4. A(n) _____ is obtained when solving a dependent system of equations.

Follow along with the two Guided Learning Video examples, and fill in the blanks on the left as you learn with the instructor. When the pencil icon appears, pause and try the Student Practice on your own.

Guided Learning Video ▶ GUIDED LEARNING VIDEO	Pause: Student Practice
1. Solve algebraically. $5x - 2y = 6$ $-10x + 4y = 1$ Use the addition method to solve the system. $5x - 2y = 6$ $-10x + 4y = 1$	2. Solve algebraically. $3x - y = 4$ $-6x + 2y = 10$

3. Solve algebraically.

$6x - 3y = 9$

$-4x + 2y = -6$

Use the addition method to solve the system.

$6x - 3y = 9$

$-4x + 2y = -6$

4. Solve algebraically.

$3x + 7y = 1$

$6x + 14y = 2$

Helpful Hint: Pause the video and write the helpful hint in your own words.

For each Active Video Lesson, when the pencil icon appears, pause and work the problem. Then press play to check your work.

Active Video Lesson 1 ✏ | **Active Video Lesson 2** ✏

5. Solve algebraically.

$2x - 6y = 2$

$-3x + 9y = -3$

6. Solve algebraically.

$-15x - 18y = 10$

$5x + 6y = -2$

Guided Learning Video Worksheet: Solve Word Problems with Systems of Equations

Text: *Beginning Algebra: Early Graphing*
Student Learning Objective 4.5.1: Use a system of equations to solve word problems.

Follow along with *Guided Learning Video 4.5.1 Solve Word Problems with Systems of Equations.*

Understanding the Big Picture: As you listen to the first part of the video, when the pencil icon appears, pause and fill in the blanks, choosing from the words listed.

organize • unknowns • system • understand • applied

1. A(n) _____ of linear equations can be used to solve _____ problems.

2. When using a system of linear equations to solve problems, it's important to _____ the question asked and _____ the given information.

3. A system of equations must be written that represents the _____ in the problem.

Follow along with the two Guided Learning Video examples, and fill in the blanks on the left as you learn with the instructor. When the pencil icon appears, pause and try the Student Practice on your own.

Guided Learning Video	Pause: Student Practice
1. The West Chop lighthouse on Martha's Vineyard has two diesel generators. Yesterday, the large one was run for 7 and the small one for 3 hours. Together, they consumed 41 gallons of fuel. Today, the large one was run for 5 hours and the small one for 4 hours. Together, they consumed 33 gallons of fuel. How many of gallons of fuel per hour does the large generator use? How many of gallons of fuel per hour does the small generator use? Use the four-step process for creating and solving a system of equations to solve this problem. **(continued)**	2. Mark purchased 11 tickets to a baseball game paying a total of 449. Infield seats cost 44 each, and outfield seats cost 37 each. How many of each type of ticket did he purchase?

1.

2.

3.

4.

3. A plumber worked for 2 hours on a project, and his helper worked 3 hours. The plumber charged $101 for that project. Later, the plumber worked for 3 hours, and his helper worked for 7 hours on another project, and the bill was $189. How much does the plumber charge per hour for his work? How much does the plumber charge per hour for his helper?

Use the four-step process for creating and solving a system of equations to solve this problem.

1.

2.

3.

4.

4. A prep school's drama club sold 650 tickets to their theatre production for a total of $4375. If orchestra seats cost $7.50 and balcony seats cost $3.50, how many of each type of seat were sold?

Helpful Hint: Pause the video and write the helpful hint in your own words.

For each Active Video Lesson, when the pencil icon appears, pause and work the problem. Then press play to check your work.

Active Video Lesson 1	Active Video Lesson 2

5. A car manufacturing plant makes compact cars and luxury cars. It takes 4 hours on the assembly line to make a compact car and 7 hours to make a luxury car. The plant makes a $1000 profit on each compact car made and $3000 on each luxury car made. The plant will be operating for 260 hours this month, and plant officials have set a profit goal of $90,000. How many of each car should be made?

6. For the concert on campus. Advance tickets cost $15 and tickets at the door cost $20. The ticket sales this year came to $6250. The activities coordinator wants to raise ticket prices next year to $17 for advance tickets and $25 for tickets at the door. She said that if exactly the same number of people attend next year, the ticket sales at these new prices will total $7375. If she is correct, how many tickets were sold in advance this year? How many tickets were sold at the door?

Name: _____   Date: _____
Instructor: _____   Section: _____

Guided Learning Video Worksheet: Multiply Algebraic Expressions with Exponents

Text: *Beginning Algebra: Early Graphing*
Student Learning Objective 5.1.1: Use the product rule to multiply exponential expressions with like bases.

Follow along with *Guided Learning Video 5.1.1, Multiply Algebraic Expressions with Exponents.*

Understanding the Big Picture: As you listen to the first part of the video, when the pencil icon appears, pause and fill in the blanks, choosing from the words listed.

variable • exponents • multiplied • coefficient • add • one • base

1. When multiplying two exponential expressions having the same base, keep the
   _____ and _____ the _____.

2. When a base doesn't have a written exponent, the exponent is understood to be equal to
   _____.

3. A numerical _____ is a number _____ with a(n)
   _____.

Follow along with the two Guided Learning Video examples, and fill in the blanks on the left as you learn with the instructor. When the pencil icon appears, pause and try the Student Practice on your own.

Guided Learning Video	**Pause: Student Practice**
1. Multiply. $(5a^2)(3a^4)$	2. Multiply. $(7x^4)(4x)$
Use the procedure for multiplying algebraic expressions.	

For example 1:

$$(\boxed{})(\boxed{})$$

$$= \boxed{} \cdot \boxed{} \cdot \boxed{} \cdot \boxed{} = \boxed{}$$

$$= \boxed{}$$

Pause: Student Practice

3. Multiply. $(2x^7)(3y^4)(6x^2)$

Use the procedure for multiplying algebraic expressions.

$(\boxed{})(\boxed{})(\boxed{})$

$= \boxed{} \cdot \boxed{} \cdot \boxed{} \cdot \boxed{} \cdot \boxed{} \cdot \boxed{} = \boxed{}$

$= \boxed{}$

4. Multiply. $(3x^2)(5x^3)(2y^2)$

Helpful Hint: Pause the video and write the helpful hint in your own words.

For each Active Video Lesson, when the pencil icon appears, pause and work the problem. Then press play to check your work.

Active Video Lesson 1

Active Video Lesson 2

5. Multiply. $(5x^3)(2x)(4x^7)$

6. Multiply. $(a^5)(3b^4)(7b^2)$

Name: _____   Date: _____

Instructor: _____   Section: _____

Guided Learning Video Worksheet: The Product Rule

Text: *Beginning Algebra: Early Graphing*
Student Learning Objective 5.1.1: Use the product rule to multiply exponential expressions with like bases.

Follow along with *Guided Learning Video 5.1.1, The Product Rule.*

Understanding the Big Picture: As you listen to the first part of the video, when the pencil icon appears, pause and fill in the blanks, choosing from the words listed.

one • variable • exponents • same • coefficient • base

1. The product rule states that when multiplying exponential expressions having the same _____, keep the _____ and add the _____.

2. The product rule only applies when the bases are the _____.

3. An exponent is understood to be equal to _____ when it is unwritten.

4. A numerical _____ is a number multiplied times a _____.

Follow along with the two Guided Learning Video examples, and fill in the blanks on the left as you learn with the instructor. When the pencil icon appears, pause and try the Student Practice on your own.

Guided Learning Video	**Pause: Student Practice**
1. Multiply.	2. Multiply.
a. $x^9 \cdot x^4$	**a.** $x^6 \cdot x$
Use the product rule to multiply.	**b.** $2^3 \cdot 2^7$
$x^9 \cdot x^4 = \boxed{} = \boxed{}$	
b. $5^3 \cdot 5^8$	
Use the product rule to multiply.	
$5^3 \cdot 5^8 = \boxed{} = \boxed{}$	

Guided Learning Video	**Pause: Student Practice** ✏

3. Multiply.

    a. $9x^5 \cdot 6x^3$

       Use the product rule to multiply.

       $9x^5 \cdot 6x^3 = \boxed{} = \boxed{}$

    b. $-y^2 \cdot 7y^6$

       Use the product rule to multiply.

       $-y^2 \cdot 7y^6 = \boxed{} = \boxed{}$

4. Multiply.

    a. $4x^5 \cdot 7x^4$

    b. $-3x^2 \cdot x^4$

Helpful Hint: Pause the video and write the helpful hint in your own words.

For each Active Video Lesson, when the pencil icon appears, pause and work the problem. Then press play to check your work.

Active Video Lesson 1 ✏	**Active Video Lesson 2** ✏
5. Multiply. $\left(-7a^4\right)\left(-8a\right)$	**6.** Multiply. $\left(-4b^8\right)\left(9b^5\right)$

Name: _____   Date: _____

Instructor: _____   Section: _____

Guided Learning Video Worksheet: Quotient Rule for Exponents

Text: *Beginning Algebra: Early Graphing*
Student Learning Objective 5.1.2: Use the quotient rule to divide exponential expressions with like bases.

Follow along with *Guided Learning Video 5.1.2, Quotient Rule for Exponents*.

Understanding the Big Picture: As you listen to the first part of the video, when the pencil icon appears, pause and fill in the blanks, choosing from the words listed.

numerator • subtracted • fractional • denominator • larger • same

1. The division of variable expressions can be written as _____ expressions.

2. The quotient rule states if the bases in the numerator and denominator are the _____, the exponents can be _____.

3. When subtracting exponents to simplify the fractional expression, if the _____ exponent is in the _____, the resultant factor will be in the numerator.

4. After subtracting exponents to simplify the fractional expression, if the resultant factor is in the _____, then the larger exponent was originally in the denominator.

Follow along with the two Guided Learning Video examples, and fill in the blanks on the left as you learn with the instructor. When the pencil icon appears, pause and try the Student Practice on your own.

Guided Learning Video	Pause: Student Practice
1. Simplify. $\dfrac{6a^{13}}{8a^5}$	2. Simplify. $\dfrac{12x^7}{18x^4}$
Simplify the coefficients and apply the quotient rule.	

$$\frac{6a^{13}}{8a^5} = \frac{\boxed{}}{\boxed{}} = \frac{\boxed{}}{\boxed{}} = \boxed{}$$

3. Simplify. $\dfrac{x^3}{x^8}$

    Apply the quotient rule.

$$\frac{x^3}{x^8} = \frac{\boxed{}}{\boxed{}} = \frac{\boxed{}}{\boxed{}}$$

4. Simplify. $\dfrac{x^9}{x^{15}}$

Helpful Hint: Pause the video and write the helpful hint in your own words.

For each Active Video Lesson, when the pencil icon appears, pause and work the problem. Then press play to check your work.

Active Video Lesson 1 ✏ | **Active Video Lesson 2** ✏

5. Simplify. $\dfrac{y^3 z^4}{y^7 z}$

6. Simplify. $\dfrac{-14y^4}{35y^9}$

Name: _____   Date: _____

Instructor: _____   Section: _____

Guided Learning Video Worksheet: The Power Rule of Exponents

Text: *Beginning Algebra: Early Graphing*
Student Learning Objective 5.1.3: Raise exponential expressions to a power.

Follow along with *Guided Learning Video 5.1.3, The Power Rule of Exponents.*

Understanding the Big Picture: As you listen to the first part of the video, when the pencil icon appears, pause and fill in the blanks, choosing from the words listed.

one • fractional • exponents • numerator • each • base

1. When raising a power to a power, keep the same _____ and multiply the
   _____.

2. If the product of two or more factors inside parentheses are raised to a power,
   _____ factor must be raised to the power.

3. When a _____ expression is raised to a power, both the _____
   and the denominator must be raised to the power.

4. A base raised to a zero power is equal to _____.

Follow along with the two Guided Learning Video examples, and fill in the blanks on the left as you learn with the instructor. When the pencil icon appears, pause and try the Student Practice on your own.

Guided Learning Video ▶ GUIDED LEARNING VIDEO	Pause: Student Practice ✏
1. Simplify. $\left(x^8\right)^6$	2. Simplify. $\left(x^5\right)^7$
Follow the rule for raising a power to a power.	
$\left(x^8\right)^6 = \boxed{} = \boxed{}$	

Guided Learning Video ▶ GUIDED LEARNING VIDEO	Pause: Student Practice ✏
3. Simplify.	**4.** Simplify.

3. Simplify.

a. $\left(-2y^5\right)^3$

Follow the rule for raising a power to a power.

$\left(-2y^5\right)^3 = \boxed{} = \boxed{}$

b. $\left(\dfrac{-7a^9b^0}{ac^4}\right)^2$

Follow the rule for raising a power to a power.

$\left(\dfrac{-7a^9b^0}{ac^4}\right)^2 = \left(\dfrac{\boxed{}}{\boxed{}}\right) = \dfrac{\boxed{}}{\boxed{}}$

4. Simplify.

a. $\left(-3x^4\right)^3$

b. $\left(\dfrac{-4x^3y^0}{xz^3}\right)^3$

Helpful Hint: Pause the video and write the helpful hint in your own words.

For each Active Video Lesson, when the pencil icon appears, pause and work the problem. Then press play to check your work.

Active Video Lesson 1 ✏	Active Video Lesson 2 ✏
5. Simplify. $\left(\dfrac{-3xy^6}{x^3z^0}\right)^2$	**6.** Simplify. $\left(-5z^6\right)^3$

Name: _____   Date: _____

Instructor: _____   Section: _____

Guided Learning Video Worksheet: Negative Exponents

Text: *Beginning Algebra: Early Graphing*
Student Learning Objective 5.2.1: Use negative exponents.

Follow along with *Guided Learning Video 5.2.1, Negative Exponents.*

Understanding the Big Picture: As you listen to the first part of the video, when the pencil icon appears, pause and fill in the blanks, choosing from the words listed.

simplify • positive • zero • reciprocal • integers • negative

1. The set of _____ consists of negative and positive whole numbers and
   _____.

2. To evaluate a numerical expression with a(n) _____ exponent, first convert the
   expression to one with a positive exponent, then _____ if possible.

3. A base raised to a negative exponent is equal to the _____ of the base raised
   to a(n) _____ exponent.

Follow along with the two Guided Learning Video examples, and fill in the blanks on the left as you learn with the instructor. When the pencil icon appears, pause and try the Student Practice on your own.

Guided Learning Video ▶ GUIDED LEARNING VIDEO	Pause: Student Practice
1. Simplify. Write the expression with no negative exponents. $\dfrac{x^{-2}y^{-3}}{z^{-4}}$	2. Simplify. Write the expression with no negative exponents. $\dfrac{x^{4}y^{-5}}{z^{-7}}$
Use properties of exponents and the definition of a negative exponent to simplify. $\dfrac{x^{-2}y^{-3}}{z^{-4}} = \dfrac{\boxed{}}{\boxed{}}$	

Guided Learning Video ▶ GUIDED LEARNING VIDEO	**Pause: Student Practice**
3. Simplify. Write the expression with no negative exponents. $\left(8a^{-6}b^7\right)^{-2}$	**4.** Simplify. Write the expression with no negative exponents. $\left(4x^5y^{-3}\right)^{-2}$

Use properties of exponents and the definition of a negative exponent to simplify.

$$\left(8a^{-6}b^7\right)^{-2} = \frac{\boxed{}}{\boxed{}}$$

Helpful Hint: Pause the video and write the helpful hint in your own words.

For each Active Video Lesson, when the pencil icon appears, pause and work the problem. Then press play to check your work.

Active Video Lesson 1	**Active Video Lesson 2**
5. Simplify. Write the expression with no negative exponents. $\dfrac{x^{-9}y^{-5}}{x^{-11}y^3}$	**6.** Simplify. Write the expression with no negative exponents. $\dfrac{a^{-3}b^{-7}}{c^{-6}}$

Name: _____   Date: _____

Instructor: _____   Section: _____

Guided Learning Video Worksheet: Change Scientific Notation to Standard Notation

Text: *Beginning Algebra: Early Graphing*
Student Learning Objective 5.2.2: Use scientific notation.

Follow along with *Guided Learning Video 5.2.2, Change Scientific Notation to Standard Notation.*

Understanding the Big Picture: As you listen to the first part of the video, when the pencil icon appears, pause and fill in the blanks, choosing from the words listed.

ten • one • positive • negative • scientific notation

1. To perform a calculation involving very large or very small numbers, _____ is often used to express the numbers.

2. If a number written in scientific notation has a _____ power of ten, the number is greater than or equal to _____.

3. If a number written in scientific notation has a _____ power of ten, the number is less than _____.

Follow along with the two Guided Learning Video examples, and fill in the blanks on the left as you learn with the instructor. When the pencil icon appears, pause and try the Student Practice on your own.

Guided Learning Video ▶ GUIDED LEARNING VIDEO	Pause: Student Practice
1. Write in standard notation. 8.3×10^3 Move the decimal point to write the number in standard notation. $8.3 \times 10^3 = \boxed{}$	2. Write in standard notation. 5.7×10^4

Guided Learning Video ▶ GUIDED LEARNING VIDEO	Pause: Student Practice ✏
3. Write in standard notation. 6.1×10^{-3} Move the decimal point to write the number in standard notation. $6.1 \times 10^{-3} = $ ⬚	**4.** Write in standard notation. 2.98×10^{-4}

Helpful Hint: Pause the video and write the helpful hint in your own words.

For each Active Video Lesson, when the pencil icon appears, pause and work the problem. Then press play to check your work.

Active Video Lesson 1 ✏	Active Video Lesson 2 ✏
5. Write in standard notation. 9.36×10^{-5}	**6.** Write in standard notation. 7.451×10^{4}

Guided Learning Video Worksheet: Add Polynomials

Text: *Beginning Algebra: Early Graphing*
Student Learning Objective 5.3.2: Add polynomials.

Follow along with *Guided Learning Video 5.3.2, Add Polynomials.*

Understanding the Big Picture: As you listen to the first part of the video, when the pencil icon appears, pause and fill in the blanks, choosing from the words listed.

like • associative • parentheses • sum • commutative

1. Sometimes, to distinguish between two or more polynomials being added, _____ are placed around the polynomials.

2. To find the sum of polynomials, first use the _____ and _____ properties of addition to rearrange terms.

3. Find the _____ of polynomials by combining _____ terms.

Follow along with the two Guided Learning Video examples, and fill in the blanks on the left as you learn with the instructor. When the pencil icon appears, pause and try the Student Practice on your own.

Guided Learning Video	Pause: Student Practice
1. Add. $\left(5x^2 + 3x - 7\right) + \left(-3x^2 - 8x\right)$	2. Add. $\left(8x^2 - 5x + 2\right) + \left(-6x^2 - 3x\right)$
Rearrange the terms so that like terms are together, and then combine the like terms. $\left(5x^2 + 3x - 7\right) + \left(-3x^2 - 8x\right)$ $=$ ☐ $=$ ☐	

3. Add.

$$\left(-6y^2 + 7y - 2\right) + \left(7y^2 - 4y + 9\right)$$

Rearrange the terms so that like terms are together, and then combine the like terms.

$$\left(-6y^2 + 7y - 2\right) + \left(7y^2 - 4y + 9\right)$$

$$= \boxed{}$$

$$= \boxed{}$$

4. Add. $\left(3y^2 + 10y - 9\right) + \left(-5y^2 - 5y + 3\right)$

Helpful Hint: Pause the video and write the helpful hint in your own words.

For each Active Video Lesson, when the pencil icon appears, pause and work the problem. Then press play to check your work.

Active Video Lesson 1 ✏ | **Active Video Lesson 2** ✏

5. Add. $\left(9z^2 + 2z - 8\right) + \left(-5z^2 - 3z + 6\right)$

6. Add. $\left(3c^2 - 2c - 4\right) + \left(-2c + 5\right)$

Name: _____   Date: _____

Instructor: _____   Section: _____

Guided Learning Video Worksheet: Subtract Polynomials

Text: *Beginning Algebra: Early Graphing*
Student Learning Objective 5.3.3: Subtract polynomials.

Follow along with *Guided Learning Video 5.3.3, Subtract Polynomials*.

Understanding the Big Picture: As you listen to the first part of the video, when the pencil icon appears, pause and fill in the blanks, choosing from the words listed.

terms • like • opposite • second • signs • add

1. To subtract a polynomial, _____ the _____ of the polynomial being subtracted.

2. To find the opposite of a polynomial enclosed in parentheses, change the _____ of all the _____ inside the parentheses.

3. When subtracting two polynomials, change the subtraction symbol to addition, change the sign of each term in the _____ polynomial to its opposite, and then combine _____ terms.

Follow along with the two Guided Learning Video examples, and fill in the blanks on the left as you learn with the instructor. When the pencil icon appears, pause and try the Student Practice on your own.

Guided Learning Video ▶ GUIDED LEARNING VIDEO	Pause: Student Practice
1. Perform the operation indicated for the expression: $(4a^2 + 3a - 6) - (a^2 + 7a - 4)$ Follow the rules for subtracting polynomials and combining like terms. $(4a^2 + 3a - 6) - (a^2 + 7a - 4)$ = [ ] = [ ]	2. Perform the operation indicated for the expression: $(7x^2 - x + 3) - (-5x^2 + 3x + 9)$

Guided Learning Video ▶ GUIDED LEARNING VIDEO	Pause: Student Practice
3. Perform the operation indicated for the expression:	**4.** Perform the operation indicated for the expression:

3. Perform the operation indicated for the expression:

$$3a - \left(a^2 + 4a\right) - 2\left(-a^2 - 5a + 1\right)$$

Follow the rules for subtracting polynomials and combining like terms.

$$3a - \left(a^2 + 4a\right) - 2\left(-a^2 - 5a + 1\right)$$

$= $ [ ]

$= $ [ ]

4. Perform the operation indicated for the expression:

$$8a - \left(6a^2 - 3a\right) - 3\left(a^2 - 4a + 2\right)$$

Helpful Hint: Pause the video and write the helpful hint in your own words.

For each Active Video Lesson, when the pencil icon appears, pause and work the problem. Then press play to check your work.

Active Video Lesson 1	**Active Video Lesson 2**
5. Perform the operation indicated for the expression:	**6.** Perform the operation indicated for the expression:
$$\left(11x^2 - 2x + 4\right) - \left(-5x^2 - 3x + 1\right)$$	$$7a - 2\left(4a^2 + 3\right) - \left(a^2 - a - 2\right)$$

Name: _____   Date: _____

Instructor: _____   Section: _____

Guided Learning Video Worksheet: Multiply a Monomial and a Binomial

Text: *Beginning Algebra: Early Graphing*
Student Learning Objective 5.4.1: Multiply a monomial by a polynomial.

Follow along with *Guided Learning Video 5.4.1, Multiply a Monomial and a Binomial.*

Understanding the Big Picture: As you listen to the first part of the video, when the pencil icon appears, pause and fill in the blanks, choosing from the words listed.

monomial • terms • whole • distributive • variable • binomial

1. Polynomials are expressions containing _____ parts with _____ number exponents only.

2. A _____ is a polynomial composed of two _____.

3. The _____ property is used for multiplying a _____ times a binomial.

Follow along with the two Guided Learning Video examples, and fill in the blanks on the left as you learn with the instructor. When the pencil icon appears, pause and try the Student Practice on your own.

Guided Learning Video ▶ GUIDED LEARNING VIDEO	Pause: Student Practice ✏️
1. Multiply. $9x^2\left(6x^3 - 7\right)$ Use the distributive property to multiply. $9x\left(6x^3 - 7\right)$ $= \boxed{} \cdot \left(\boxed{}\right) + \boxed{} \cdot \left(\boxed{}\right)$ $= \boxed{}$	2. Multiply. $4x^3\left(7x^2 - 8\right)$

3. Multiply. $(8y - 5)(-6y^3)$

Use the distributive property to multiply.

$(8y - 5)(-6y^3)$

$(\boxed{})(\boxed{})$

$= \boxed{} \cdot (\boxed{}) + \boxed{} \cdot (\boxed{})$

$= \boxed{}$

4. Multiply. $(6x^2 - 3)(-2x^2)$

Helpful Hint: Pause the video and write the helpful hint in your own words.

For each Active Video Lesson, when the pencil icon appears, pause and work the problem. Then press play to check your work.

Active Video Lesson 1 | **Active Video Lesson 2**

5. Multiply. $(7x^3 - 9x)(8x)$

6. Multiply. $(7x^3)(5x^3 + 4x^2)$

Guided Learning Video Worksheet: Multiply a Monomial Times a Polynomial

Text: *Beginning Algebra: Early Graphing*
Student Learning Objective 5.4.1: Multiply a monomial by a polynomial.

Follow along with *Guided Learning Video 5.4.1, Multiply a Monomial Times a Polynomial.*

Understanding the Big Picture: As you listen to the first part of the video, when the pencil icon appears, pause and fill in the blanks, choosing from the words listed.

distributive • multiply • trinomials • three • binomials • one

1. Three different types of polynomials are monomials, _____, and
   _____.

2. A monomial is a polynomial with _____ term and a trinomial is a polynomial with
   _____ terms.

3. When using the _____ property, the signs of the terms must be properly
   identified in order to _____ correctly.

Follow along with the two Guided Learning Video examples, and fill in the blanks on the left as you learn with the instructor. When the pencil icon appears, pause and try the Student Practice on your own.

Guided Learning Video	Pause: Student Practice
1. Multiply. $-6x(8x^2 - x + 9)$	2. Multiply. $-3x(8x^3 + 9x + 2)$
Use the distributive property to multiply.	
$-6x(8x^2 - x + 9)$	
$= \boxed{}() - \boxed{}() - \boxed{}()$	
$= \boxed{}$	

3. Multiply. $(7m^2 + 5m - 2)(-4m^3)$

Use the distributive property to multiply.

$(7m^2 + 5m - 2)(-4m^3)$

$= \boxed{}\Big|\boxed{}$

$= \boxed{}() - \boxed{}() - \boxed{}()$

$= \boxed{}$

4. Multiply. $(4x^2 + 7x - 1)(-5x)$

Helpful Hint: Pause the video and write the helpful hint in your own words.

For each Active Video Lesson, when the pencil icon appears, pause and work the problem. Then press play to check your work.

Active Video Lesson 1 | **Active Video Lesson 2**

5. Multiply. $(9a^2 - 7b + 3)(-8a^3)$

6. Multiply. $-5y^2(6y + 11z - 4)$

Guided Learning Video Worksheet: Multiplying Binomials Using FOIL

Text: *Beginning Algebra: Early Graphing*
Student Learning Objective 5.4.2: Multiply two binomials.

Follow along with *Guided Learning Video 5.4.2, Multiplying Binomials Using FOIL.*

Understanding the Big Picture: As you listen to the first part of the video, when the pencil icon appears, pause and fill in the blanks, choosing from the words listed.

first • distributive • last • inner • terms • outer

1. Multiplying binomials requires the use of the _____ property.

2. The acronym FOIL helps identify the _____ to multiply when multiplying a binomial times a binomial.

3. In the FOIL method, F means multiply the _____ terms, O means multiply the _____ terms, I means multiply the _____ terms, and L means multiply the _____ terms.

Follow along with the two Guided Learning Video examples, and fill in the blanks on the left as you learn with the instructor. When the pencil icon appears, pause and try the Student Practice on your own.

Guided Learning Video	Pause: Student Practice
1. Multiply. $(x+6)(x+7)$ Use the FOIL method to multiply, then combine like terms. $(x+6)(x+7)$ $=\square(\ )+\square(\ )+\square(\ )+\square(\ )$ $=\boxed{}$ $=\boxed{}$	2. Multiply. $(x+5)(4x+7)$

3. Multiply. $(y-3)(2y+5)$

Use the FOIL method to multiply, then combine like terms.

$(y-3)(2y+5)$

$= \square(\quad) + \square(\quad) + \square(\quad) + \square(\quad)$

$= \boxed{}$

$= \boxed{}$

4. Multiply. $(3x-7)(x+2)$

Helpful Hint: Pause the video and write the helpful hint in your own words.

For each Active Video Lesson, when the pencil icon appears, pause and work the problem. Then press play to check your work.

Active Video Lesson 1

5. Multiply. $(x+4)(3x-9)$

Active Video Lesson 2

6. Multiply. $(y-9)(y-8)$

Guided Learning Video Worksheet: Multiply Binomials: $(a+b)(a-b)$

Text: *Beginning Algebra: Early Graphing*
Student Learning Objective 5.5.1: Multiply binomials of the type $(a+b)(a-b)$.

Follow along with *Guided Learning Video 5.5.1, Multiply Binomials:* $(a+b)(a-b)$.

Understanding the Big Picture: As you listen to the first part of the video, when the pencil icon appears, pause and fill in the blanks, choosing from the words listed.

simplify • outer • zero • combining • products • last

1. The memory device FOIL stands for "first," "_____," "inner," and
   "_____."

2. After multiplying two binomials, _____ if possible by _____ like terms.

3. When multiplying binomials that are the sum and difference of the same terms, the sum of the inner and outer _____ is _____.

Follow along with the two Guided Learning Video examples, and fill in the blanks on the left as you learn with the instructor. When the pencil icon appears, pause and try the Student Practice on your own.

Guided Learning Video ▶ GUIDED LEARNING VIDEO	Pause: Student Practice
1. Multiply. $(9x+4)(9x-4)$ Use the formula for multiplying the sum and difference of two binomials with the same terms. $(9x+4)(9x-4)$ $=\boxed{}=\boxed{}$	2. Multiply. $(8x-5)(8x+5)$

Guided Learning Video ▶ GUIDED LEARNING VIDEO	**Pause: Student Practice** ✎
3. Multiply. $(8r+7s)(8r-7s)$	4. Multiply. $(11x+3y)(11x-3y)$

Use the formula for multiplying the sum and difference of two binomials with the same terms.

$(8r+7s)(8r-7s)$

$=$ [   ] $=$ [   ]

Helpful Hint: Pause the video and write the helpful hint in your own words.

For each Active Video Lesson, when the pencil icon appears, pause and work the problem. Then press play to check your work.

Active Video Lesson 1 ✎	**Active Video Lesson 2** ✎
5. Multiply. $(6m-5n)(6m+5n)$	6. Multiply. $(7y+3)(7y-3)$

Guided Learning Video Worksheet: Multiply Binomials: $(a+b)^2$ **and** $(a-b)^2$

Text: *Beginning Algebra: Early Graphing*

Student Learning Objective 5.5.2: Multiply binomials of the type $(a+b)^2$ **and** $(a-b)^2$.

Follow along with *Guided Learning Video 5.5.2, Multiply Binomials:* $(a+b)^2$ *and* $(a-b)^2$.

Understanding the Big Picture: As you listen to the first part of the video, when the pencil icon appears, pause and fill in the blanks, choosing from the words listed.

double • binomial • squares • product • trinomial • middle

1. When a binomial of the form $(a+b)$ or $(a-b)$ is squared, the product is a

   _____ expression.

2. The first and last terms of the trinomial product are the _____ of the first and
   second terms of the _____.

3. The trinomial's _____ term is _____ the _____ of the
   two terms of the binomial.

Follow along with the two Guided Learning Video examples, and fill in the blanks on the left as you learn with the instructor. When the pencil icon appears, pause and try the Student Practice on your own.

Guided Learning Video ⓖ	**Pause: Student Practice**
1. Multiply. $(6x+7)^2$ Use the formula for squaring a binomial to find the product. $(6x+7)^2$ = [ ] = [ ]	2. Multiply. $(3x+5)^2$

---

3. Multiply. $(11a - 4b)^2$

Use the formula for squaring a binomial to find the product.

$(11a - 4b)^2$

$= \boxed{}$

$= \boxed{}$

4. Multiply. $(12x - 15y)^2$

Helpful Hint: Pause the video and write the helpful hint in your own words.

For each Active Video Lesson, when the pencil icon appears, pause and work the problem. Then press play to check your work.

Active Video Lesson 1 ✏ | **Active Video Lesson 2** ✏

---

5. Multiply. $(9r - 8s)^2$

6. Multiply. $(10y + 3)^2$

Guided Learning Video Worksheet: Multiply Polynomials with More Than Two Terms

Text: *Beginning Algebra: Early Graphing*
Student Learning Objective 5.5.3: Multiply polynomials with more than two terms.

Follow along with *Guided Learning Video 5.5.3, Multiply Polynomials with More Than Two Terms*.

Understanding the Big Picture: As you listen to the first part of the video, when the pencil icon appears, pause and fill in the blanks, choosing from the words listed.

top • partial • vertically • distributive • combine • longer

1. The _____ property can be used to multiply polynomials.

2. When multiplying polynomials _____, the _____ polynomial is written at the _____.

3. To calculate the final product _____ the _____ products.

Follow along with the two Guided Learning Video examples, and fill in the blanks on the left as you learn with the instructor. When the pencil icon appears, pause and try the Student Practice on your own.

Guided Learning Video ▶ GUIDED LEARNING VIDEO	Pause: Student Practice ✏️
1. **a.** Multiply vertically. $$(2x^3 - 5x^2 + 4x)$$ $$\times \quad (x^2 + 2x - 3)$$ Multiply each term of the top polynomial by each term of the bottom polynomial. $$(2x^3 - 5x^2 + 4x)$$ $$\times \quad (x^2 + 2x - 3)$$ [ ] [ ] [ ] [ ] **(continued)**	2. **a.** Multiply vertically. $$(x^2 + 5x - 3)$$ $$\times \quad (2x + 1)$$ **b.** Multiply horizontally. $$(3x^2 + x - 2)(x^2 - x + 1)$$

Guided Learning Video ▶ GUIDED LEARNING VIDEO	**Pause: Student Practice**

b. Multiply horizontally.

$$(x^2 - 3x - 4)(3x^2 + 2x - 10)$$

Multiply each term of the second polynomial by each term of the first polynomial.

$$(x^2 - 3x - 4)(3x^2 + 2x - 10)$$

$$= \boxed{}$$

$$= \boxed{}$$

3. Multiply. $(x-3)(4x+1)(x+2)$

Multiply the first two binomials, and then multiply the result times the third binomial.

$$(x-3)(4x+1)(x+2)$$

$$= \boxed{}\,\boxed{}$$

$$= \boxed{}$$

4. Multiply. $(x+3)(x+5)(x-3)$

Helpful Hint: Pause the video and write the helpful hint in your own words.

For each Active Video Lesson, when the pencil icon appears, pause and work the problem. Then press play to check your work.

Active Video Lesson 1	**Active Video Lesson 2**
5. Multiply. $(2x+5)(x-1)(x+3)$	**6.** Multiply. $(3x^2 - x + 2)(5x^2 + 2x - 1)$

Name: _____   Date: _____

Instructor: _____   Section: _____

Guided Learning Video Worksheet: Divide a Polynomial by a Monomial

Text: *Beginning Algebra: Early Graphing*
Student Learning Objective 5.6.1: Divide a polynomial by a monomial.

Follow along with *Guided Learning Video 5.6.1, Divide a Polynomial by a Monomial.*

Understanding the Big Picture: As you listen to the first part of the video, when the pencil icon appears, pause and fill in the blanks, choosing from the words listed.

simplify • fractions • sum • monomial • divisor

1. The easiest type of polynomial division occurs when the _____ is a _____.

2. When performing the division of a polynomial by a monomial, write the expression as the _____ of individual _____.

3. After writing individual _____, _____ each if possible to express the final quotient.

Follow along with the two Guided Learning Video examples, and fill in the blanks on the left as you learn with the instructor. When the pencil icon appears, pause and try the Student Practice on your own.

Guided Learning Video ▶ GUIDED LEARNING VIDEO	**Pause: Student Practice**
1. Divide. $\dfrac{16x^5 + 8x^3 - 20x^2}{4x}$	2. Divide. $\dfrac{21x^5 - 15x^3 + 9x^2}{3x}$
Use the procedure for dividing a polynomial by a monomial.	
$\dfrac{16x^5 + 8x^3 - 20x^2}{4x}$	
$= \dfrac{\boxed{}}{\boxed{}} + \dfrac{\boxed{}}{\boxed{}} - \dfrac{\boxed{}}{\boxed{}}$	
$= \boxed{}$	

3. Divide. $\dfrac{35y^4 - 45y^3 + 15y^2}{5y^2}$

4. Divide. $\dfrac{48x^7 + 66x^5 - 18x^3}{6x^3}$

Use the procedure for dividing a polynomial by a monomial.

$$\dfrac{35y^4 - 45y^3 + 15y^2}{5y^2}$$

$$= \dfrac{\boxed{}}{\boxed{}} + \dfrac{\boxed{}}{\boxed{}} - \dfrac{\boxed{}}{\boxed{}}$$

$$= \boxed{}$$

Helpful Hint: Pause the video and write the helpful hint in your own words.

For each Active Video Lesson, when the pencil icon appears, pause and work the problem. Then press play to check your work.

| **Active Video Lesson 1** ✏ | **Active Video Lesson 2** ✏ |

5. Divide. $\dfrac{63y^7 + 72y^5 - 54y^3}{9y^3}$

6. Divide. $\dfrac{48w^6 - 64w^4 + 56w^2}{8w}$

Guided Learning Video Worksheet: Divide a Polynomial by a Binomial

Text: *Beginning Algebra: Early Graphing*
Student Learning Objective 5.6.2: Divide a polynomial by a binomial.

Follow along with *Guided Learning Video 5.6.2, Divide a Polynomial by a Binomial.*

Understanding the Big Picture: As you listen to the first part of the video, when the pencil icon appears, pause and fill in the blanks, choosing from the words listed.

subtract • descending • multiply • long • zero • less

1. Division of a polynomial by a binomial is similar to _____ division in arithmetic.

2. Write the terms of the polynomial and the binomial in _____ exponential order, filling in a _____ for any missing terms.

3. In the procedure for dividing polynomials by a binomial, continue to divide, _____, and _____ until the degree of the remainder is _____ than the degree of the divisor.

Follow along with the two Guided Learning Video examples, and fill in the blanks on the left as you learn with the instructor. When the pencil icon appears, pause and try the Student Practice on your own.

Guided Learning Video ▶ GUIDED LEARNING VIDEO	**Pause: Student Practice** ✏
1. Divide. $4x+3\overline{)8x^2-14x-15}$ Use the procedure for dividing a polynomial by a binomial to determine the quotient. $4x+3\overline{)8x^2-14x-15}$	2. Divide. $2x+1\overline{)2x^2+7x+3}$

3. Divide. $x+2\overline{)2x^3 - 3x + 2}$

Use the procedure for dividing a polynomial by a binomial to determine the quotient.

$$x+2\overline{)2x^3 - 3x + 2}$$

4. Divide. $2x-1\overline{)8x^3 + 8x + 5}$

Helpful Hint: Pause the video and write the helpful hint in your own words.

For each Active Video Lesson, when the pencil icon appears, pause and work the problem. Then press play to check your work.

Active Video Lesson 1 ✏️ | **Active Video Lesson 2** ✏️

5. Divide. $2x-3\overline{)10x^3 - 13x^2 + x - 4}$

6. Divide. $3x-1\overline{)21x^2 + 5x - 4}$

Guided Learning Video Worksheet: Factor Out the Greatest Common Factor from a Polynomial

Text: *Beginning Algebra: Early Graphing*
Student Learning Objective 6.1.1: Factor polynomials whose terms contain a common factor.

Follow along with *Guided Learning Video 6.1.1, Factor Out the Greatest Common Factor from a Polynomial.*

Understanding the Big Picture: As you listen to the first part of the video, when the pencil icon appears, pause and fill in the blanks, choosing from the words listed.

smallest • coefficient • integer • common • exponent

1. The abbreviation GCF means "greatest _____ factor."

2. Determine the greatest common numerical factor of an expression by identifying the largest _____ that divides evenly into the _____ of each term.

3. Determine the greatest common variable factor of an expression by identifying which variable(s) are common to all terms and then identifying the _____ _____ of the common variable(s).

Follow along with the two Guided Learning Video examples, and fill in the blanks on the left as you learn with the instructor. When the pencil icon appears, pause and try the Student Practice on your own.

Guided Learning Video ▶ GUIDED LEARNING VIDEO	**Pause: Student Practice**
1. Factor the expression: $15x^2 + 20x - 10$ Use the procedure for factoring a polynomial to factor the expression. $15x^2 + 20x - 10$ $= \Box \cdot \Box + \Box \cdot \Box - \Box \cdot \Box$ $= \Box (\underline{\qquad\qquad})$	2. Factor the expression: $28x^2 + 14x - 35$

3. Factor the expression:
$16ab^3 - 24a^2b^2$

Use the procedure for factoring a polynomial to factor the expression.

$16ab^3 - 24a^2b^2$

$$= \boxed{} \left[\frac{\boxed{}}{\boxed{}} - \frac{\boxed{}}{\boxed{}} \right]$$

$$= \boxed{} \left(\boxed{} \right)$$

4. Factor the expression:
$20x^2y^3 - 45xy^4$

Helpful Hint: Pause the video and write the helpful hint in your own words.

For each Active Video Lesson, when the pencil icon appears, pause and work the problem. Then press play to check your work.

Active Video Lesson 1 | **Active Video Lesson 2**

5. Factor the expression:
$35x^3y + 21x^2y^3$

6. Factor the expression:
$18a - 27b + 36$

Guided Learning Video Worksheet: Factor by Grouping

Text: *Beginning Algebra: Early Graphing*
Student Learning Objective 6.2.1: Factor expressions with four terms by grouping.

Follow along with *Guided Learning Video 6.2.1, Factor by Grouping.*

Understanding the Big Picture: As you listen to the first part of the video, when the pencil icon appears, pause and fill in the blanks, choosing from the words listed.

different • expression • parentheses • first • binomial • variable

1. A common factor of a polynomial can be a number, a(n) _____, or an algebraic _____.

2. Sometimes, the common factor of a polynomial can be a(n) _____ enclosed by _____.

3. To factor by grouping, remove a common factor from the _____ two terms and then a(n) _____ common factor from the second two terms.

Follow along with the two Guided Learning Video examples, and fill in the blanks on the left as you learn with the instructor. When the pencil icon appears, pause and try the Student Practice on your own.

Guided Learning Video ▶ GUIDED LEARNING VIDEO	Pause: Student Practice
1. Factor. $2x^2 + 8x + 3x + 12$ Factor by grouping. $2x^2 + 8x + 3x + 12$ = [_____] = [_____]	2. Factor. $3x^2 - 21x + 4x - 28$

3. Factor. $9xy - 14 - 21y + 6x$

Rearrange the polynomial so the first two terms have a common factor and the last two terms have a common factor. Then, factor by grouping.

$9xy - 14 - 21y + 6x$

= [   ]

= [   ]

= [   ]

4. Factor. $10x^2 + 5xy - 6x - 3y$

Helpful Hint: Pause the video and write the helpful hint in your own words.

For each Active Video Lesson, when the pencil icon appears, pause and work the problem. Then press play to check your work.

Active Video Lesson 1 ✏ | **Active Video Lesson 2** ✏

5. Factor. $2ab - 15 + 5a - 6b$

6. Factor. $21mx - 3x - 7my + y$

Name: _____   Date: _____
Instructor: _____   Section: _____

Guided Learning Video Worksheet: Factor Trinomials: $x^2 + bx + c$

Text: *Beginning Algebra: Early Graphing*
Student Learning Objective 6.3.1: Factor polynomials of the form $x^2 + bx + c$.

Follow along with *Guided Learning Video 6.3.1, Factor Trinomials:* $x^2 + bx + c$.

Understanding the Big Picture: As you listen to the first part of the video, when the pencil icon appears, pause and fill in the blanks, choosing from the words listed.

last • middle • error • product • sum • trial

1. The _____ and _____ process is sometimes used to factor a trinomial.

2. The _____ of the last terms of the binomial factors should equal the
   _____ term of the trinomial.

3. The _____ of the last terms of the binomial factors should equal the coefficient of the
   trinomial's _____ term.

Follow along with the two Guided Learning Video examples, and fill in the blanks on the left as you learn with the instructor. When the pencil icon appears, pause and try the Student Practice on your own.

Guided Learning Video ▶ GUIDED LEARNING VIDEO	**Pause: Student Practice**
1. Factor. $x^2 + 9x + 14$ Use the guidelines for factoring trinomials of the form $x^2 + bx + c$. $\boxed{} + \boxed{} = 9$, and $\boxed{} \cdot \boxed{} = 14$ $x^2 + 9x + 14$ $= \boxed{()()}$	2. Factor. $x^2 - 8x + 12$

3. Factor. $x^2 - 4x - 45$

Use the guidelines for factoring trinomials of the form $x^2 + bx + c$.

$\boxed{} + \boxed{} = -4$, and

$\boxed{} \cdot \boxed{} = -45$

$x^2 - 4x - 45$

$= \boxed{()()}$

4. Factor. $x^2 - 7x - 60$

Helpful Hint: Pause the video and write the helpful hint in your own words.

For each Active Video Lesson, when the pencil icon appears, pause and work the problem. Then press play to check your work.

Active Video Lesson 1 ✏ | **Active Video Lesson 2** ✏

5. Factor. $x^2 + 3x - 54$

6. Factor. $y^2 - 13y + 42$

Guided Learning Video Worksheet: Factor Polynomials That Have a Common Factor and a Factor of the Form $x^2 + bx + c$

Text: *Beginning Algebra: Early Graphing*
Student Learning Objective 6.3.2: Factor polynomials that have a common factor and a factor of the form $x^2 + bx + c$.

Follow along with *Guided Learning Video 6.3.2, Factor Polynomials That Have a Common Factor and a Factor of the Form $x^2 + bx + c$.*

Understanding the Big Picture: As you listen to the first part of the video, when the pencil icon appears, pause and fill in the blanks, choosing from the words listed.

factor • binomials • greatest • product • common

1. Sometimes, all terms of a trinomial share a _____ factor.

2. When factoring a trinomial, first look for and remove the _____ common _____.

3. After removing the _____ common _____, if possible, factor the remaining trinomial as the _____ of two _____.

Follow along with the two Guided Learning Video examples, and fill in the blanks on the left as you learn with the instructor. When the pencil icon appears, pause and try the Student Practice on your own.

Guided Learning Video ▶ GUIDED LEARNING VIDEO	Pause: Student Practice
1. Factor. $3x^2 + 18x + 24$ Look for and remove the greatest common factor, then factor the remaining polynomial. $3x^2 + 18x + 24$ $= \boxed{}(\qquad\qquad)$ $= \boxed{}(\qquad)(\qquad)$	2. Factor. $3x^2 - 15x - 18$

3. Factor. $5y^2 - 5y - 100$

Look for and remove the greatest common factor, then factor the remaining polynomial.

$5y^2 - 5y - 100$

$= \boxed{}\left(\right)$

$= \boxed{}\left(\right)\left(\right)$

4. Factor. $4x^2 - 28x + 40$

Helpful Hint: Pause the video and write the helpful hint in your own words.

For each Active Video Lesson, when the pencil icon appears, pause and work the problem. Then press play to check your work.

Active Video Lesson 1 ✎ | **Active Video Lesson 2** ✎

5. Factor. $2x^2 + 12x - 54$

6. Factor. $7x^2 - 35x + 42$

Guided Learning Video Worksheet: Factor Trinomials: $ax^2 + bx + c$ Using the Grouping Method

Text: *Beginning Algebra: Early Graphing*

Student Learning Objective 6.4.2: Factor a trinomial of the form $ax^2 + bx + c$ by the grouping method.

Follow along with *Guided Learning Video 6.4.2, Factor Trinomials: $ax^2 + bx + c$ Using the Grouping Method.*

Understanding the Big Picture: As you listen to the first part of the video, when the pencil icon appears, pause and fill in the blanks, choosing from the words listed.

sum • four • greatest • product • grouping

1. To factor a trinomial of the form $ax^2 + bx + c$, write it as an equivalent expression with _____ terms and factor it by _____.

2. When factoring by this method, first obtain the _____ number, which is the _____ of a times c.

3. Rewrite the bx term as the _____ of the two numbers whose _____ is the _____ number.

Follow along with the two Guided Learning Video examples, and fill in the blanks on the left as you learn with the instructor. When the pencil icon appears, pause and try the Student Practice on your own.

Guided Learning Video ▶ GUIDED LEARNING VIDEO	**Pause: Student Practice**
1. Factor. $4x^2 + 11x + 6$	2. Factor. $2x^2 + 11x + 12$

Use the grouping number method to factor the trinomial.

$a \cdot c = \boxed{}$

$\boxed{} \cdot \boxed{} = \boxed{}$

$\boxed{} + \boxed{} = \boxed{}$

$4x^2 + 11x + 6 = \boxed{}$

$= \boxed{}$

$= \boxed{}$

3. Factor. $6x^2 + 7x - 5$	**4.** Factor. $6x^2 + 13x - 5$

Use the grouping number method to factor the trinomial.

$a \cdot c = \boxed{}$

$\boxed{} \cdot \boxed{} = \boxed{}$

$\boxed{} + \boxed{} = \boxed{}$

$6x^2 + 7x - 5 = \boxed{}$

$= \boxed{}$

$= \boxed{}$

Helpful Hint: Pause the video and write the helpful hint in your own words.

For each Active Video Lesson, when the pencil icon appears, pause and work the problem. Then press play to check your work.

Active Video Lesson 1 ✏	**Active Video Lesson 2** ✏
5. Factor. $15x^2 - x - 2$	**6.** Factor. $8x^2 - 14x + 3$

Name: _____   Date: _____

Instructor: _____   Section: _____

Guided Learning Video Worksheet: Factor Trinomials: $ax^2 + bx + c$ with a Greatest Common Factor

Text: *Beginning Algebra: Early Graphing*

Student Learning Objective 6.4.3: Factor a trinomial of the form $ax^2 + bx + c$ after a common factor has been factored out of each term.

Follow along with *Guided Learning Video 6.4.3, Factor Trinomials: $ax^2 + bx + c$ with a Greatest Common Factor.*

Understanding the Big Picture: As you listen to the first part of the video, when the pencil icon appears, pause and fill in the blanks, choosing from the words listed.

binomial • common • error • grouping • trial

1. When factoring trinomials of the form $ax^2 + bx + c$, first remove the greatest _____ factor.

2. Trinomials of the form $ax^2 + bx + c$ can be factored by using either the _____ and _____ method, or the _____ method.

3. Examine the final answer to see if a common factor is present in any _____ in a set of parentheses.

Follow along with the two Guided Learning Video examples, and fill in the blanks on the left as you learn with the instructor. When the pencil icon appears, pause and try the Student Practice on your own.

Guided Learning Video ▶ GUIDED LEARNING VIDEO	**Pause: Student Practice** ✏
1. Factor. $12x^2 + 26x + 10$	2. Factor. $12x^2 + 21x - 6$
Factor out the greatest common factor, then factor the remaining trinomial.	
$12x^2 + 26x + 10$	
$= \boxed{}(\boxed{})$	
$= \boxed{}()()$	

3. Factor. $24x^2 + 30x - 9$

Factor out the greatest common factor, then factor the remaining trinomial.

$24x^2 + 30x - 9$

$= \boxed{}(\boxed{})$

$= \boxed{}()()$

4. Factor. $36x^2 - 66x + 18$

Helpful Hint: Pause the video and write the helpful hint in your own words.

For each Active Video Lesson, when the pencil icon appears, pause and work the problem. Then press play to check your work.

Active Video Lesson 1 | **Active Video Lesson 2**

5. Factor. $30x^2 + 5x - 5$

6. Factor. $40x^2 - 28x + 4$

Guided Learning Video Worksheet: Factor the Difference of Two Squares

Text: *Beginning Algebra: Early Graphing*
Student Learning Objective 6.5.1: Recognize and factor expressions of the type $a^2 - b^2$ (difference of two squares).

Follow along with *Guided Learning Video 6.5.1, Factor the Difference of Two Squares*.

Understanding the Big Picture: As you listen to the first part of the video, when the pencil icon appears, pause and fill in the blanks, choosing from the words listed.

negative • sum • squares • difference • last

1. One of the special cases of factoring is called the _____ of two
   _____.

2. The difference of two squares can be factored into the _____ and _____
   of those values that were squared to equal the original terms of the expression.

3. The difference of two squares formula only works if the _____ term is
   _____.

Follow along with the two Guided Learning Video examples, and fill in the blanks on the left as you learn with the instructor. When the pencil icon appears, pause and try the Student Practice on your own.

Guided Learning Video	Pause: Student Practice
1. Factor. $16x^2 - 9$ Use the formula for the difference of two squares to factor the expression. $16x^2 - 9$ $= ($ $)($ $)$	2. Factor. $64x^2 - 9$

Guided Learning Video ▶ GUIDED LEARNING VIDEO	**Pause: Student Practice** ✎
3. Factor. $49x^2 - 25y^2$	**4.** Factor. $121x^2 - 225y^2$
Use the formula for the difference of two squares to factor the expression.	
$49x^2 - 25y^2$	
$= (\qquad)(\qquad)$	

Helpful Hint: Pause the video and write the helpful hint in your own words.

For each Active Video Lesson, when the pencil icon appears, pause and work the problem. Then press play to check your work.

Active Video Lesson 1 ✎	**Active Video Lesson 2** ✎
5. Factor. $100x^2 - 81y^2$	**6.** Factor. $36m^2 - 1$

Guided Learning Video Worksheet: Factor Perfect Square Trinomials

Text: *Beginning Algebra: Early Graphing*
Student Learning Objective 6.5.2: Recognize and factor expressions of the type
$$a^2 + 2ab + b^2 \text{ (perfect square trinomial).}$$

Follow along with *Guided Learning Video 6.5.2, Factor Perfect Square Trinomials.*

Understanding the Big Picture: As you listen to the first part of the video, when the pencil icon appears, pause and fill in the blanks, choosing from the words listed.

first • trinomial • twice • last • perfect • product

1. A second special case of factoring is called a _____ square
   _____.

2. A perfect square trinomial has _____ and _____ terms that are perfect squares.

3. The middle term of a perfect square trinomial is _____ the _____ of the square roots of the first and last terms.

Follow along with the two Guided Learning Video examples, and fill in the blanks on the left as you learn with the instructor. When the pencil icon appears, pause and try the Student Practice on your own.

Guided Learning Video ▶ GUIDED LEARNING VIDEO	Pause: Student Practice ✏
1. Factor. $x^2 + 4x + 4$ Use the formula for a perfect square trinomial to factor the expression. $x^2 + 4x + 4$ $= (\qquad)(\qquad)$ $= (\qquad)^{\square}$	2. Factor. $81x^2 + 12x + 1$

Guided Learning Video	**Pause: Student Practice**
3. Factor. $16x^2 - 40x + 25$	**4.** Factor. $4x^2 - 60x + 225$

Use the formula for a perfect square trinomial to factor the expression.

$16x^2 - 40x + 25$

$= ($ $)($ $)$

$= ($ $)^{\square}$

Helpful Hint: Pause the video and write the helpful hint in your own words.

For each Active Video Lesson, when the pencil icon appears, pause and work the problem. Then press play to check your work.

Active Video Lesson 1	**Active Video Lesson 2**
5. Factor. $9y^2 - 42y + 49$	**6.** Factor. $x^2 + 12x + 36$

Guided Learning Video Worksheet: Solve Quadratic Equations by Factoring

Text: *Beginning Algebra: Early Graphing*
Student Learning Objective 6.7.1: Solve quadratic equations by factoring.

Follow along with *Guided Learning Video 6.7.1, Solve Quadratic Equations by Factoring*.

Understanding the Big Picture: As you listen to the first part of the video, when the pencil icon appears, pause and fill in the blanks, choosing from the words listed.

factoring • two • zero • standard • solutions • quadratic

1. A _____ equation is a polynomial equation in one variable with a highest degree variable term of degree _____.

2. The _____ form of a quadratic equation is $ax^2 + bx + c = 0$, where a, b, and c are real numbers and $a \neq 0$.

3. Many quadratic equations have two real number _____, and can be found using _____ methods and the _____ factor property.

Follow along with the two Guided Learning Video examples, and fill in the blanks on the left as you learn with the instructor. When the pencil icon appears, pause and try the Student Practice on your own.

Guided Learning Video ▶ GUIDED LEARNING VIDEO	Pause: Student Practice
1. Solve. $4x^2 + 11x + 6 = 0$	2. Solve. $x^2 - 5x - 24 = 0$
Follow the guidelines for solving a quadratic equation to determine the solution.	
$4x^2 + 11x + 6 = 0$	
$(\quad\quad)(\quad\quad) = 0$	
$(\quad\quad) = 0$ or $(\quad\quad) = 0$	
$\boxed{\quad = \quad}$ $\boxed{\quad = \quad}$	

3. Solve. $x^2 = 15 + 2x$

Follow the guidelines for solving a quadratic equation to determine the solution.

$x^2 = 15 + 2x$

$\boxed{}$

$()() = 0$

$() = 0$ or $() = 0$

$\boxed{ = }$ $\boxed{ = }$

4. Solve. $2x^2 = 11x - 12$

Helpful Hint: Pause the video and write the helpful hint in your own words.

For each Active Video Lesson, when the pencil icon appears, pause and work the problem. Then press play to check your work.

Active Video Lesson 1 ✏ | **Active Video Lesson 2** ✏

5. Solve. $x^2 = 5x + 14$

6. Solve. $8x^2 - 10x - 3 = 0$

Guided Learning Video Worksheet: Using Quadratic Equations to Solve Applied Problems

Text: *Beginning Algebra: Early Graphing*
Student Learning Objective 6.7.2: Use quadratic equations to solve applied problems.

Follow along with *Guided Learning Video 6.7.2, Using Quadratic Equations to Solve Applied Problems.*

Understanding the Big Picture: As you listen to the first part of the video, when the pencil icon appears, pause and fill in the blanks, choosing from the words listed.

elimination • roots • geometry • quadratic • zero

1. Certain types of word problems such as _____ problems lead to _____ equations.

2. After factoring the quadratic equation, each factor is set equal to _____ and both equations are solved to find the solutions or _____.

3. The conditions of a problem sometimes leads to the _____ of one of the equation's solutions or _____.

Follow along with the two Guided Learning Video examples, and fill in the blanks on the left as you learn with the instructor. When the pencil icon appears, pause and try the Student Practice on your own.

Guided Learning Video ▶ GUIDED LEARNING VIDEO	**Pause: Student Practice** ✏️
1. Allen has a small rectangular garden in his backyard. The length of the garden is 2 feet shorter than triple the width. The garden has an area of 65 square feet. Find the length and width of the garden. Write expressions for each dimension. Then write an equation representing the situation, and solve it to determine the problem's answer.	2. The length of a rectangle is 3 inches longer than double the width. The area of the rectangle is 90 square inches. Find the length and width of the rectangle.

[_____]

[_____]

[_____]

[_____]

(_____) = 0 or (_____) = 0

[_____ = _____]   [_____ = _____]

Guided Learning Video ▶ GUIDED LEARNING VIDEO	Pause: Student Practice

3. The top of a local cable television tower has several small triangular reflectors. The area of each is 25 square inches. The altitude of each triangle is 5 inches longer than the base. Find the altitude and the base of one of the triangles.

Follow the guidelines for solving a quadratic equation to determine the solution.

(   ) = 0 or (   ) = 0

=		=

4. A triangular shape has an area of 35 square feet. The triangle's base is 3 feet longer than the altitude. Find the measure of the triangle's altitude and base.

Helpful Hint: Pause the video and write the helpful hint in your own words.

For each Active Video Lesson, when the pencil icon appears, pause and work the problem. Then press play to check your work.

Active Video Lesson 1	Active Video Lesson 2

5. A racing sailboat has a triangular sail. Find the base and altitude of a triangular sail that has an area of 33 square meters, and has an altitude that is 1 meter shorter than twice the base.

6. A quilt is partly made up of large squares and smaller squares. The side of a larger square is 1 inch shorter than triple the side of a smaller square, and the area of a larger square is 21 square inches greater than the area of a smaller square. Find the dimensions of each square.

Name: _____ Date: _____

Instructor: _____ Section: _____

Guided Learning Video Worksheet: Simplifying Rational Expressions by Factoring

Text: *Beginning Algebra: Early Graphing*
Student Learning Objective 7.1.1: Simplify rational expressions by factoring.

Follow along with *Guided Learning Video 7.1.1, Simplifying Rational Expressions by Factoring.*

Understanding the Big Picture: As you listen to the first part of the video, when the pencil icon appears, pause and fill in the blanks, choosing from the words listed.

simplified • zero • rational • expression • integer

1. A _____ number is a number that can be written as a(n) _____ divided by a non-zero _____.

2. A rational _____ is a polynomial divided by another polynomial where the denominator is not equal to _____.

3. A rational expression is _____ when all common factors are divided out from the numerator and denominator.

Follow along with the two Guided Learning Video examples, and fill in the blanks on the left as you learn with the instructor. When the pencil icon appears, pause and try the Student Practice on your own.

Guided Learning Video ▶ GUIDED LEARNING VIDEO	Pause: Student Practice ✏
1. Simplify. $\dfrac{x^2-3x+2}{x^2+2x-8}$	2. Simplify. $\dfrac{x^2-7x+10}{x^2-3x-10}$
Factor the numerator and denominator and divide out any common factors. $\dfrac{x^2-3x+2}{x^2+2x-8}$ $=\dfrac{\boxed{}}{\boxed{}}$ $=\dfrac{\boxed{}}{\boxed{}}$	

3. Simplify. $\dfrac{a^2 - b^2}{3a^2 - 2ab - 5b^2}$

Factor the numerator and denominator and divide out any common factors.

$$\dfrac{a^2 - b^2}{3a^2 - 2ab - 5b^2}$$

$$= \dfrac{\boxed{}}{\boxed{}}$$

$$= \dfrac{\boxed{}}{\boxed{}}$$

4. Simplify. $\dfrac{x^2 - 9}{x^2 + 6x + 9}$

Helpful Hint: Pause the video and write the helpful hint in your own words.

For each Active Video Lesson, when the pencil icon appears, pause and work the problem. Then press play to check your work.

Active Video Lesson 1

5. Simplify. $\dfrac{16a^2 - 9b^2}{8a^2 - 2ab - 3b^2}$

Active Video Lesson 2

6. Simplify. $\dfrac{2y^2 + 3y - 5}{2y^2 + y - 10}$

Name: _____   Date: _____

Instructor: _____   Section: _____

Guided Learning Video Worksheet: Multiplying Rational Expressions

Text: *Beginning Algebra: Early Graphing*
Student Learning Objective 7.2.1: Multiply rational expressions.

Follow along with *Guided Learning Video 7.2.1, Multiplying Rational Expressions.*

Understanding the Big Picture: As you listen to the first part of the video, when the pencil icon appears, pause and fill in the blanks, choosing from the words listed.

simplify • factor • rational • divide • fractions • integer

1. Multiplication of _____ expressions follows the same rules as multiplication of _____ fractions.

2. When multiplying fractions, it's easier to _____ all numerators and denominators first, and then _____ out any common factors.

3. Use the basic rule of _____ to _____ rational expressions.

Follow along with the two Guided Learning Video examples, and fill in the blanks on the left as you learn with the instructor. When the pencil icon appears, pause and try the Student Practice on your own.

Guided Learning Video ⏵ GUIDED LEARNING VIDEO	**Pause: Student Practice** ✏️
1. Multiply. $\dfrac{x^2-4x-5}{2x^2+7x+3} \cdot \dfrac{x^2-9}{x^2-8x+15}$	2. Multiply. $\dfrac{x^2-4}{x^2-5x+6} \cdot \dfrac{x^2-1}{x^2+3x+2}$

Factor all numerators and denominators, divide out any common factors, and multiply.

$$\frac{x^2-4x-5}{2x^2+7x+3} \cdot \frac{x^2-9}{x^2-8x+15}$$

$$= \frac{\boxed{}}{\boxed{}} \cdot \frac{\boxed{}}{\boxed{}}$$

$$= \frac{\boxed{}}{\boxed{}}$$

3. Multiply. $\dfrac{5x^2-10x-15}{x^3+x^2} \cdot \dfrac{x^3-x^2}{x^2-x-6}$

Factor all numerators and denominators, divide out any common factors, and multiply.

$\dfrac{5x^2-10x-15}{x^3+x^2} \cdot \dfrac{x^3-x}{x^2-x-6}$

$= \dfrac{\boxed{}}{\boxed{}} \cdot \dfrac{\boxed{}}{\boxed{}}$

$= \dfrac{\boxed{}}{\boxed{}} \cdot \dfrac{\boxed{}}{\boxed{}}$

$= \dfrac{\boxed{}}{\boxed{}}$

4. Multiply. $\dfrac{5x-10}{x^2-25} \cdot \dfrac{x^2-4x-5}{x^2+2x-8}$

Helpful Hint: Pause the video and write the helpful hint in your own words.

For each Active Video Lesson, when the pencil icon appears, pause and work the problem. Then press play to check your work.

5. Multiply. $\dfrac{3a^4+3a^3}{6x-18} \cdot \dfrac{2x^2-2x-12}{a^3-3a^2-4a}$

6. Multiply.

$$\dfrac{x^2-2x-35}{x^2-1} \cdot \dfrac{6x^3+6x^2-12x}{3x^2+15x}$$

Guided Learning Video Worksheet: Divide Rational Expressions

Text: *Beginning Algebra: Early Graphing*
Student Learning Objective 7.2.2: Divide rational expressions.

Follow along with *Guided Learning Video 7.2.2, Divide Rational Expressions.*

Understanding the Big Picture: As you listen to the first part of the video, when the pencil icon appears, pause and fill in the blanks, choosing from the words listed.

one • divisor • reciprocal • product • multiplying • invert

1. Dividing a first fraction by a second fraction is equal to _____ the first fraction times the _____ of the second fraction.

2. Two numbers are reciprocals of each other if their _____ equals _____.

3. When dividing rational expressions, first _____ the _____, and then express the quotient as a(n) _____.

Follow along with the two Guided Learning Video examples, and fill in the blanks on the left as you learn with the instructor. When the pencil icon appears, pause and try the Student Practice on your own.

Guided Learning Video ▶ GUIDED LEARNING VIDEO	Pause: Student Practice ✎
1. Divide. $\dfrac{2x-2y}{x^2+10xy+21y^2} \div \dfrac{x^2-xy}{x^3-9xy^2}$	2. Divide. $\dfrac{5x-10y}{x^2+6xy+8y^2} \div \dfrac{x^2-4y^2}{x^2+7xy+12y^2}$

Invert the divisor and multiply, dividing out any common factors first.

$$\dfrac{2x-2y}{x^2+10xy+21y^2} \div \dfrac{x^2-xy}{x^3-9xy^2}$$

$$= \dfrac{\boxed{}}{\boxed{}} \cdot \dfrac{\boxed{}}{\boxed{}}$$

$$= \dfrac{\boxed{}}{\boxed{}} \cdot \dfrac{\boxed{}}{\boxed{}}$$

$$= \dfrac{\boxed{}}{\boxed{}}$$

Pause: Student Practice

3. Divide. $\dfrac{5x^2 - 10x - 15}{x^2 - 6x - 7} \div 5x - 15$

Invert the divisor and multiply, dividing out any common factors first.

$\dfrac{5x^2 - 10x - 15}{x^2 - 6x - 7} \div 5x - 15$

$= \dfrac{\boxed{}}{\boxed{}} \cdot \dfrac{\boxed{}}{\boxed{}}$

$= \dfrac{\boxed{}}{\boxed{}} \cdot \dfrac{\boxed{}}{\boxed{}}$

$= \dfrac{\boxed{}}{\boxed{}}$

4. Divide. $\dfrac{4x^2 - 32x + 60}{x^2 - 9x + 18} \div 4x - 20$

Helpful Hint: Pause the video and write the helpful hint in your own words.

For each Active Video Lesson, when the pencil icon appears, pause and work the problem. Then press play to check your work.

Active Video Lesson 1

5. Divide. $\dfrac{2a^2 - 3a - 2}{a^2 + 5a + 6} \div \dfrac{a^2 - 5a + 6}{a^2 - 9}$

Active Video Lesson 2

6. Divide. $\dfrac{x^2 - 1}{x + 3} \div 8 + 8x$

Name: _____   Date: _____

Instructor: _____   Section: _____

Guided Learning Video Worksheet: Adding & Subtracting Rational Expressions with Different Denominators

Text: *Beginning Algebra: Early Graphing*
Student Learning Objective 7.3.3: Add and subtract rational expressions with different denominators.

Follow along with *Guided Learning Video 7.3.3, Adding & Subtracting Rational Expressions with Different Denominators.*

Understanding the Big Picture: As you listen to the first part of the video, when the pencil icon appears, pause and fill in the blanks, choosing from the words listed.

common • factor • different • greatest • denominators

1. When adding fractions, the _____ must be the same.

2. To add or subtract two rational expressions with _____ denominators, change them to equivalent rational expressions with a least _____ denominator.

3. For rational expressions, the LCD must contain each _____ that appears in any denominator.

4. If a _____ is repeated, the LCD must contain that _____ the _____ number of times it appears in any one denominator.

Follow along with the two Guided Learning Video examples, and fill in the blanks on the left as you learn with the instructor. When the pencil icon appears, pause and try the Student Practice on your own.

Guided Learning Video ▶ GUIDED LEARNING VIDEO	Pause: Student Practice
1. Add. $\dfrac{5}{x} + \dfrac{6}{xy}$	2. Add. $\dfrac{3}{x} + \dfrac{5}{2xy}$

Rewrite the fractions with the least common denominator, then add the fractions and simplify if necessary.

$$\frac{5}{x} + \frac{6}{xy} = \frac{\boxed{}}{\boxed{}} \cdot \frac{\boxed{}}{\boxed{}} + \frac{\boxed{}}{\boxed{}}$$

$$= \frac{\boxed{}}{\boxed{}} + \frac{\boxed{}}{\boxed{}} = \frac{\boxed{}}{\boxed{}}$$

3. Subtract. $\dfrac{2a}{2ab+b^2} - \dfrac{3a}{4a^2-b^2}$

4. Subtract. $\dfrac{x}{3x-6y} - \dfrac{2xy}{x^2-4y^2}$

Rewrite the fractions with the least common denominator, then subtract the fractions and simplify if necessary.

$$\dfrac{2a}{2ab+b^2} - \dfrac{3a}{4a^2-b^2}$$

$$= \dfrac{\square}{\square} \cdot \dfrac{\square}{\square} - \dfrac{\square}{\square} \cdot \dfrac{\square}{\square}$$

$$= \dfrac{\square}{\square} - \dfrac{\square}{\square} = \dfrac{\square}{\square}$$

$$= \dfrac{\square}{\square}$$

Helpful Hint: Pause the video and write the helpful hint in your own words.

For each Active Video Lesson, when the pencil icon appears, pause and work the problem. Then press play to check your work.

Active Video Lesson 1
| **Active Video Lesson 2**

5. Add. $\dfrac{2}{x^2+5x+6} + \dfrac{5}{x^2+7x+12}$

6. Subtract. $\dfrac{7}{(3x-6)} - \dfrac{2x}{(x^2-4)}$

Guided Learning Video Worksheet: Simplifying Complex Rational Expressions: Method 1—Adding or Subtracting in the Numerator and Denominator

Text: *Beginning Algebra: Early Graphing*
Student Learning Objective 7.4.1: Simplify complex rational expressions by adding or subtracting in the numerator and denominator.

Follow along with *Guided Learning Video 7.4.1, Simplifying Complex Rational Expressions: Method 1—Adding or subtracting in the Numerator and Denominator.*

Understanding the Big Picture: As you listen to the first part of the video, when the pencil icon appears, pause and fill in the blanks, choosing from the words listed.

both • dividing • fraction • combining • least • rational

1. A complex _____ expression is also called a complex _____.

2. A complex rational expression is composed of a numerator and denominator and has at _____ one rational expression in the numerator, the denominator or in _____.

3. One method for simplifying a complex rational expression involves _____ quantities to obtain single fractions in the numerator and denominator, and then _____ the numerator by the denominator.

Follow along with the two Guided Learning Video examples, and fill in the blanks on the left as you learn with the instructor. When the pencil icon appears, pause and try the Student Practice on your own.

Guided Learning Video	Pause: Student Practice
1. Simplify. $\dfrac{3+\dfrac{5}{a}}{\dfrac{4}{a}+\dfrac{1}{a^2}}$	2. Simplify. $\dfrac{\dfrac{1}{x}+4}{\dfrac{2}{x^2}-\dfrac{1}{x}}$
Simplify the numerator and the denominator. Then, divide the numerator by the denominator, and simplify the final answer.	
$\dfrac{3+\dfrac{5}{a}}{\dfrac{4}{a}+\dfrac{1}{a^2}} =$	

3. Simplify. $\dfrac{\dfrac{3}{x+4}-\dfrac{2}{x-4}}{\dfrac{5}{x^2-16}+\dfrac{1}{x+4}}$

4. Simplify. $\dfrac{\dfrac{2}{x-3}-\dfrac{1}{x+3}}{\dfrac{4}{x^2-9}+\dfrac{1}{x-3}}$

Simplify the numerator and the denominator. Then, divide the numerator by the denominator, and simplify the final answer.

$$\dfrac{\dfrac{3}{x+4}-\dfrac{2}{x-4}}{\dfrac{5}{x^2-16}+\dfrac{1}{x+4}} =$$

Helpful Hint: Pause the video and write the helpful hint in your own words.

For each Active Video Lesson, when the pencil icon appears, pause and work the problem. Then press play to check your work.

Active Video Lesson 1

Active Video Lesson 2

5. Simplify. $\dfrac{\dfrac{2x}{x-3}-\dfrac{x^2}{x^2-2x-3}}{\dfrac{7}{x-3}+\dfrac{4}{x+1}}$

6. Simplify. $\dfrac{\dfrac{5}{b}-\dfrac{6}{c}}{\dfrac{4}{a^2}+\dfrac{1}{a}}$

**Guided Learning Video Worksheet: Simplifying Complex Rational Expressions:
Method 2—Using the Least Common Denominator**

Text: *Beginning Algebra: Early Graphing*
Student Learning Objective 7.4.2: Simplify complex rational expressions using the LCD.

Follow along with *Guided Learning Video 7.4.2, Simplifying Complex Rational Expressions:
Method 2—Using the Least Common Denominator.*

**Understanding the Big Picture: As you listen to the first part of the video, when the pencil
icon appears, pause and fill in the blanks, choosing from the words listed.**

both • common • fraction • multiplying • least • rational

1. A complex _____ expression is also called a complex _____.

2. A complex rational expression is composed of a numerator and denominator and has at
   _____ one rational expression in the numerator, the denominator or in _____.

3. Simplifying a complex rational expression can be accomplished by _____
   each term of the numerator and denominator by the lowest _____
   denominator of all the individual fractions.

**Follow along with the two Guided Learning Video examples, and fill in the blanks on the
left as you learn with the instructor. When the pencil icon appears, pause and try the
Student Practice on your own.**

Guided Learning Video ▶ GUIDED LEARNING VIDEO	**Pause: Student Practice** ✏️
1. Simplify. $\dfrac{\dfrac{9}{a}-\dfrac{4}{a^2 b}}{\dfrac{1}{ab}-2}$	2. Simplify. $\dfrac{\dfrac{5}{x}+\dfrac{1}{y}}{\dfrac{3}{xy}-4}$
Multiply all terms in the numerator and the denominator by the LCD, and simplify the result. $\dfrac{\dfrac{9}{a}-\dfrac{4}{a^2 b}}{\dfrac{1}{ab}-2}=$	

Guided Learning Video ▶ GUIDED LEARNING VIDEO	Pause: Student Practice ✏
3. Simplify. $\dfrac{\dfrac{7}{x}}{\dfrac{4}{x-2}-\dfrac{3}{x^2-2x}}$	4. Simplify. $\dfrac{\dfrac{3}{x}}{\dfrac{7}{x-6}-\dfrac{5}{x^2-6x}}$

Multiply all terms in the numerator and the denominator by the LCD, and simplify the result.

$$\frac{\dfrac{7}{x}}{\dfrac{4}{x-2}-\dfrac{3}{x^2-2x}}=$$

Helpful Hint: Pause the video and write the helpful hint in your own words.

For each Active Video Lesson, when the pencil icon appears, pause and work the problem. Then press play to check your work.

Active Video Lesson 1 ✏	Active Video Lesson 2 ✏
5. Simplify. $\dfrac{\dfrac{5}{y}}{\dfrac{2}{y^2+3y}+\dfrac{7}{y+3}}$	6. Simplify. $\dfrac{\dfrac{5}{st}+\dfrac{3}{st^2}}{\dfrac{4}{t}-\dfrac{1}{s}}$

**Guided Learning Video Worksheet: Rational Expressions:
Solve Equations That Have Solutions**

Text: *Beginning Algebra: Early Graphing*
Student Learning Objective 7.5.1: Solve equations involving rational expressions that
have solutions.

Follow along with *Guided Learning Video 7.5.1, Rational Expressions: Solve Equations That
Have Solutions.*

**Understanding the Big Picture: As you listen to the first part of the video, when the pencil
icon appears, pause and fill in the blanks, choosing from the words listed.**

polynomial • fractions • denominators • multiply • expressions • rational

1. A _____ equation is an equation having one or more rational
    _____ as terms.

2. To solve a rational equation, _____ each side of the equation by the LCD
    of all the _____ in the equation.

3. The _____ of rational equations will often contain _____
    expressions.

**Follow along with the two Guided Learning Video examples, and fill in the blanks on the
left as you learn with the instructor. When the pencil icon appears, pause and try the
Student Practice on your own.**

Guided Learning Video ▶ GUIDED LEARNING VIDEO	**Pause: Student Practice** ✐
1. Solve. $\dfrac{5}{x}+\dfrac{1}{2}=\dfrac{7}{x}$	2. Solve. $\dfrac{3}{x}+\dfrac{1}{4}=\dfrac{6}{x}$

Use the procedure for solving an
equation with rational expressions.

$$\frac{5}{x}+\frac{1}{2}=\frac{7}{x}$$

$$\boxed{}\left(\frac{}{}\right)+\boxed{}\left(\frac{}{}\right)=\boxed{}\left(\frac{}{}\right)$$

$$\boxed{}=\boxed{}$$

$$\boxed{}=\boxed{}$$

Guided Learning Video ▶ GUIDED LEARNING VIDEO | **Pause: Student Practice** ✏️

---

3. Solve. $\dfrac{4}{x-2}+\dfrac{3}{x+1}=\dfrac{x+16}{x^2-x-2}$

| **4.** Simplify. $\dfrac{2}{x-3}+\dfrac{1}{x+4}=\dfrac{x+1}{x^2+x-12}$

Use the procedure for solving an equation with rational expressions.

$$\dfrac{4}{x-2}+\dfrac{3}{x+1}=\dfrac{x+16}{x^2-x-2}$$

$$\boxed{}\left(\dfrac{}{}\right)+\boxed{}\left(\dfrac{}{}\right)=\boxed{}\left(\dfrac{}{}\right)$$

$$\boxed{}=\boxed{}$$

$$\boxed{}=\boxed{}$$

$$\boxed{}=\boxed{}$$

$$\boxed{}=\boxed{}$$

Helpful Hint: Pause the video and write the helpful hint in your own words.

For each Active Video Lesson, when the pencil icon appears, pause and work the problem. Then press play to check your work.

Active Video Lesson 1 ✏️ | **Active Video Lesson 2** ✏️

---

5. Solve. $\dfrac{x-3}{x-2}=\dfrac{2x^2-15}{x^2+x-6}-\dfrac{x+1}{x+3}$

| **6.** Solve. $\dfrac{7}{3x}+\dfrac{5}{6}=\dfrac{x+1}{x}$

238

Copyright © 2017 Pearson Education, Inc.

Guided Learning Video Worksheet: Identifying Rational Equations with No Solution

Text: *Beginning Algebra: Early Graphing*
Student Learning Objective 7.5.2: Determine whether an equation involving rational expressions has no solution.

Follow along with *Guided Learning Video 7.5.2, Identifying Rational Equations with No Solution.*

Understanding the Big Picture: As you listen to the first part of the video, when the pencil icon appears, pause and fill in the blanks, choosing from the words listed.

zero • fractions • extraneous • multiply • contradiction • eliminated

1. To solve a rational equation, _____ each side of the equation by the LCD of all the _____ in the equation.

2. Some rational equations have no solution because all variable terms are _____ and a(n) _____ results.

3. When the solution to a rational equation leads to division by _____ , the equation has no solution, and the apparent solution is called _____ .

Follow along with the two Guided Learning Video examples, and fill in the blanks on the left as you learn with the instructor. When the pencil icon appears, pause and try the Student Practice on your own.

Guided Learning Video ▶ GUIDED LEARNING VIDEO	**Pause: Student Practice** ✏️
1. Solve. $\dfrac{3}{a-1}+\dfrac{4}{a+5}=\dfrac{7a+5}{a^2+4a-5}$	2. Solve. $\dfrac{2}{x+1}-\dfrac{1}{x-3}=\dfrac{x+4}{x^2-2x-3}$

Use the procedure for solving a rational equation and check the solution.

$$\frac{3}{a-1}+\frac{4}{a+5}=\frac{7a+5}{a^2+4a-5}$$

$$\boxed{}\left(\dfrac{}{}\right)+\boxed{}\left(\dfrac{}{}\right)=\boxed{}\left(\dfrac{}{}\right)$$

$$\boxed{}=\boxed{}$$

$$\boxed{}=\boxed{}$$

Check: $\boxed{}=\boxed{}$

$$\boxed{}=\boxed{}$$

3. Solve. $\dfrac{4}{y+6}+\dfrac{1}{y-3}=\dfrac{9}{y^2+3y-18}$

Use the procedure for solving a rational equation and check the solution.

$$\dfrac{4}{y+6}+\dfrac{1}{y-3}=\dfrac{9}{y^2+3y-18}$$

$$\boxed{}\left(\dfrac{}{}\right)+\boxed{}\left(\dfrac{}{}\right)=\boxed{}\left(\dfrac{}{}\right)$$

$$\boxed{}=\boxed{}$$

$$\boxed{}=\boxed{}$$

Check: $\boxed{}=\boxed{}$

$$\boxed{}=\boxed{}$$

4. Solve. $5-\dfrac{3}{3+x}=\dfrac{x}{x+3}$

Helpful Hint: Pause the video and write the helpful hint in your own words.

For each Active Video Lesson, when the pencil icon appears, pause and work the problem. Then press play to check your work.

Active Video Lesson 1 ✎ | **Active Video Lesson 2** ✎

5. Solve. $\dfrac{12}{x^2-9}+\dfrac{3}{x+3}=\dfrac{2}{x-3}$

6. Solve. $\dfrac{2}{y+5}+\dfrac{4}{y-3}=\dfrac{6y+7}{y^2+2y-15}$

Guided Learning Video Worksheet: Solving Proportions

Text: *Beginning Algebra: Early Graphing*
Student Learning Objective 7.6.1: Solve problems involving ratio and proportion.

Follow along with *Guided Learning Video 7.6.1, Solving Proportions.*

Understanding the Big Picture: As you listen to the first part of the video, when the pencil icon appears, pause and fill in the blanks, choosing from the words listed.

equation • unknown • positions • cross • denominators

1. To solve a proportion, begin by _____ multiplying numerators times _____.

2. Create a(n) _____ by setting the cross products equal to each other.

3. The letter n is often used to represent the _____ quantity.

4. When setting up the proportion, the same units should be in the same _____ in each of the fractions.

Follow along with the two Guided Learning Video examples, and fill in the blanks on the left as you learn with the instructor. When the pencil icon appears, pause and try the Student Practice on your own.

Guided Learning Video ▶ GUIDED LEARNING VIDEO	Pause: Student Practice ✏
1. Find the value of n: $\dfrac{13}{3} = \dfrac{26}{n}$	2. Find the value of n: $\dfrac{17}{4} = \dfrac{51}{n}$

Find the cross products, then divide to solve for n.

$$\frac{13}{3} = \frac{26}{n}$$

☐ = ☐   Check:

☐ = ☐ $\dfrac{\square}{\square} = \dfrac{\square}{\square}$

☐ = ☐

☐ = ☐

3. If the current dose of an antibiotic for a 180-pound man is 15 milligrams, what is the correct dose for a 120-pound woman?

   Set up the proportion, find the cross products, then divide to solve for *n*.

   $$\frac{\boxed{}}{\boxed{}} = \frac{\boxed{}}{\boxed{}}$$

   $$\boxed{} = \boxed{}$$ Check:

   $$\boxed{} = \boxed{}$$ $$\frac{\boxed{}}{\boxed{}} = \frac{\boxed{}}{\boxed{}}$$

   $$\boxed{} = \boxed{}$$ $$\boxed{} = \boxed{}$$

4. If 3 grams of saline need to be added to a 15-liter solution, how many grams need to be added to a 20-liter solution?

Helpful Hint: Pause the video and write the helpful hint in your own words.

For each Active Video Lesson, when the pencil icon appears, pause and work the problem. Then press play to check your work.

Active Video Lesson 1 ✏ | **Active Video Lesson 2** ✏

5. Find the value of *n*: $\dfrac{0.5}{n} = \dfrac{3}{4.2}$

6. Find the value of the unknown amount, in dollars, of: 54 dollars per 18 square feet is equal to how much per 15 square feet?

Guided Learning Video Worksheet: Solve Word Problems: Similar Triangles

Text: *Beginning Algebra: Early Graphing*
Student Learning Objective 7.6.2: Solve problems involving similar triangles.

Follow along with *Guided Learning Video 7.6.2, Solve Word Problems: Similar Triangles.*

Understanding the Big Picture: As you listen to the first part of the video, when the pencil icon appears, pause and fill in the blanks, choosing from the words listed.

angles • proportional • equal • corresponding • different

1. Similar triangles are triangles having the same shape but with _____ measures for their _____ sides.

2. In similar triangles, the _____ sides are _____.

3. The degree measures of the corresponding _____ of similar triangles are _____.

Follow along with the two Guided Learning Video examples, and fill in the blanks on the left as you learn with the instructor. When the pencil icon appears, pause and try the Student Practice on your own.

Guided Learning Video ▶ GUIDED LEARNING VIDEO	Pause: Student Practice
1. Ramp A is 50 feet long and rises up 35 feet. Ramp B, at the same angle, rises up 28 feet. How long is ramp B? Use similar triangles and a proportion equation to solve for the unknown. $$\frac{\boxed{}}{\boxed{}} = \frac{\boxed{}}{\boxed{}}$$ $\boxed{} = \boxed{}$ $\boxed{} = \boxed{}$	2. A ramp 42 feet long rises 3 feet. At the same angle, a second ramp rises 2.5 . How long is the second ramp?

3. A woman who is 5.5 feet tall casts a shadow that is 4 feet long. At the same time of day, a tree casts a shadow that is 50 feet long. How tall is the tree?

   Use similar triangles and a proportion equation to solve for the unknown.

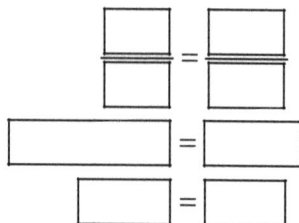

   $$\frac{\boxed{}}{\boxed{}} = \frac{\boxed{}}{\boxed{}}$$

   $$\boxed{} = \boxed{}$$

   $$\boxed{} = \boxed{}$$

4. A man who is 6 feet tall casts a shadow that is 4 feet long. At the same time of day, a flagpole casts a shadow that is 12 feet long. How tall is the flagpole?

Helpful Hint: Pause the video and write the helpful hint in your own words.

For each Active Video Lesson, when the pencil icon appears, pause and work the problem. Then press play to check your work.

Active Video Lesson 1 ✎ | **Active Video Lesson 2** ✎

5. A wire anchoring a television tower to the ground measures 85 meters from the tower to the ground anchor pin. The wire is secured 60 meters up on the tower. If a second wire is secured 582 meters up on the tower and is extended from the tower at the same angle as the first wire, how long would the second wire need to be to reach an anchor pin on the ground? Round the answer to the nearest meter.

6. A slide at the park is 20 feet long and rises up 12 feet. A toddler slide beside it at the same angle rises up 6 feet. How long is the toddler slide?

Guided Learning Video Worksheet: Square Roots

Text: *Beginning Algebra: Early Graphing*
Student Learning Objective 8.1.1: Evaluate the square root of a perfect square.

Follow along with *Guided Learning Video 8.1.1, Square Roots.*

Understanding the Big Picture: As you listen to the first part of the video, when the pencil icon appears, pause and fill in the blanks, choosing from the words listed.

zero • integer • radicand • principal • perfect • nonnegative

1. The expression inside a radical is called a _____.

2. A radicand is said to be a _____ square because its square root is a(n)
   _____.

3. A(n) _____ square root or _____ is also called the
   _____ square root.

Follow along with the two Guided Learning Video examples, and fill in the blanks on the left as you learn with the instructor. When the pencil icon appears, pause and try the Student Practice on your own.

Guided Learning Video ▶ GUIDED LEARNING VIDEO	Pause: Student Practice ✏️
1. Find the square root. $-\sqrt{36}$	2. Find the square root. $-\sqrt{81}$
Find the square root.	
$-\sqrt{36} = \boxed{}$	

Pause: Student Practice

3. Find the square root. $\sqrt{\dfrac{49}{121}}$

4. Find the square root. $\sqrt{\dfrac{16}{225}}$

Find the principal square root.

$$\sqrt{\dfrac{49}{121}} = \boxed{}$$

Helpful Hint: Pause the video and write the helpful hint in your own words.

For each Active Video Lesson, when the pencil icon appears, pause and work the problem. Then press play to check your work.

Active Video Lesson 1

5. Find the square root. $-\sqrt{64}$

Active Video Lesson 2

6. Find the square root. $\sqrt{\dfrac{25}{144}}$

Name: _____   Date: _____

Instructor: _____   Section: _____

Guided Learning Video Worksheet: Radical Expressions: Simplify Perfect Squares

Text: *Beginning Algebra: Early Graphing*
Student Learning Objective 8.2.1: Simplify a radical expression with a radicand that is a perfect square.

Follow along with *Guided Learning Video 8.2.1, Radical Expressions: Simplify Perfect Squares.*

Understanding the Big Picture: As you listen to the first part of the video, when the pencil icon appears, pause and fill in the blanks, choosing from the words listed.

variable • multiplication • zero • base • positive • nonnegative

1. If a radicand is written as a square of a nonnegative number, the square root is the nonnegative _____.

2. If a _____ is under a radical sign, it is assumed to represent a _____ number or _____.

3. The _____ rule for square roots is given as $\sqrt{a} \cdot \sqrt{b} = \sqrt{ab}$, where a and b are _____ numbers.

Follow along with the two Guided Learning Video examples, and fill in the blanks on the left as you learn with the instructor. When the pencil icon appears, pause and try the Student Practice on your own.

Guided Learning Video ▶ GUIDED LEARNING VIDEO	Pause: Student Practice 🖊
1. Find the square root. $\sqrt{x^{22}}$ Use the power rule of exponents to rewrite the radicand and simplify. $\sqrt{x^{22}} = \boxed{}$	2. Find the square root. $\sqrt{x^{32}}$

Guided Learning Video ▶ GUIDED LEARNING VIDEO	**Pause: Student Practice** ✏
3. Find the square root. $\sqrt{81a^{10}b^2}$ Use the multiplication rule for square roots and simplify. $\sqrt{81a^{10}b^2} = $ [ ]	**4.** Find the square root. $\sqrt{49x^6y^2}$

Helpful Hint: Pause the video and write the helpful hint in your own words.

For each Active Video Lesson, when the pencil icon appears, pause and work the problem. Then press play to check your work.

Active Video Lesson 1 ✏	**Active Video Lesson 2** ✏
5. Find the square root. $\sqrt{121x^{12}y^{18}}$	**6.** Find the square root. $\sqrt{25c^6}$

Guided Learning Video Worksheet: Adding and Subtracting Radical Expressions.

Text: *Beginning Algebra: Early Graphing*
Student Learning Objective 8.3.2: Add and subtract radical expressions that must first be simplified.

Follow along with *Guided Learning Video 8.3.2, Adding and Subtracting Radical Expressions.*

Understanding the Big Picture: As you listen to the first part of the video, when the pencil icon appears, pause and fill in the blanks, choosing from the words listed.

radicands • same • expressions • perfect • simplified • numbers

1. To add or subtract square root radicals, the _____ or the algebraic _____ under the radical signs must be the _____.

2. Sometimes radicals must be _____ so they are rewritten as like radicals and can be combined.

3. In order to simplify radicals, look for the _____ square factors of the _____.

Follow along with the two Guided Learning Video examples, and fill in the blanks on the left as you learn with the instructor. When the pencil icon appears, pause and try the Student Practice on your own.

Guided Learning Video ▶ GUIDED LEARNING VIDEO	Pause: Student Practice ✏️
1. Combine. $\sqrt{18b} - \sqrt{50b} + \sqrt{12b}$	2. Combine. $\sqrt{48x} - \sqrt{40x} + \sqrt{75x}$

Look for perfect square factors to simplify the radicands, then combine the like radicals.

$\sqrt{18b} - \sqrt{50b} + \sqrt{12b}$

$=$ ▭

$=$ ▭

$=$ ▭

3. Combine. $5\sqrt{28} - 3\sqrt{7} + 6\sqrt{63}$

Look for perfect square factors to simplify the radicands, then combine the like radicals.

$5\sqrt{28} - 3\sqrt{7} + 6\sqrt{63}$

$=$ [   ]

$=$ [   ]

$=$ [   ]

4. Combine. $3\sqrt{18} - 6\sqrt{27} + 4\sqrt{128}$

Helpful Hint: Pause the video and write the helpful hint in your own words.

For each Active Video Lesson, when the pencil icon appears, pause and work the problem. Then press play to check your work.

Active Video Lesson 1

5. Combine. $7\sqrt{45} - 3\sqrt{20} - 4\sqrt{5}$

Active Video Lesson 2

6. Combine. $\sqrt{27x} + \sqrt{50x} - \sqrt{8x}$

Guided Learning Video Worksheet: Radical Expressions: Multiply Monomials

Text: *Beginning Algebra: Early Graphing*
Student Learning Objective 8.4.1: Multiply monomial radical expressions.

Follow along with *Guided Learning Video 8.4.1, Radical Expressions: Multiply Monomials.*

Understanding the Big Picture: As you listen to the first part of the video, when the pencil icon appears, pause and fill in the blanks, choosing from the words listed.

radicands • simplest • product • simplify • coefficients

1. To multiply square root radicals, multiply the _____, and then _____ the product if necessary.

2. If necessary, first multiply the _____, then multiply the radicals.

3. The direction to "multiply" square root radical expressions means to express the _____ in _____ form.

Follow along with the two Guided Learning Video examples, and fill in the blanks on the left as you learn with the instructor. When the pencil icon appears, pause and try the Student Practice on your own.

Guided Learning Video ▶ GUIDED LEARNING VIDEO	**Pause: Student Practice** ✏️
1. Multiply. $\sqrt{10x} \cdot \sqrt{2}$ Use the rule for multiplying radicals. $\sqrt{10x} \cdot \sqrt{2}$ = [　　　　] = [　　　　] = [　　　　]	2. Multiply. $\sqrt{3} \cdot \sqrt{30x}$

3. Multiply. $\left(5\sqrt{3y}\right)\left(2y\sqrt{6y}\right)$

Use the rule for multiplying radicals.

$\left(5\sqrt{3y}\right)\left(2y\sqrt{6y}\right)$

$=$ [   ]

$=$ [   ]

$=$ [   ]

4. Multiply. $\left(2x\sqrt{6x}\right)\left(3\sqrt{2x}\right)$

Helpful Hint: Pause the video and write the helpful hint in your own words.

For each Active Video Lesson, when the pencil icon appears, pause and work the problem. Then press play to check your work.

Active Video Lesson 1 ✏ | **Active Video Lesson 2** ✏

5. Multiply. $\left(4b\sqrt{10b}\right)\left(2\sqrt{5b}\right)$

6. Multiply. $\left(\sqrt{3z}\right)\left(\sqrt{15}\right)$

Name: _____   Date: _____
Instructor: _____   Section: _____

Guided Learning Video Worksheet: Radical Expressions: Multiply a Monomial by a Polynomial

Text: *Beginning Algebra: Early Graphing*
Student Learning Objective 8.4.2: Multiply a monomial radical expression by a polynomial.

Follow along with *Guided Learning Video 8.4.2, Radical Expressions: Multiply a Monomial by a Polynomial.*

Understanding the Big Picture: As you listen to the first part of the video, when the pencil icon appears, pause and fill in the blanks, choosing from the words listed.

nonnegative • two • distributive • multiplication • binomial

1. A _____ consists of _____ terms.

2. The _____ property is used to multiply a binomial times another factor.

3. The _____ rule of radicals states that for any _____ real number a, $\sqrt{a} \cdot \sqrt{a} = a$.

Follow along with the two Guided Learning Video examples, and fill in the blanks on the left as you learn with the instructor. When the pencil icon appears, pause and try the Student Practice on your own.

Guided Learning Video ▶ GUIDED LEARNING VIDEO	Pause: Student Practice
1. Multiply and simplify. $\sqrt{3}\left(6\sqrt{5} - 2\sqrt{6}\right)$	2. Multiply and simplify. $\sqrt{5}\left(2\sqrt{10} - 7\sqrt{3}\right)$

Use the multiplication rule of radicals to simplify the expression.

$\sqrt{3}\left(6\sqrt{5} - 2\sqrt{6}\right)$

$= \boxed{}$

$= \boxed{}$

$= \boxed{}$

Pause: Student Practice

---

3. Multiply and simplify.

$$2\sqrt{y}\left(\sqrt{5y} - 6\sqrt{3}\right)$$

Use the multiplication rule of radicals to simplify the expression.

$$2\sqrt{y}\left(\sqrt{5y} - 6\sqrt{3}\right)$$

$$= \boxed{}$$

$$= \boxed{}$$

$$= \boxed{}$$

4. Multiply and simplify.

$$6\sqrt{x}\left(4\sqrt{2x} + 3\sqrt{5}\right)$$

Helpful Hint: Pause the video and write the helpful hint in your own words.

For each Active Video Lesson, when the pencil icon appears, pause and work the problem. Then press play to check your work.

Active Video Lesson 1

Active Video Lesson 2

---

5. Multiply and simplify.

$$3\sqrt{x}\left(3\sqrt{6} + 8\sqrt{7x}\right)$$

6. Multiply and simplify.

$$\sqrt{2}\left(3\sqrt{10} - 9\sqrt{7}\right)$$

Name: _____   Date: _____

Instructor: _____   Section: _____

Guided Learning Video Worksheet: Radical Expressions: Multiply Two Polynomials

Text: *Beginning Algebra: Early Graphing*
Student Learning Objective 8.4.3: Multiply two polynomial radical expressions.

Follow along with *Guided Learning Video 8.4.3, Radical Expressions: Multiply Two Polynomials.*

Understanding the Big Picture: As you listen to the first part of the video, when the pencil icon appears, pause and fill in the blanks, choosing from the words listed.

simplified • inner • binomials • radical • multiply • outer

1. The FOIL method is used to _____ two _____.

2. Two binomial _____ expressions can also be multiplied using FOIL and then _____ to equal the final product.

3. When using the FOIL method, multiply the first terms of the binomials, then the _____ terms and the _____ terms, and finally the last terms.

Follow along with the two Guided Learning Video examples, and fill in the blanks on the left as you learn with the instructor. When the pencil icon appears, pause and try the Student Practice on your own.

Guided Learning Video ▶ GUIDED LEARNING VIDEO	**Pause: Student Practice**
1. Multiply. a. $\left(\sqrt{3}+7\right)\left(\sqrt{3}-2\right)$ b. $\left(2\sqrt{5}+\sqrt{7}\right)\left(\sqrt{5}-\sqrt{7}\right)$	2. Multiply. a. $\left(\sqrt{2}-6\right)\left(\sqrt{2}+5\right)$ b. $\left(3\sqrt{2}-\sqrt{5}\right)\left(\sqrt{2}-\sqrt{5}\right)$

Use the FOIL method to multiply.

a. $\left(\sqrt{3}+7\right)\left(\sqrt{3}-2\right)$

= [ ]

= [ ]

b. $\left(2\sqrt{5}+\sqrt{7}\right)\left(\sqrt{5}-\sqrt{7}\right)$

= [ ]

= [ ]

3. Multiply. $\left(3\sqrt{5} - 2\sqrt{10}\right)^2$

Rewrite the expression as the product of binomials and use the FOIL method to multiply.

$\left(3\sqrt{5} - 2\sqrt{10}\right)^2$

$=$ [ ]

$=$ [ ]

$=$ [ ]

4. Multiply. $\left(\sqrt{7} - 2\sqrt{5}\right)^2$

Helpful Hint: Pause the video and write the helpful hint in your own words.

For each Active Video Lesson, when the pencil icon appears, pause and work the problem. Then press play to check your work.

Active Video Lesson 1 | **Active Video Lesson 2**

5. Multiply. $\left(2\sqrt{6} - \sqrt{3}\right)^2$

6. Multiply. $\left(4\sqrt{2} - \sqrt{5}\right)\left(3\sqrt{2} + \sqrt{5}\right)$

Guided Learning Video Worksheet: Rationalizing Denominators

Text: *Beginning Algebra: Early Graphing*
Student Learning Objective 8.5.2: Rationalize the denominator of a fraction with a square root in the denominator.

Follow along with *Guided Learning Video 8.5.2, Rationalizing Denominators*.

Understanding the Big Picture: As you listen to the first part of the video, when the pencil icon appears, pause and fill in the blanks, choosing from the words listed.

rationalize • denominator • perfect • integer • smallest • radical

1. When calculating with fractions containing radicals, it is advantageous to have a(n) _____ in the _____.

2. If a fraction has a(n) _____ in its denominator, _____ the denominator.

3. Multiply by the _____ possible radical that leads to a square root of a _____ square in the denominator.

Follow along with the two Guided Learning Video examples, and fill in the blanks on the left as you learn with the instructor. When the pencil icon appears, pause and try the Student Practice on your own.

Guided Learning Video (▶) GUIDED LEARNING VIDEO	Pause: Student Practice ✏️
1. Simplify. a. $\dfrac{\sqrt{5}}{\sqrt{12}}$ b. $\dfrac{9y}{\sqrt{y^5}}$ Rationalize the denominators and simplify the resulting fractions. a. $\dfrac{\sqrt{5}}{\sqrt{12}} \cdot \dfrac{\square}{\square} = \dfrac{\square}{\square} = \dfrac{\square}{\square}$ b. $\dfrac{9y}{\sqrt{y^5}} \cdot \dfrac{\square}{\square} = \dfrac{\square}{\square} = \dfrac{\square}{\square}$	2. Simplify. a. $\dfrac{\sqrt{3}}{\sqrt{7}}$ b. $\dfrac{6x}{\sqrt{x^3}}$

3. Simplify. $\dfrac{\sqrt{7}}{\sqrt{45y}}$

First simplify the denominator, then rationalize the denominator and simplify the resulting fraction.

$$\frac{\sqrt{7}}{\sqrt{45y}} \cdot \frac{\square}{\square} = \frac{\square}{\square} = \frac{\square}{\square}$$

4. Simplify. $\dfrac{\sqrt{2}}{\sqrt{27x}}$

Helpful Hint: Pause the video and write the helpful hint in your own words.

For each Active Video Lesson, when the pencil icon appears, pause and work the problem. Then press play to check your work.

Active Video Lesson 1 ✏ | **Active Video Lesson 2** ✏

5. Simplify. $\dfrac{\sqrt{5}}{\sqrt{18x}}$

6. Simplify. $\dfrac{6x}{\sqrt{x^7}}$

Name: _____   Date: _____

Instructor: _____   Section: _____

Guided Learning Video Worksheet: Rationalizing Denominators Using Conjugates

Text: *Beginning Algebra: Early Graphing*
Student Learning Objective 8.5.3: Rationalize the denominator of a fraction with a binomial denominator containing at least one square root.

Follow along with *Guided Learning Video 8.5.3, Rationalizing Denominator Using Conjugates.*

Understanding the Big Picture: As you listen to the first part of the video, when the pencil icon appears, pause and fill in the blanks, choosing from the words listed.

square • eliminated • opposite • binomial • conjugate • difference

1. When working with a fraction with a(n) _____ expression containing radicals in the denominator, the radicals must sometimes be _____.

2. When binomials with opposite second terms are multiplied, the result is the _____ of two perfect _____ terms.

3. The _____ of a binomial is formed by expressing the second term of the original binomial as its _____.

Follow along with the two Guided Learning Video examples, and fill in the blanks on the left as you learn with the instructor. When the pencil icon appears, pause and try the Student Practice on your own.

Guided Learning Video (▶) GUIDED LEARNING VIDEO	**Pause: Student Practice** ✎
1. Simplify. $\dfrac{3x}{\sqrt{5}+2}$	2. Simplify. $\dfrac{2x}{\sqrt{3}-1}$
Multiply the numerator and the denominator by the conjugate of the denominator and simplify. $\dfrac{3x}{\sqrt{5}+2} \cdot \dfrac{\Box}{\Box} = \dfrac{\Box}{\Box} = \dfrac{\Box}{\Box}$	

3. Simplify. $\dfrac{\sqrt{2}+3}{\sqrt{2}-5}$

Multiply the numerator and the denominator by the conjugate of the denominator and simplify.

$$\dfrac{\sqrt{2}+3}{\sqrt{2}-5} \cdot \dfrac{\boxed{}}{\boxed{}} = \dfrac{\boxed{}}{\boxed{}} = \dfrac{\boxed{}}{\boxed{}}$$

4. Simplify. $\dfrac{\sqrt{x}+1}{\sqrt{x}-2}$

Helpful Hint: Pause the video and write the helpful hint in your own words.

For each Active Video Lesson, when the pencil icon appears, pause and work the problem. Then press play to check your work.

Active Video Lesson 1 | **Active Video Lesson 2**

5. Simplify. $\dfrac{\sqrt{3}+4}{5+\sqrt{3}}$

6. Simplify. $\dfrac{\sqrt{2}}{\sqrt{7}-\sqrt{2}}$

Guided Learning Video Worksheet: Solve Radical Equations

Text: *Beginning Algebra: Early Graphing*
Student Learning Objective 8.6.2: Solve radical equations.

Follow along with *Guided Learning Video 8.6.2, Solve Radical Equations.*

Understanding the Big Picture: As you listen to the first part of the video, when the pencil icon appears, pause and fill in the blanks, choosing from the words listed.

squaring • extraneous • isolate • variable • no • original • radicands

1. A radical equation is an equation with a(n) _____ in one or more of the _____.

2. For some square root radical equations, _____ the radical before _____ each side of the equation.

3. If the apparent solution doesn't satisfy the _____ equation, the solution is called an _____ root, and there is _____ solution to the equation.

Follow along with the two Guided Learning Video examples, and fill in the blanks on the left as you learn with the instructor. When the pencil icon appears, pause and try the Student Practice on your own.

Guided Learning Video	Pause: Student Practice
1. Solve and check. $\sqrt{2x+6} = x-1$	2. Solve and check. $\sqrt{2x+9} = x+3$
Isolate the radical and solve the equation.	

$$\sqrt{2x+6} = x-1$$

```
┌─────────────────────┐
│                     │
└─────────────────────┘
┌─────────────────────┐
│                     │
└─────────────────────┘
┌─────────────────────┐
│                     │
└─────────────────────┘
┌─────────────────────┐
│                     │
└─────────────────────┘
┌─────────────────────┐
│                     │
└─────────────────────┘
```

Check:
```
┌─────────────────────┐
│                     │
└─────────────────────┘
┌─────────────────────┐
│                     │
└─────────────────────┘
┌─────────────────────┐
│                     │
└─────────────────────┘
```

3. Solve and check. $\sqrt{3x+13}+3=2x$

Isolate the radical and solve the equation.

$$\sqrt{3x+13}+3=2x$$

Check:

4. Solve and check. $\sqrt{2x+3}-x=2$

Helpful Hint: Pause the video and write the helpful hint in your own words.

For each Active Video Lesson, when the pencil icon appears, pause and work the problem. Then press play to check your work.

Active Video Lesson 1 | **Active Video Lesson 2**

5. Solve and check. $5+\sqrt{x+15}=x$

6. Solve and check. $x-5=\sqrt{x+7}$

Name: _____   Date: _____

Instructor: _____   Section: _____

Guided Learning Video Worksheet: Word Problems Involving Direct Variation

Text: *Beginning Algebra: Early Graphing*
Student Learning Objective 8.7.1: Solve problems involving direct variation.

Follow along with *Guided Learning Video 8.7.1, Word Problems Involving Direct Variation.*

Understanding the Big Picture: As you listen to the first part of the video, when the pencil icon appears, pause and fill in the blanks, choosing from the words listed.

known • constant • equation • variation • quantities

1. Relationships between two measureable _____ are often seen in
   _____ problems.

2. In direct variation problems, the _____ of _____ is
   represented by the variable k.

3. The direct variation _____ expresses the relationship between the
   _____ values and the constant of variation.

Follow along with the two Guided Learning Video examples, and fill in the blanks on the left as you learn with the instructor. When the pencil icon appears, pause and try the Student Practice on your own.

Guided Learning Video ▶ GUIDED LEARNING VIDEO	**Pause: Student Practice** ✏
1. If y varies directly as x, and $y = 16$ when $x = 12$, find the value of y when $x = 36$. Use the procedure for solving direct variation problems. $y = k \cdot x$ ☐ ☐ ☐ ☐ ☐	2. If y varies directly as x, and $y = 28.5$ when $x = 3$, find the value of y when $x = 5.5$.

| 3. The volume of water flowing from a hose varies directly as the square of the radius of the hose. 48 gallons flows from a hose of radius 2 centimeters into a barrel. In the same period of time, how many gallons would flow from the hose if it had a radius of 4 centimeters?

Use the procedure for solving direct variation problems.

$V = k \cdot r^2$ | 4. The volume of water flowing from a hose varies directly as the square of the radius of the hose. 26 gallons flows from a hose of radius 1.25 centimeters into a barrel. In the same period of time, how many gallons would flow from the hose if it had a radius of 1.75 centimeters? |

Helpful Hint: Pause the video and write the helpful hint in your own words.

For each Active Video Lesson, when the pencil icon appears, pause and work the problem. Then press play to check your work.

Active Video Lesson 1	Active Video Lesson 2
5. In a certain class of racing cars, the maximum speed varies directly as the square root of the horsepower of the engine. If a car with 225 horsepower can achieve a maximum speed of 135 mph, what speed could it achieve with 196 horsepower?	6. If y varies directly as the cube of x, and $y = 56$ when $x = 2$, find the value of y when $x = 3$.

Name: _____   Date: _____

Instructor: _____   Section: _____

Guided Learning Video Worksheet: Word Problems Involving Inverse Variation

Text: *Beginning Algebra: Early Graphing*
Student Learning Objective 8.7.2: Solve problems involving inverse variation.

Follow along with *Guided Learning Video 8.7.2, Word Problems Involving Inverse Variation*.

Understanding the Big Picture: As you listen to the first part of the video, when the pencil icon appears, pause and fill in the blanks, choosing from the words listed.

inverse • constant • reciprocal • variation • inversely

1. Two variables vary _____ when one variable is a constant multiple of the _____ of the other.

2. In inverse variation problems, the _____ of _____ is represented by k.

3. The form of the _____ variation equation is $y = \dfrac{k}{x}$.

Follow along with the two Guided Learning Video examples, and fill in the blanks on the left as you learn with the instructor. When the pencil icon appears, pause and try the Student Practice on your own.

Guided Learning Video ▶ GUIDED LEARNING VIDEO	Pause: Student Practice
1. If y varies inversely as x, and $y = 12$ when $x = 7$, find the value of y when $x = \dfrac{2}{3}$. Use the procedure for solving inverse variation problems. $y = \dfrac{k}{x}$ ⬚ ⬚ ⬚ ⬚ ⬚	2. If y varies inversely as x, and $y = 62$ when $x = 0.75$, find the value of y when $x = 1.2$.

265

3. The number of characters that fit on a line of print varies inversely with the font size. If an average of 102 characters with font size 10 fit on a line of print, on average, how many characters with size 12 will fit on the same-size line of print?

Use the procedure for solving inverse variation problems.

$$c = \frac{k}{f}$$

4. The amount of time, in minutes, it takes ice to melt in water varies inversely as the temperature of the water into which the ice is placed. If ice placed in 50°F takes 3 minutes to melt, how long does it take for ice placed in 60°F water to melt?

Helpful Hint: Pause the video and write the helpful hint in your own words.

For each Active Video Lesson, when the pencil icon appears, pause and work the problem. Then press play to check your work.

Active Video Lesson 1 ✏ | **Active Video Lesson 2** ✏

5. The illumination of a light varies inversely as the square of the distance from the source. The illumination measures 16 candlepower when a certain light is 5 meters away. Find the illumination when the light is 10 meters away.

6. Over the last three years, a calculator company found that the sales of its calculators varies inversely as the price of the calculator. One year, 120,000 calculators were sold at $30 each. How many calculators were sold the next year when each calculator was $24?

Name: _____   Date: _____

Instructor: _____   Section: _____

Guided Learning Video Worksheet: Quadratic Equations: Solve $ax^2 + bx = 0$ by Factoring

Text: *Beginning Algebra: Early Graphing*

Student Learning Objective 9.1.2: Solve quadratic equations of the form $ax^2 + bx = 0$ by factoring.

Follow along with *Guided Learning Video 9.1.2, Quadratic Equations: Solve $ax^2 + bx = 0$ by Factoring.*

Understanding the Big Picture: As you listen to the first part of the video, when the pencil icon appears, pause and fill in the blanks, choosing from the words listed.

numerical • degree • greatest • quadratic • factoring

1. A _____ equation is a polynomial equation of _____ two.

2. When solving quadratic equations of the form $ax^2 + bx = 0$, begin by _____ out x and any common _____ factor.

3. When factoring, always remove the _____ common factor.

Follow along with the two Guided Learning Video examples, and fill in the blanks on the left as you learn with the instructor. When the pencil icon appears, pause and try the Student Practice on your own.

Guided Learning Video ▶ GUIDED LEARNING VIDEO	Pause: Student Practice
1. Solve $2x^2 - 5x = 0$.	2. Solve $7x^2 + x = 0$.

Factor out the greatest common factor, and set each factor to zero to solve the equation.

$2x^2 - 5x = 0$

```
┌─────────────────────┐
│                     │
└─────────────────────┘

┌──────────┐   ┌──────────┐
│          │   │          │
└──────────┘   └──────────┘

┌──────────┐   ┌──────────┐
│          │   │          │
└──────────┘   └──────────┘
```

Guided Learning Video ▶ GUIDED LEARNING VIDEO	**Pause: Student Practice**
3. Solve $6x^2 - 3x + 5 = x + 5$. Rewrite the equation in standard form, then factor out the greatest common factor, and set each factor to zero to solve the equation. $6x^2 - 3x + 5 = x + 5$	**4.** Solve $5x^2 - 9x + 8 = 6x + 8$.

Helpful Hint: Pause the video and write the helpful hint in your own words.

For each Active Video Lesson, when the pencil icon appears, pause and work the problem. Then press play to check your work.

Active Video Lesson 1	**Active Video Lesson 2**
5. Solve $12x^2 - 7x + 3 = 8x + 3$.	**6.** Solve $4x^2 - 3x = 0$.

Guided Learning Video Worksheet: Solving Quadratic Equations by the Square Root Property

Text: *Beginning Algebra: Early Graphing*
Student Learning Objective 9.2.1: Solve quadratic equations using the square root property.

Follow along with *Guided Learning Video 9.2.1, Solving Quadratic Equations by the Square Root Property*.

Understanding the Big Picture: As you listen to the first part of the video, when the pencil icon appears, pause and fill in the blanks, choosing from the words listed.

square • standard • form • real • quadratic

1. Equations of the _____ $ax^2 + bx + c = 0$ when a, b, and c are real numbers and $a \neq 0$ are called _____ equations.

2. The _____ form of a quadratic equation is $ax^2 + bx + c = 0$.

3. The _____ root property states that if $x^2 = a$, then $x = \pm\sqrt{a}$ for all nonnegative _____ numbers a.

Follow along with the two Guided Learning Video examples, and fill in the blanks on the left as you learn with the instructor. When the pencil icon appears, pause and try the Student Practice on your own.

Guided Learning Video	Pause: Student Practice
1. Solve $x^2 = 40$. Use the square root property to solve the equation. $x^2 = 40$ ▭ ▭ ▭	2. Solve $x^2 = 128$.

3. Solve $2x^2 - 7 = 91$.

Use the square root property to solve the equation.

$2x^2 - 7 = 91$

$$\boxed{}$$

$$\boxed{}$$

$$\boxed{}$$

$$\boxed{}$$

4. Solve $4x^2 - 5 = 31$.

Helpful Hint: Pause the video and write the helpful hint in your own words.

For each Active Video Lesson, when the pencil icon appears, pause and work the problem. Then press play to check your work.

Active Video Lesson 1

Active Video Lesson 2

5. Solve $3x^2 - 7 = 13 - 2x^2$.

6. Solve $x^2 = 24$.

Guided Learning Video Worksheet: Solve Quadratic Equations by Completing the Square

Text: *Beginning Algebra: Early Graphing*
Student Learning Objective 9.2.2: Solve quadratic equations by completing the square.

Follow along with *Guided Learning Video 9.2.2, Solve Quadratic Equations by Completing the Square.*

Understanding the Big Picture: As you listen to the first part of the video, when the pencil icon appears, pause and fill in the blanks, choosing from the words listed.

trinomial • square • completing • perfect • factorable • constant

1. For quadratic equations that are not _____, the method of
    _____ the square is sometimes used to solve the equation.

2. When completing the square, the quadratic equation is changed so that one side of the
    equation is a _____ square _____, and the opposite side is a
    constant.

3. In a perfect square trinomial, the _____ term is the _____ of
    one-half of the coefficient of the linear term.

Follow along with the two Guided Learning Video examples, and fill in the blanks on the left as you learn with the instructor. When the pencil icon appears, pause and try the Student Practice on your own.

Guided Learning Video ▶ GUIDED LEARNING VIDEO	Pause: Student Practice
1. Solve by completing the square. $x^2 - 6x = 1$	2. Solve by completing the square. $x^2 - 8x = 3$

Use the procedure for completing the square to solve the equation.

$x^2 - 6x = 1$

[ ]

[ ]

[ ]

[ ]

3. Solve by completing the square.
$9x^2 - 6x - 4 = 0$

Use the procedure for completing the square to solve the equation.

$9x^2 - 6x - 4 = 0$

4. Solve by completing the square.
$2x^2 - 3x - 20 = 0$

Helpful Hint: Pause the video and write the helpful hint in your own words.

For each Active Video Lesson, when the pencil icon appears, pause and work the problem. Then press play to check your work.

Active Video Lesson 1 ✏ | **Active Video Lesson 2** ✏

5. Solve by completing the square.
$x^2 + 5x - 4 = 0$

6. Solve by completing the square.
$x^2 + 10x + 1 = 0$

Name: _____   Date: _____
Instructor: _____   Section: _____

Guided Learning Video Worksheet: Solve Quadratic Equations Using the Quadratic Formula

Text: *Beginning Algebra: Early Graphing*
Student Learning Objective 9.3.1: Solve a quadratic equation using the quadratic formula.

Follow along with *Guided Learning Video 9.3.1, Solve Quadratic Equations Using the Quadratic Formula.*

Understanding the Big Picture: As you listen to the first part of the video, when the pencil icon appears, pause and fill in the blanks, choosing from the words listed.

substituted • standard • square • identified • formula • quadratic

1. Completing the _____ of $ax^2 + bx + c = 0$ leads to the quadratic

    _____ .

2. To solve a quadratic equation using the _____ formula, the equation must
    be written in the _____ form $ax^2 + bx + c = 0$.

3. The values of a, b, and c must be _____ from the standard form and
    _____ into the quadratic formula to determine the equation's solution.

Follow along with the two Guided Learning Video examples, and fill in the blanks on the left as you learn with the instructor. When the pencil icon appears, pause and try the Student Practice on your own.

Guided Learning Video ▶ GUIDED LEARNING VIDEO	Pause: Student Practice
1. Solve using the quadratic formula. $7x^2 + 2x - 5 = 0$. Identify and substitute the values of a, b, and c into the quadratic formula, then evaluate and simplify to compute the solution. $7x^2 + 2x - 5 = 0$ $a = \boxed{}, b = \boxed{}, c = \boxed{}$ $x = \dfrac{\boxed{}}{\boxed{}} = \dfrac{\boxed{}}{\boxed{}}$ $x = \boxed{}$ $x = \boxed{}$	2. Solve using the quadratic formula. $x^2 + 3x - 10 = 0$.

3. Solve using the quadratic formula.
$$\frac{x}{2}+1=\frac{5}{4}x^2.$$

Identify and substitute the values of a, b, and c into the quadratic formula, then evaluate and simplify to compute the solution.

$$\frac{x}{2}+1=\frac{5}{4}x^2$$

$a=\boxed{}, b=\boxed{}, c=\boxed{}$

$x=\dfrac{\boxed{}}{\boxed{}}=\dfrac{\boxed{}}{\boxed{}}$

$x=\boxed{}$ $x=\boxed{}$

4. Solve using the quadratic formula.
$$x^2=1-\frac{2}{5}x.$$

Helpful Hint: Pause the video and write the helpful hint in your own words.

For each Active Video Lesson, when the pencil icon appears, pause and work the problem. Then press play to check your work.

Active Video Lesson 1 | **Active Video Lesson 2**

5. Solve using the quadratic formula.
$$\frac{4}{7}x^2=x-\frac{3}{7}.$$

6. Solve using the quadratic formula.
$$3x^2-2x-4=0.$$

Guided Learning Video Worksheet: Determining Whether a Quadratic Equation Has No Real Solutions

Text: *Beginning Algebra: Early Graphing*
Student Learning Objective 9.3.3: Determine whether a quadratic equation has no real solutions.

Follow along with *Guided Learning Video 9.3.3, Determining Whether a Quadratic Equation Has No Real Solutions*.

Understanding the Big Picture: As you listen to the first part of the video, when the pencil icon appears, pause and fill in the blanks, choosing from the words listed.

zero • no • discriminant • negative • two • radical

1. In the quadratic equation, the expression under the _____ sign is called the _____.

2. If the discriminant is a _____ number, there is _____ real solution to the quadratic equation.

3. When the discriminant equals a positive number, there will be _____ real number solutions.

4. When the discriminant equals _____, there is one real number solution.

Follow along with the two Guided Learning Video examples, and fill in the blanks on the left as you learn with the instructor. When the pencil icon appears, pause and try the Student Practice on your own.

Guided Learning Video ▶ GUIDED LEARNING VIDEO	**Pause: Student Practice** ✏
1. Determine whether $5x^2 + 3x + 1 = 0$ has real number solutions. Evaluate the discriminant to determine if the equation has real number solutions. $5x^2 + 3x + 1 = 0$ $a = \boxed{}, b = \boxed{}, c = \boxed{}$ $b^2 - 4ac = \boxed{} = \boxed{}$ $b^2 - 4ac = \boxed{}$	2. Determine whether $2x^2 + x + 7 = 0$ has real number solutions.

3. Determine whether $x^2 + 2 = 4x$ has real number solutions.

Evaluate the discriminant to determine if the equation has real number solutions.

$x^2 + 2 = 4x$

$a = \boxed{}, b = \boxed{}, c = \boxed{}$

$b^2 - 4ac = \boxed{} = \boxed{}$

$b^2 - 4ac = \boxed{}$

4. Determine whether $2x^2 + 3 = 5x$ has real number solutions.

Helpful Hint: Pause the video and write the helpful hint in your own words.

For each Active Video Lesson, when the pencil icon appears, pause and work the problem. Then press play to check your work.

Active Video Lesson 1 | **Active Video Lesson 2**

5. Determine whether $3x^2 = 2x - 3$ has real number solutions.

6. Determine whether $2x^2 - 5x + 1 = 0$ has real number solutions.

Guided Learning Video Worksheet: Graph Quadratic Equations Using the Vertex Formula

Text: *Beginning Algebra: Early Graphing*

Student Learning Objective 9.4.2: Graph equations of the form $y = ax^2 + bx + c$ using the vertex formula.

Follow along with *Guided Learning Video 9.4.2, Graph Quadratic Equations Using the Vertex Formula.*

Understanding the Big Picture: As you listen to the first part of the video, when the pencil icon appears, pause and fill in the blanks, choosing from the words listed.

negative • upward • parabola • vertex • *x*-coordinate

1. When graphing a(n) _____, it is important to know the coordinates of the _____.

2. The _____ of the vertex of a parabola described by
   $y = ax^2 + bx + c$ is computed using the formula $x = \dfrac{-b}{2a}$.

3. In the form $y = ax^2 + bx + c$, when a is positive, the parabola opens _____,
   and when a is _____, the parabola opens downward.

Follow along with the two Guided Learning Video examples, and fill in the blanks on the left as you learn with the instructor. When the pencil icon appears, pause and try the Student Practice on your own.

Guided Learning Video ▶ GUIDED LEARNING VIDEO	Pause: Student Practice ✏
1. Determine the vertex and *x*-intercepts, then sketch the graph. $y = x^2 - 6x + 8$ Use the vertex formula to find the *x*-coordinate of the vertex, and then use the quadratic equation to find the corresponding *y*-coordinate. Finally, solve the equation to find the *x*-intercept(s). $x = \dfrac{-b}{2a} = \dfrac{\boxed{}}{\boxed{}}$ **(continued)**	2. Determine the vertex and *x*-intercepts, then sketch the graph. $y = x^2 - 4x + 3$

$0 = x^2 - 6x + 8$

[ ]

[ ]

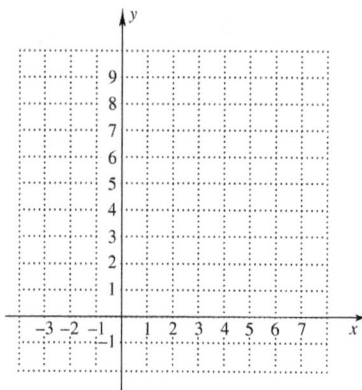

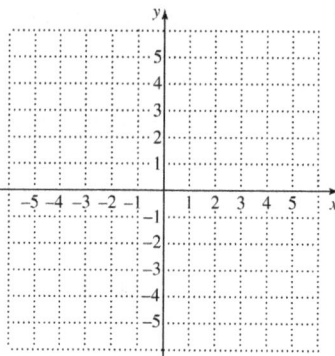

3. A toy rocket is launched and its distance h, in feet, from the ground is given by the equation $h = -16t^2 + 48t + 3$, where t is measured in seconds. Graph the equation $h = -16t^2 + 48t + 3$. What is the greatest height of the rocket? After how many seconds will the rocket hit the ground?

   Determine the vertex, t-intercept, and the h-intercept.

   $$t = \frac{-b}{2a} = \frac{\boxed{}}{\boxed{}}$$

   $$(0, h) = -16\left(\boxed{}\right)^2 + 48\left(\boxed{}\right) + 3 = \left(0, \boxed{}\right)$$

   $0 = -16t^2 + 48t + 3$

   [ ]

   [ ]

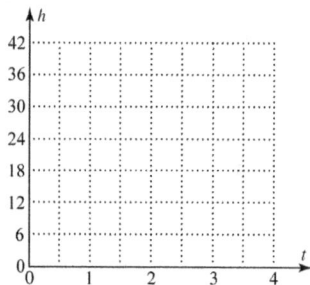

4. A ball is thrown upward and its distance h, in feet, from the ground is given by $h = -16t^2 + 32t + 6$ where t is measured in seconds. Graph the equation $h = -16t^2 + 32t + 6$. What is the greatest height of the ball? After how many seconds will the ball hit the ground?

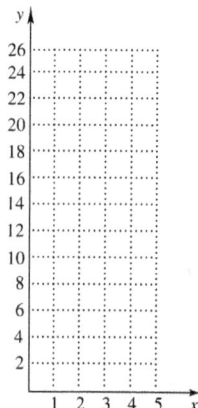

Helpful Hint: Pause the video and write the helpful hint in your own words.

Name: _____   Date: _____

Instructor: _____   Section: _____

For each Active Video Lesson, when the pencil icon appears, pause and work the problem. Then press play to check your work.

Active Video Lesson 1	Active Video Lesson 2
5. A person throws a ball into the air and its distance h, in feet, from the ground is given by the equation $h = -16t^2 + 32t + 7$, where t is measured in seconds. Graph the equation $h = -16t^2 + 32t + 7$. What is the greatest height of the ball? After how many seconds will the ball hit the ground?	6. Determine the vertex and x-intercepts, then sketch the graph. $y = x^2 - 8x + 12$

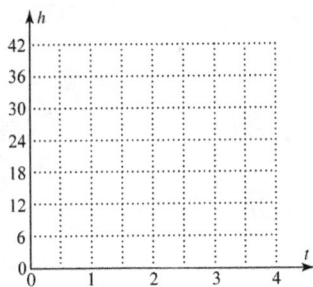

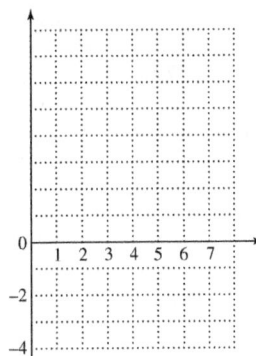

Answers
Guided Learning Video Worksheets

Student Learning Objective 0.1.2 Answers
Understanding the Big Picture: 1. equivalent 2. reduced, factor 3. divide

Guided Learning Video: 1a. $\dfrac{3}{7}$ b. $\dfrac{4}{5}$ 3. $\dfrac{2}{5}$

Student Practice: 2a. $\dfrac{3}{5}$ b. $\dfrac{5}{9}$ 4. $\dfrac{7}{10}$

Active Video Lesson: 5. $\dfrac{2}{5}$ 6. $\dfrac{8}{13}$

Student Learning Objective 0.1.3A Answers
Understanding the Big Picture: 1. proper 2. improper 3. zero, mixed

Guided Learning Video: 1. $\dfrac{59}{6}$ 3. $\dfrac{119}{8}$

Student Practice: 2. $\dfrac{23}{5}$ 4. $\dfrac{155}{9}$

Active Video Lesson: 5. $\dfrac{137}{7}$ 6. $\dfrac{58}{7}$

Student Learning Objective 0.1.3B Answers
Understanding the Big Picture: 1. divide, numerator, denominator 2. quotient 3. remainder

Guided Learning Video: 1. $3\dfrac{3}{4}$ 3. $7\dfrac{5}{8}$

Student Practice: 2. $3\dfrac{5}{6}$ 4. $8\dfrac{1}{9}$

Active Video Lesson: 5. $6\dfrac{4}{9}$ 6. $8\dfrac{2}{3}$

Student Learning Objective 0.1.4 Answers
Understanding the Big Picture: 1. equivalent 2. value 3. building

Guided Learning Video: 1. $\dfrac{20}{36}$ 3. $\dfrac{32}{56}$

Student Practice: 2. $\dfrac{24}{40}$ 4. $\dfrac{63}{72}$

Active Video Lesson: 5. $\dfrac{21}{45}$ 6. $\dfrac{55}{60}$

Student Learning Objective 0.2.2 Answers
Understanding the Big Picture: 1. compare 2. prime 3. greatest, one
Guided Learning Video: 1. 36 3. 54
Student Practice: 2. 75 4. 72
Active Video Lesson: 5. 30 6. 72

Student Learning Objective 0.2.4 Answers

Understanding the Big Picture: 1. least 2. fractions, whole numbers

Guided Learning Video: 1. $4\frac{14}{15}$ 3. $1\frac{5}{6}$

Student Practice: 2. $2\frac{9}{20}$ 4. $5\frac{11}{15}$

Active Video Lesson: 5. $1\frac{13}{20}$ 6. $1\frac{3}{4}$

Student Learning Objective 0.3.1A Answers

Understanding the Big Picture: 1. multiplication 2. numerators 3. denominators

Guided Learning Video: 1. $\frac{8}{63}$ 3. $\frac{3}{10}$

Student Practice: 2. $\frac{10}{99}$ 4. $\frac{3}{20}$

Active Video Lesson: 5. $\frac{2}{7}$ 6. $\frac{3}{8}$

Student Learning Objective 0.3.1B Answers

Understanding the Big Picture: 1. improper 2. divide 3. mixed number

Guided Learning Video: 1. $14\frac{2}{3}$ 3. $27\frac{1}{3}$

Student Practice: 2. $16\frac{4}{5}$ 4. $31\frac{1}{3}$

Active Video Lesson: 5. $11\frac{2}{3}$ square inches 6. $8\frac{1}{2}$

Student Learning Objective 0.3.2A Answers

Understanding the Big Picture: 1. reciprocal 2. inverted 3. second

Guided Learning Video: 1. $\frac{63}{80}$ 3. $\frac{5}{6}$

Student Practice: 2. $\frac{28}{55}$ 4. $\frac{10}{13}$

Active Video Lesson: 5. $\frac{3}{4}$ 6. $\frac{9}{10}$

Student Learning Objective 0.3.2B Answers

Understanding the Big Picture: 1. improper 2. multiplication

Guided Learning Video: 1. $1\frac{1}{3}$ 3. $\frac{2}{3}$

Student Practice: 2. $\frac{15}{22}$ 4. $\frac{2}{5}$

Active Video Lesson: 5. $1\frac{1}{2}$ 6. $\frac{22}{25}$

Student Learning Objective 0.4.2A Answers

Understanding the Big Picture: 1. decimal 2. numerator, equal, zeros 3. delete

Guided Learning Video: 1. 6.59 3. 0.078

VWA-2

Student Practice: 2. 4.06 4. 0.045
Active Video Lesson: 5. 8.003 6. 0.07

Student Learning Objective 0.4.2B Answers
Understanding the Big Picture: 1. numerator, denominator 2. terminating 3. repeat, repeating
Guided Learning Video: 1. 0.875 3. $0.\overline{4}$
Student Practice: 2. 0.375 4. $0.8\overline{3}$
Active Video Lesson: 5. $0.41\overline{6}$ 6. 0.5375

Student Learning Objective 0.4.3 Answers
Understanding the Big Picture: 1. whole, point 2. and, point 3. last 4. numerator
Guided Learning Video: 1. $6\frac{317}{1000}$ 3. $\frac{3}{4}$
Student Practice: 2. $4\frac{629}{1000}$ 4. $\frac{8}{25}$
Active Video Lesson: 5. $\frac{27}{200}$ 6. $4\frac{29}{100}$

Student Learning Objective 0.4.4A Answers
Understanding the Big Picture: 1. fractions 2. lining 3. same, sum
Guided Learning Video: 1. 11.8 3. 34.509
Student Practice: 2. 13.3 4. 50.406
Active Video Lesson: 5. 44.838 6. $954.38

Student Learning Objective 0.4.4B Answers
Understanding the Big Picture: 1. borrow 2. zeros, right, same 3. difference
Guided Learning Video: 1a. 11.6 b. 18.59 3. 8.231
Student Practice: 2a. 21.2 b. 51.97 4. 16.408
Active Video Lesson: 5. 14.955 6. 48.06

Student Learning Objective 0.4.5 Answers
Understanding the Big Picture: 1. whole 2. total 3. equal 4. zeros, left
Guided Learning Video: 1. 0.056 3. 247.987
Student Practice: 2. 0.024 4. 150.65
Active Video Lesson: 5. 0.036 6. 72.768

Student Learning Objective 0.4.6 Answers
Understanding the Big Picture: 1. divisor, right 2. dividend, divisor 3. quotient, dividend
Guided Learning Video: 1. 1.79 3. 3.15
Student Practice: 2. 1.53 4. 2.65
Active Video Lesson: 5. 2.35 6. 4.5

Student Learning Objective 0.4.7 Answers
Understanding the Big Picture: 1. multiplication 2. zeros, places, right 3. less
Guided Learning Video: 1. $8,174,000$ 3. $91,600$
Student Practice: 2. $667,300$ 4. 4950
Active Video Lesson: 5. $1,700$ 6. $62,800\,\text{m}$

Student Learning Objective 0.5.1 Answers
Understanding the Big Picture: 1. percent 2. ratios, denominators 3. parts
Guided Learning Video: 1. 72.6%   3. 580%
Student Practice: 2. 53.9%   4. 310%
Active Video Lesson: 5. 290% 6. 16.5%

Student Learning Objective 0.5.3 Answers
Understanding the Big Picture: 1. amount, percent, base 2. multiplication 3. percent, decimal
Guided Learning Video: 1. 83.2   3. 2450 freshman
Student Practice: 2. 328.5   4. 5750 ballots
Active Video Lesson: 5. 510 students 6. 12,700 people

Student Learning Objective 0.5.4 Answers
Understanding the Big Picture: 1. amount, percent, base 2. divide, amount, base 3. two, right
Guided Learning Video: 1. 25%   3. 8%
Student Practice: 2. 60%   4. 6%
Active Video Lesson: 5. 80% 6. 40%

Student Learning Objective 0.5.5 Answers
Understanding the Big Picture: 1. place 2. less 3. increase, one
Guided Learning Video: 1. 14.26   3. 0.703
Student Practice: 2. 63.54   4. 0.095
Active Video Lesson: 5. $9.87   6. 3.2

Student Learning Objective 0.6.1 Answers
Understanding the Big Picture: 1. organize, plan, blueprint 2. understand, calculate 3. check
Guided Learning Video: 1. current salary of $31,200 is greater than $28,000   3. $137
Student Practice: 2. second job offer's salary $46,800 is greater than $45,760   4. $1987
Active Video Lesson: 5. $2401 6. 15 miles per gallon

Student Learning Objective 1.1.3A Answers
Understanding the Big Picture: 1. positive, direction 2. absolute, distance, zero 3. vertical
4. positive
Guided Learning Video: 1. 21   3. −16
Student Practice: 2. 14   4. −33
Active Video Lesson: 5. 53   6. 0

Student Learning Objective 1.1.3B Answers
Understanding the Big Picture: 1. absolute values 2. adding, same, common 3. negative
Guided Learning Video: 1. −108   3. −1770
Student Practice: 2. −101   4. −1160
Active Video Lesson: 5. −1980   6. −444